Transport of Fluids in Nanoporous Materials

Transport of Fluids in Nanoporous Materials

Special Issue Editors

Suresh K. Bhatia
David Nicholson
Xuechao Gao
Guozhao Ji

MDPI • Basel • Beijing • Wuhan • Barcelona • Belgrade

MDPI

Special Issue Editors

Suresh K. Bhatia	David Nicholson	Xuechao Gao
The University of Queensland	The University of Queensland	Nanjing Tech University
Australia	Australia	China

Guozhao Ji
Dalian University of Technology
China

Editorial Office
MDPI
St. Alban-Anlage 66
4052 Basel, Switzerland

This is a reprint of articles from the Special Issue published online in the open access journal *Processes* (ISSN 2227-9717) from 2018 to 2019 (available at: https://www.mdpi.com/journal/processes/special_issues/transport_fluids)

For citation purposes, cite each article independently as indicated on the article page online and as indicated below:

LastName, A.A.; LastName, B.B.; LastName, C.C. Article Title. *Journal Name* **Year**, *Article Number, Page Range.*

ISBN 978-3-03897-529-8 (Pbk)
ISBN 978-3-03897-530-4 (PDF)

Contents

About the Special Issue Editors

Suresh K. Bhatia is Professor of Chemical Engineering at the University of Queensland in Australia. His main research interests are in the adsorption and transport of fluids in nanoporous materials and in heterogeneous reaction engineering, where he has authored over 250 refereed journal publications and four book chapters. He has received numerous awards for his outstanding contributions to the subject of chemical engineering, including the Shanti Swarup Bhatnagar Prize from the Government of India, the ExxonMobil Award for excellence from the Institution of Chemical Engineers, and the Australian Professorial Fellowship from the Australian Research Council. He is an elected Fellow of two major academies—the Australian Academy of Technological Sciences and Engineering, and the Indian Academy of Sciences—and of the Institution of Chemical Engineers. He is an editorial board member of the open-access journal *Processes*, the international journal *Molecular Simulation*, and of the journal *Advanced Porous Materials*.

David Nicholson is an Honorary Visiting Professor at the University of Queensland in Australia and holds a DSc from the University of London. His main research interests are in the theory and simulation of interfacial systems and in the use of molecular simulation in predicting transport properties of fluids. He has co-authored over 300 publications, including a book on the theory and simulation of adsorption and papers in learned journals, and was an elected member of the international adsorption society from 2001–2005.

Xuechao Gao is an Associate Professor of Chemical Engineering at Nanjing Tech University in China. His main research activities are about the experimental characterization and numerical modelling of the mass transfer process of fluids in nanoporous materials, such as inorganic membranes, zeolites, and carbons. He has co-authored over 25 refereed journal papers.

Guozhao Ji is an Associate Professor of Environmental Science and Technology at Dalian University of Technology in China. His research interests are modelling of gas transport in microporous material, gas separation by inorganic membrane, high temperature CO_2 capture, solid waste gasification, and computational fluid dynamic application in chemical engineering processes. He has authored over 35 refereed journal publications and one book chapter.

Preface to "Transport of Fluids in Nanoporous Materials"

The transport of fluids in nanoporous materias is central to their numerous conventional and emerging applications. Recent advances in the synthesis of ordered nanomaterials with well-defined pore channels has enabled an improved understanding of the mechanisms affecting transport in confined spaces, and this Special Issue aims to highlight the current research trends related to this topic. This Special Issue collects 14 papers from leading scholars in this field and demonstrates the diverse behaviors of fluid molecules in different processes. In addition to featuring state-of-the art reviews and research in diverse topics, this collection demonstrates the versatility of the area, ranging from fundamental theories to practical applications of confined fluids.

Suresh K. Bhatia, David Nicholson, Xuechao Gao, Guozhao Ji
Special Issue Editors

processes

MDPI

Editorial

Special Issue on "Transport of Fluids in Nanoporous Materials"

Xuechao Gao [1], Guozhao Ji [2], Suresh K. Bhatia [3,*] and David Nicholson [3]

[1] State Key Laboratory of Materials-Oriented Chemical Engineering, College of Chemical Engineering, Nanjing Tech University, 5 Xinmofan Road, Nanjing 210009, China; xuechao.gao@njtech.edu.cn
[2] School of Environmental Science and Technology, Dalian University of Technology, Dalian 116024, China; guozhaoji@dlut.edu.cn
[3] School of Chemical Engineering, the University of Queensland, Brisbane, QLD 4072, Australia; d.nicholson@uq.edu.au
* Correspondence: s.bhatia@uq.edu.au; Tel.: +61-7-3365-4263

Received: 26 November 2018; Accepted: 26 November 2018; Published: 1 January 2019

Understanding the transport behavior of fluid molecules in confined spaces is central to the design of innovative processes involving porous materials and is indispensable to the correlation of process behavior with the material structure and properties typically used for structural characterizations such as pore dimension, surface texture, and tortuosity. The interest in fluid transport in nanopores dates back a century, when Martin Knudsen [1] performed experiments on gaseous transport in macroporous glass tubes several microns in diameter and derived the well-known Knudsen diffusion equation for the transport of rarefied gases. Since then, the Knudsen model has been widely used to predict gas transport coefficients in confined spaces and has had a significant impact on technological developments. The Knudsen model was improved by Smolouchowski [2], who showed that the mode of scattering from a solid surface can radically modify flux rates. Subsequently, it was recognized that the Knudsen model suffers from a significant weakness, especially in channels of nanoscale cross sections, because the potential energy field arising from fluid–solid interactions distorts the linear trajectories assumed in classical gas kinetic theory. As a result, this model and others such as the Dusty Gas model [3], which is applicable to dense gas systems, fail at the nanoscale where the impact of fluid–solid interactions is dominant. Moreover, in dense fluids at the nanoscale, it has become clear that that these interactions mean that fluid density is not uniform, which implies that there must be a spatial dependence in viscosity. The recognition that transport in nanoporous materials differs in many respects from macroscale transport has spawned new research in the last decade, particularly to facilitate the development of new technologies at the nanoscale with enhanced efficiency.

The selective nature of transport in nanoporous materials has resulted in numerous new separation processes that have replaced energetically less efficient conventional technologies. Zeolite-and carbon-based membranes have been used to purify azeotropic and saline solutions, for which membrane-based pervaporation and desalination techniques have been developed, exploiting the large surface area and adsorptive capacity of nanoporous materials. Pervaporation involves hydrophilic/hydrophobic surfaces and a phase-change in a liquid, whilst in desalination, as well as in any phase change, the hydration interaction of ions with water molecules plays a central role in determining the rejection ratio. In these and all separations in general, fluid–solid interactions strongly influence adsorption capacity and permeability and therefore the process efficiency. The transport coefficients associated with such systems can be estimated from the analysis of the trajectories of fluid molecules from computer simulations; however, structural non-idealities, lack of reliable models for the solid-fluid interactions, and unduly large computational requirements impose limitations in actual practice. Consequently, the wide variety of structural complexities and interaction forces make it necessary to combine experiments on permeation and transport with adsorption isotherms and molecular dynamics simulations in order to probe the movement of fluids/ions in confined systems under realistic conditions.

This Special Issue on "Transport of Fluids in Nanoporous Materials" of *Processes* collects the recent work of leading researchers in a single forum, and the contents cover a variety of theoretical studies and experimental applications, focusing on the transport of fluids in nanoporous materials. Despite the interdisciplinary nature of the different applications involved, there is a common characteristic of fluid/ion transport in nanopores connecting the areas together which we seek to capture in this issue. We believe that the advances described by the contributors have significantly helped accomplish this target. Besides the research articles, the issue features a number of reviews, covering a range of topics, which highlight the versatility of the area. For instance, Kärger et al. [4] discuss the direct measurement of transport coefficients for guest fluid molecules in complex nanoporous materials where the signal produced by the H atom is employed to predict diffusivity. Since many fluid molecules contain hydrogen, this approach is very versatile. As an attractive technique, molecular dynamics modeling has the ability to access spatial and temporal resolutions that are difficult to attain in experimental studies. Murad et al. [5] review molecular dynamics techniques to examine intramembrane transport in reverse osmosis (RO), ion exchange, and gas separation. It is comprehensively demonstrated how molecular dynamics simulations provide deep insight into the physiochemical behavior of many such membrane-based applications and aid in more efficient process design and optimization. The review article of Daivis and Todd [6] analyzes the failure of traditional Navier–Stokes theory when applied to transport at the molecular scale in nanofluidics where several fundamental phenomena such as slip, spin-angular momentum coupling, non-local response, and density inhomogeneity require consideration. Theories accounting for these effects are provided and are discussed to improve the accuracy of nanoscale transport modelling. The review of Monsalve-Bravo and Bhatia [7] focuses on models for mixed matrix membranes, an emerging nanomaterial now widely being investigated for gas separations. Such membranes combine the favorable selectivity properties of polymers with high flux capabilities of zeolites or other adsorbents to achieve high separation efficiency, and their modelling is critical to membrane design and optimization. At a more applied level, the review of Zhu and coworkers [8] discusses the synthesis and potential applications of superparamagnetic Nano-Fe_3O_4, an important nanoscale material with desirable properties that are conducive to its application in a number of areas including magnetic resonance imaging, biosensors, and drug delivery. In another application-focused review, Nguyen and Lee [9] discuss the separation of lithium, an important material for battery electrodes, ceramic glass, alloys, dying, and a host of other applications.

Adsorption and separation is one of the key application areas of nanoporous materials where they provide enhanced efficiency, and several articles address problems related to such applications. Gu and coworkers [10] describe experimental measurements of ethanol dehydration by cation-treated zeolite membranes, in which the pore channels were decorated by ion-exchange. The treated membranes achieved the desired pore size, cation charges, and hydrophilicity, thereby significantly enhancing the separation factor. Pore size distribution measurement is an important issue in nanoporous materials and becomes challenging when the pores are in the ultra-microporous range. Ji et al. [11] describe a novel method for determining such pore size distributions based on the measurement of transport parameters and interpreting these through effective medium theory while using a molecular level theory for single pore transport. Qing Liu and coworkers [12] discuss the synthesis of carbon aerogels for CO_2 capture, an application now considered one of the significant challenges of our time. In other adsorption related work, Xue, Ju, and coworkers [13] discuss the synthesis of ion sieves for Li ion adsorption, while Liang, Zou, and Li [14] report the effect of different cyclic stress paths on the damage and permeability changes in gas bearing coal seams. These are all issues of importance to technologies related to our energy future. Of course, there are a host of other areas where nanoscale materials play a role, and some examples are discussed in several important contributions. Liu, Ba, and coworkers [15] applied microfluidics theory to microdroplet dosing for cell culture on a chip to meet the demand for narrow diffusion distances, controllable pulse dosing, and to lessen the impact to cells. The established mathematical model could analyze the rhodamine mass fraction distribution, pressure field, and velocity field around the microdroplet and cell surfaces. Good accuracy and controllability of the cell dosing pulse time and maximum drug mass fraction on cell surfaces is achieved, and the drug effect on cells analyzed with more precision, especially for neuronal cell dosing. Ye and coworkers [16] discuss a new porous-medium-based burner, which can achieve high energy efficiency, and present its numerical simulation. Combustion in porous media is a subject of long-term interest and continues to attract attention because of its potential for reducing emissions and improving combustion efficiency of low-grade combustible pollutants. Finally, Liu and Zhang [17] present an application for carbon nanotubes as a dispersant in foamed concrete to improve its mechanical properties.

The above papers demonstrate the versatility and technical importance of the area of fluid transport in nanoporous materials, ranging from the formulation of fundamental theory to practical applications. Although the basic principles of fluid transport in the nanoscale are fairly well understood, the articles address outstanding challenges related to fluid transport in different areas in terms of both application and theoretical perspectives, and much remains to be explored in the future. With the enormous variety and number of the applications currently under development, we feel confident for the longevity and future of this subject.

We thank all the contributors and the Editor-in-Chief, Michael A. Henson, for their enthusiastic support of the Special Issue, as well as the editorial staff of *Processes* for their efforts.

Xuechao Gao
Guozhao Ji
David Nicholson
Suresh K. Bhatia
Guest Editors

Funding: There is no funding support.

Conflicts of Interest: The authors declare no conflict of interest.

References

1. Knudsen, M. Molecular Effusion and Transpiration. *Nature* **1909**, *80*, 491–492. [CrossRef]
2. Smoluchowski, M. Zur kinetischen Theorie der Transpiration und Diffusion verdünnter Gase. *Ann. Phys.* **1910**, *338*, 1559–1570. [CrossRef]
3. Evans, R.; Watson, G.; Mason, E. Gaseous Diffusion in Porous Media. II. Effect of Pressure Gradients. *J. Chem. Phys.* **1962**, *36*, 1894–1902. [CrossRef]
4. Kärger, J.; Freude, D.; Haase, J. Diffusion in Nanoporous Materials: Novel Insights by Combining MAS and PFG NMR. *Processes* **2018**, *6*, 147. [CrossRef]
5. Hinkle, K., Wang, X.; Gu, X.; Jameson, C.; Murad, S. Computational Molecular Modeling of Transport Processes in Nanoporous Membranes. *Processes* **2018**, *6*, 124. [CrossRef]
6. Daivis, P.; Todd, B. Challenges in Nanofluidics—Beyond Navier–Stokes at the Molecular Scale. *Processes* **2018**, *6*, 144. [CrossRef]
7. Monsalve-Bravo, G.; Bhatia, S. Modeling Permeation through Mixed-Matrix Membranes: A Review. *Processes* **2018**, *6*, 172. [CrossRef]
8. Zhan, H.; Bian, Y.; Yuan, Q.; Ren, B.; Hursthouse, A.; Zhu, G. Preparation and Potential Applications of Super Paramagnetic Nano-Fe_3O_4. *Processes* **2018**, *6*, 33. [CrossRef]
9. Nguyen, T.; Lee, M. A Review on the Separation of Lithium Ion from Leach Liquors of Primary and Secondary Resources by Solvent Extraction with Commercial Extractants. *Processes* **2018**, *6*, 55. [CrossRef]
10. Gao, X.; Gao, B.; Wang, X.; Shi, R.; Ur Rehman, R.; Gu, X. The Influence of Cation Treatments on the Pervaporation Dehydration of NaA Zeolite Membranes Prepared on Hollow Fibers. *Processes* **2018**, *6*, 70. [CrossRef]

11. Ji, G.; Gao, X.; Smart, S.; Bhatia, S.; Wang, G.; Hooman, K.; da Costa, J. Estimation of Pore Size Distribution of Amorphous Silica-Based Membrane by the Activation Energies of Gas Permeation. *Processes* **2018**, *6*, 239. [CrossRef]
12. He, P.; Qian, X.; Fei, Z.; Liu, Q.; Zhang, Z.; Chen, X.; Tang, J.; Cui, M.; Qiao, X. Structure Manipulation of Carbon Aerogels by Managing Solution Concentration of Precursor and Its Application for CO_2 Capture. *Processes* **2018**, *6*, 35. [CrossRef]
13. Yang, F.; Chen, S.; Shi, C.; Xue, F.; Zhang, X.; Ju, S.; Xing, W. A Facile Synthesis of Hexagonal Spinel λ-MnO_2 Ion-Sieves for Highly Selective Li^+ Adsorption. *Processes* **2018**, *6*, 59. [CrossRef]
14. Li, Q.; Liang, Y.; Zou, Q. Seepage and Damage Evolution Characteristics of Gas-Bearing Coal under Different Cyclic Loading–Unloading Stress Paths. *Processes* **2018**, *6*, 190. [CrossRef]
15. Wang, Z.; Liu, K.; Ning, J.; Chen, S.; Hao, M.; Wang, D.; Mei, Q.; Ba, Y.; Ba, D. Effects of Pulse Interval and Dosing Flux on Cells Varying the Relative Velocity of Micro Droplets and Culture Solution. *Processes* **2018**, *6*, 119. [CrossRef]
16. Jia, Z.; Ye, Q.; Wang, H.; Li, H.; Shi, S. Numerical Simulation of a New Porous Medium Burner with Two Sections and Double Decks. *Processes* **2018**, *6*, 185. [CrossRef]
17. Zhang, J.; Liu, X. Dispersion Performance of Carbon Nanotubes on Ultra-Light Foamed Concrete. *Processes* **2018**, *6*, 194. [CrossRef]

processes

MDPI

Review

Diffusion in Nanoporous Materials: Novel Insights by Combining MAS and PFG NMR

Jörg Kärger * ⓘ, **Dieter Freude** ⓘ and **Jürgen Haase** ⓘ

Fakultät für Physik und Geowissenschaften, Universität Leipzig, Linnéstraße 5, 04103 Leipzig, Germany; freude@uni-leipzig.de (D.F.); j.haase@physik.uni-leipzig.de (J.H.)
* Correspondence: kaerger@uni-leipzig.de; Tel.: +49-341-973-2502

Received: 6 August 2018; Accepted: 21 August 2018; Published: 1 September 2018

Abstract: Pulsed field gradient (PFG) nuclear magnetic resonance (NMR) allows recording of molecular diffusion paths (notably, the probability distribution of molecular displacements over typically micrometers, covered during an observation time of typically milliseconds) and has thus proven to serve as a most versatile means for the in-depth study of mass transfer in complex materials. This is particularly true with nanoporous host materials, where PFG NMR enabled the first direct measurement of intracrystalline diffusivities of guest molecules. Spatial resolution, i.e., the minimum diffusion path length experimentally observable, is limited by the time interval over which the pulsed field gradients may be applied. In "conventional" PFG NMR measurements, this time interval is determined by a characteristic quantity of the host-guest system under study, the so-called transverse nuclear magnetic relaxation time. This leads, notably when considering systems with low molecular mobilities, to severe restrictions in the applicability of PFG NMR. These restrictions may partially be released by performing PFG NMR measurements in combination with "magic-angle spinning" (MAS) of the NMR sample tube. The present review introduces the fundamentals of this technique and illustrates, via a number of recent cases, the gain in information thus attainable. Examples include diffusion measurements with nanoporous host-guest systems of low intrinsic mobility and selective diffusion measurement in multicomponent systems.

Keywords: NMR; PFG; MAS; diffusion; adsorption; hierarchical host materials

---

1. Introduction

Diffusion, i.e., the irregular movement of the elements of a given entity in nature, technology, or society, is an essentially omnipresent phenomenon [1] and may often be found to decide about the performance of the systems under study. This is, in particular, true for nanoporous host-guest materials where the performance, i.e., the gain in value-added products by matter upgrading via separation [2,3] or conversion [4,5], can never be faster than allowed by the rate of mass transfer [6–8]. Measurement of the rate of mass transfer in nanoporous materials, however, is complicated by the small size of the individual crystallites (or particles). It was only with the introduction of the pulsed field gradient (PFG) technique of nuclear magnetic resonance (NMR) that the direct measurement of intracrystalline diffusivities has become possible [9–11]. The information provided by PFG NMR in its broadest significance is the probability distribution of molecular displacements, referred to as the mean propagator [12–14]. It does, thus, notably, include the intracrystalline self-diffusivity D, resulting via the Einstein relation [8,15] from the time dependence of the mean square displacement,

$$\langle r^2(t) \rangle = 6Dt \tag{1}$$

during the observation time t, i.e., from the squared width (the variance) of the propagator. As a prerequisite of such measurement, the root-mean-square displacement must be much smaller than the size of the crystals under study so that, during the observation time (typically in the range of milliseconds), the diffusion paths (typically of the order of micrometers) may be implied to remain unaffected by any significant interference with the crystal surface. The displacements must, simultaneously, be large enough for giving rise to a diffusion-related attenuation of the NMR signal.

The first requirement is seen to be easily fulfilled as soon as the material under study is accessible with sufficiently large crystal sizes. Accessibility of sufficiently large zeolite crystallites [16] did thus prove to be a very fortunate pre-condition for performance of the very first PFG NMR measurements with zeolites [9,17]. In view of Equation (1), the second pre-condition might also appear to be easily obeyed by simply choosing sufficiently long observation times. PFG NMR observation times, however, cannot be chosen to be arbitrarily large. They are rather limited by the influence of transverse nuclear magnetic relaxation which, via Equation (1), also sets a limit on the mean molecular displacements, depending on the given diffusivities (which, in PFG NMR studies, typically cover a range from 10^{-14} m^2 s^{-1} to 10^{-8} m^2 s^{-1}). The minimum displacement still observable by PFG NMR is proportional to the amplitude of the field gradient pulses and to the duration over which the field gradients may be applied. While the maximum gradient amplitude is a key parameter of the given device, the maximum width of the field gradients is determined by a characteristic quantity of the system, namely by the transverse nuclear magnetic relaxation time of the molecules under study. The here-important value T_2^{echo} can be obtained by measurements of the Hahn echo decay in dependence on the pulse distance between the two radio frequency pulses. Notably, in systems of low mobility, where gradients with particularly large widths were needed for recording particularly small displacements, rapid transverse nuclear magnetic relaxation prohibits, as a rule, their application.

Novel access towards the application of larger field gradient pulse widths has been provided by the recent combination of PFG NMR measurement with the application of magic-angle spinning (MAS) [18–22]. Enhancement of the transverse relaxation time $T_2^{MAS\ echo}$ with respect to T_2^{echo} upon MAS allows longer gradient pulse widths and is accompanied by a reduction in NMR line width so that MAS PFG NMR offers, as a second advantage, distinction between different components and, hence, the option of selective diffusion measurement in mixtures where conventional PFG NMR would fail. The advantages of MAS PFG NMR are purchased, however, with a decrease in the amplitude of the field gradient pulses applied, as a simple consequence of the reduction in space available for the PFG NMR coils, brought about by the presence of the MAS NMR rotor. Application of MAS PFG NMR should always be accompanied, therefore, by thoughtful balancing of pros and cons.

We are introducing this novel field of diffusion measurement with a short summary of the experimental procedure and the physical background in Section 2. Showcases of the application of MAS PFG NMR are presented in Section 3. They include selective diffusion measurements with mixtures of hydrocarbons in microporous materials, notably zeolites and metal–organic frameworks (MOFs) (Section 3.1) and in mesoporous silica gel (Section 3.2). Section 3.3 deals with the application of MAS PFG NMR for investigating the diffusion properties of nematic liquid crystals under confinement. Section 3.4 illustrates the potential of MAS PFG NMR for tracing and characterizing the diffusion pathways of water molecules in zeolite X. In Section 3.5, the self-diffusion coefficients from MAS PFG NMR are compared with tracer diffusion coefficients which were derived from impedance spectroscopy by the Nernst-Einstein equation and provide a model for proton mobility in functionalized mesoporous materials. The paper concludes with a summary of pros and cons and a view into promising future applications.

2. Experimental Procedure

Before describing the measurement procedure commonly used in MAS PFG NMR in more detail, we are going to briefly recollect the measuring principle of PFG NMR in its most straightforward variant (for more extensive presentations see, e.g., [8,14,23,24]). Its fundamentals can be easily rationalized within the frame of the classical interpretation of nuclear magnetism which is based on the understanding that a nuclear spin (in the cases here considered in general protons, i.e., the nuclei of hydrogen, ^{1}H) possesses both a magnetic and a mechanic momentum. Nuclear spins perform, therefore, within a magnetic field, a processional motion (i.e., they rotate around the direction of the magnetic field) with an angular frequency vector of the Larmor frequency

$$\boldsymbol{\omega}_L = -\gamma \boldsymbol{B}_0 \tag{2}$$

where the magnetic induction, $\boldsymbol{B}_0$, stands for the intensity of the external magnetic field in the z direction, and γ denotes the gyromagnetic ratio. Chemical shift reference materials of all NMR isotopes were fixed by the IUPAC (International Union of Pure and Applied Chemistry) convention in 2001 [25]. In PFG NMR, a properly chosen pulse sequence gives rise to a preferential orientation of the individual nuclear magnetic moments with a component perpendicular to the direction of the constant magnetic field. Just as each individual spins, their vector sum also performs a rotational motion about the direction of the magnetic field. This rotating (nuclear) magnetization induces a voltage in a transverse coil surrounding the sample which is recorded as the NMR signal.

Diffusion measurement by PFG NMR is based on the application of a strong additional z-gradient field $B_{add} = gz$, superimposed upon the constant external one over a short time interval δ. The thus-created spreading in the local magnetic field and, hence, via Equation (2), in the rotational frequencies of the local magnetizations gives rise to a spreading in their orientation and, hence, to the decrease in their vector sum, i.e., in total magnetization, with the NMR signal fading away. With a second, identical field gradient pulse, properly placed within the PFG NMR pulse program after a certain time interval (in the PFG NMR literature generally referred to as the observation/diffusion time Δ), one is able to counteract this process by creating a phase shift in exactly the opposite direction. Correspondingly, all phase shifts are eliminated by this second field gradient pulse if all molecules have kept their positions. Molecules, however, which have been shifted (over a distance z) in the field gradient direction undergo a phase shift $\gamma gz\delta$ and contribute, correspondingly, with only the cosine of this shift to the overall signal. The attenuation of the NMR signal intensity $S(m,t)$ under the influence of diffusion and the field gradients applied is thus easily seen to be given by the relation [10,12]

$$S(m,t) = S(0,t) \int_{-\infty}^{\infty} \cos(mz) P(z,t) dz, \tag{3}$$

where $m = \gamma g\delta$, with gradient intensity g and gradient pulse duration δ, has been introduced as a measure of the intensity of the field gradient pulses and t stands for the observation time, Δ, of the PFG NMR experiment. $P(z,t)$ is the mean propagator [12,26], referred to already in the introduction. It denotes the probability (density) that, during the observation time t, an arbitrarily selected molecule within the sample (contributing to the observed NMR signal) is shifted over a distance z in the direction of the applied field gradient. With the notation of Equation (3) it has, further on, been implied that molecular displacements occurring during the field gradient pulses are negligibly small in comparison with the displacements in the interval between the two gradient pulses. For normal diffusion in an infinitely extended medium, the mean propagator is easily found to be given by a Gaussian [8,15,27]

$$P(z,t) = (4\pi Dt)^{-1/2} \exp\left[-z^2/(4Dt)\right] \tag{4}$$

with D denoting the self-diffusivity. Inserting Equation (4) into Equation (3) yields

$$\psi = \frac{S(m,t)}{S(0,t)} = \exp\left(-\gamma^2 g^2 \delta^2 Dt\right) = \exp\left(-\frac{1}{6}\gamma^2 g^2 \delta^2 \langle r^2(t)\rangle\right) \tag{5}$$

where, with the latter equality, we have made use of Equation (1). For the PFG NMR signal attenuation we have, moreover, introduced the common notation ψ. The time and space scales relevant for Equations (3) and (5) are typically of the order of milliseconds and micrometers. Measurement of particularly small displacements $\langle r^2(t)\rangle^{1/2}$ is seen to require particularly large pulsed field gradient intensities $g\delta$. Since the amplitude g of the field gradient pulses is limited by the constructional details of the PFG NMR probe, the maximum value of the pulse width δ decides the minimum displacements accessible by PFG NMR. In conventional PFG NMR, however, the pulse width δ is limited by the relaxation time T_2^{echo} of transverse magnetization. It is this component of nuclear magnetization from which, with Equation (2), the space-dependent phase spreading and, as a consequence, signal attenuation by molecular displacements has been shown to originate. This decay in transverse magnetization, however, is notably slowed down for samples sufficiently quickly rotating, with a spinning axis oriented under an angle of $\theta_{mas} = \text{arc} \cos 3^{-1/2} \approx 54.7°$ with reference to the external magnetic field.

This option of enhancing the time interval over which magnetic field gradients may be applied is exploited in MAS PFG NMR [19,20,22,28–30]. Figure 1 introduces the experimental arrangement and the pulse program used in the measurement.

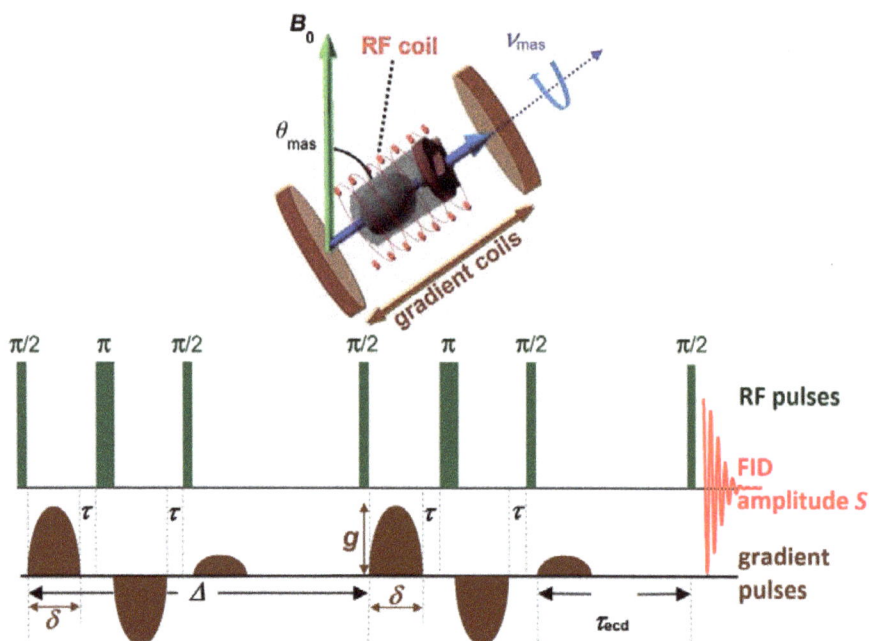

Figure 1. A representation of the MAS (magic angle spinning) design with two gradient coils on the top and on the bottom of the MAS stator in a high-resolution wide-bore MAS NMR (nuclear magnetic resonance) probe at the top. Radio frequency (RF) and gradient pulse scheme of the MAS pulsed field gradient (PFG) NMR experiment is shown below. Parameters are diffusion time Δ and gradient pulse width δ. The gradient pulse amplitude is denoted as g, the eddy current delay as τ_{ecd}, and the inter-gradient delay as τ. Two weak spoiler gradient pulses average undesirable coherences [20].

The arrangement of the NMR sample tube containing the nanoporous host material and the guest molecules is shown in the top of Figure 1. We note the "magic-angle" of about 54.7° between the spinning axis and the direction of the magnetic field. The direction of the spinning axis coincides with that of the field gradient so that the local magnetic field within the sample (and thus, with Equation (2), the local rate of rotational motion) remains unaffected by sample rotation. Molecular displacements recorded by analyzing the signal attenuation under the influence of the gradient pulses may thus indeed be attributed to diffusion phenomena within the sample.

The pulse sequence shown in the bottom of Figure 1 includes a number of differences in comparison with the basic version of PFG NMR as initially introduced. We note that the field gradient pulses are of sinusoidal, rather than of rectangular shape. This facilitates switching of the current used for generation of the field gradients and diminishes the occurrence of eddy currents in the radio frequency (RF) coil, which might interfere with the NMR signal. Serving the same purpose, pairs of opposing field gradient pulses (generated by opposing currents) rather than single ones are applied. The RF "π" pulse appearing in between such a pair gives rise to a rotation of all spins by 180° so that the "effective" gradients acting on the spins are identical. The initial π/2 pulse is recognized as the starting point of the experiment when the equilibrium magnetization showing in the direction of the constant magnetic field B_0 is turned, by 90°, into the plane perpendicular to B_0. We note that with the last π/2 pulse, magnetization is once again turned into the plane perpendicular to B_0, giving rise to the NMR signal S (the initial value of the "free induction decay"). Signal attenuation for the pulse sequence shown in Figure 1 is given by the relation [20]

$$\psi = \frac{S(m,t)}{S(0,t)} = \exp\left(-\frac{16}{\pi^2}\gamma^2 g^2 \delta^2 Dt\right) = \exp\left(-\frac{8}{3\pi^2}\gamma^2 g^2 \delta^2 \langle r^2(t)\rangle\right). \tag{6}$$

The time t also includes, in addition to Δ, corrections due to finite pulse widths. The meaning of the gradient pulse width δ (see Figure 1) is for sine-shaped alternating pulses and, hence, changed in comparison with the basic experiment with two rectangular field gradient pulses, giving rise to a slightly different pre-factor in the exponents in Equations (5) and (6).

By combining the application of field gradient pulses with fast sample spinning, the time interval δ over which magnetic field gradients may be applied can be notably enhanced, in some cases over several orders of magnitude. In this way, molecular displacements which are too small to be observable by conventional PFG NMR become accessible by direct observation. This possibility is provided by the dramatic enhancement of the transverse nuclear magnetic relaxation time under the conditions of magic-angle spinning, $T_2^{\text{MAS echo}}$, in comparison with T_2^{echo}, the value observed without MAS. Enhancement of the transverse nuclear magnetic relaxation time leads, simultaneously, to decreasing line widths so that different chemical compounds can be distinguished on the basis of their NMR spectra.

Magnetic field gradient pulses in conventional PFG NMR can today be operated with values above 20 T/m. Schlayer et al. [31], e.g., achieved a value of 37 T/m with a 100 A power supply. The design with two gradient coils on the top and on the bottom of the MAS stator of a Bruker probe reaches about 0.5 T/m with a 10 A power supply. Narrow-bore MAS designs without gradient coils which are located in the imaging gradient tube of a wide-bore magnet achieve, with three 60 A power supplies, about 2 T/m.

In spite of these low gradient intensities, application of MAS PFG NMR proves to be the method of choice in quite a number of situations including, notably, selective multicomponent diffusion measurement in such systems where the spectra recorded by conventional PFG NMR fail in providing the resolution necessary for differentiating between various guest components. The subsequent section is going to highlight these advantages with a number of show cases.

While, as a matter of course, gravity in a non-rotating sample coincides with that in our natural surroundings, it becomes some hundred-thousand-fold higher in a rotating sample at 10 kHz. It is worthwhile mentioning, therefore, that molecular microdynamics within the spinning sample tubes remains essentially unaffected by this high rotational frequency, since intermolecular forces are much larger than the influence of gravity. Thus it could be observed by ^{11}B MAS NMR spectroscopy [32] that the parameters of crystallization upon super-gravity downgrade by less than one order of magnitude. For adsorption or catalysis in porous materials, we do not know of any report about the influence of supergravity by MAS. In cases with availability of both PFG NMR and MAS PFG NMR results, like for *n*-alkanes adsorbed in silicalite-1 [33], no significant difference could be found. It is true, however, that operation with a high-velocity gas stream for ensuring high-speed sample rotation up to tens of kHz gives rise to a temperature difference between the "rotor" (with the fused sample tube) and the bearing air (whose temperature is accessible to direct measurement). This requires a separate temperature calibration. It is commonly based on measurement of the ^{207}Pb MAS NMR signal of Pb(NO$_3$)$_2$ within the sample tube whose temperature dependence is well known [34–36]. Temperatures within rotor and sample were thus found to exceed the temperature in the stator by about 10 K for a 4 mm rotor spinning at 10 kHz, with the option of a slight temperature gradient versus the sample.

3. Diffusion Measurement by MAS PFG NMR

3.1. Mixture Diffusion in Microporous Materials

Figure 2 introduces the potential of MAS PFG NMR for investigating mixture diffusion with a particularly simple case. On considering a 1:1 molar mixture of ethene and ethane (each with a loading of two molecules per cage), one benefits from the exceptional situation that in either compound all hydrogens are chemically equivalent. They are, therefore, "shielded" by identical "electronic clouds" so that the shift in the local field (and, hence, via Equation (2), in the frequency of the NMR signal) in comparison with the externally applied one is identical. Both molecules therefore give rise to different signals. These two lines are, as to be seen on the left in Figure 2, well separated from each other so that one may easily record the attenuation of the intensity (that is, the area under the line (the "band")) of each of them as a function of the gradient intensity. The logarithmic representation of this attenuation as a function of the squared gradient intensity in Figure 2 is found to be in nice agreement with the expected behavior as predicted by Equation (6). The resulting diffusivities are 1.21×10^{-10} m^2 s^{-1} for ethene and 0.27×10^{-10} m^2 s^{-1} for ethane.

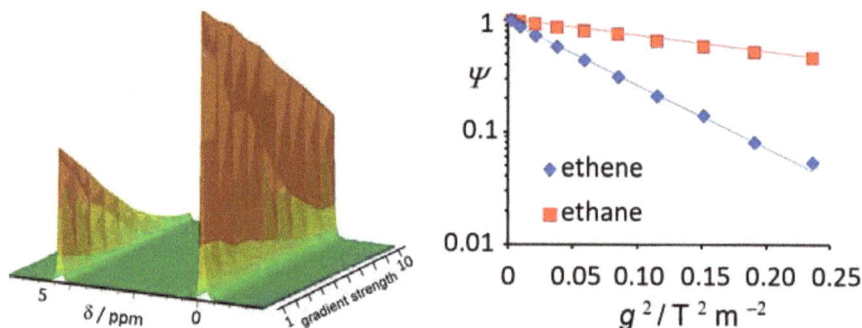

Figure 2. On the left is a 2D presentation of the ^{1}H MAS PFG NMR signal decay with 10 steps from 0.05 to 0.5 T/m linearly increasing strength of the gradient pulses for a ZIF-8 sample loaded with two ethene and two ethane molecules per cavity measured at 363 K. The logarithmic decay as a function of the squared gradient magnitude is presented on the right [37].

Given the similarity in the molecular critical diameters (0.39 nm for ethene, 0.40 nm for ethane [38]) the substantial difference in the diffusivities is quite remarkable. It can, however, be associated with the fact that these diameters significantly exceed the diameter of about 0.34 nm of the "windows" between adjacent cages as determined by X-ray diffraction analysis [39]. Under such conditions, already minor changes in size of the guest molecules may indeed be expected to dramatically affect their diffusion properties. With "windows" smaller than the diffusing molecules, diffusion is only possible in sufficiently flexible host lattices. This, however, is among the main features of metal-organic frameworks (MOFs) quite in general and of ZIF-8 as one of its representatives in particular. High resistivity and durability in comparison with other MOFs have made it an interesting topic of research [40,41]. Studies include extensive diffusion measurements based on microimaging via IR microscopy [42]. For ethene and ethane the thus-obtained results were found to be in nice agreement with the diffusivities obtained by MAS PFG NMR [38]. IR microimaging did, moreover, allow an extension of the measurements to longer chains and, thus, to lower diffusivities.

As a common feature of all these systems, molecular jumps through the windows between adjacent cages can be considered as the rate-controlling step in molecular propagation. By adopting classical transition state theory (TST, with the jump through the window as the "activated state" [43–45]), the concentration dependence of the jump rates in such systems and, hence, of the self-diffusivity may be shown to be proportional to the ratio $p(c)/c$ between the guest pressure in the surrounding atmosphere and the guest concentration under equilibrium [46]. This prediction has been confirmed in [46] with the diffusivity data obtained by recording molecular uptake and release of a large spectrum of guest molecules (ethene, ethane, propene, propane, methanol, and ethanol) on ZIF-8 [38]. Such measurements benefit from the fact that in uptake and release measurements there exists, essentially, no lower limit on the accessible diffusivities. Moreover, pressure variation in the surrounding atmosphere allows a straightforward and most accurate variation of the guest concentration. Such a possibility does not exist in MAS PFG NMR measurements. Although also here, in the course of sample preparation, very accurate guest concentrations may be attained, these concentrations tend to decrease during sample fusing, as the indispensable last step in sample preparation for enabling sample spinning during the measurement. Although it is true that guest concentrations may be determined quite accurately from their signal intensity, there is an unavoidable (and, with factors up to 2, quite substantial) scattering in the attained concentrations, which significantly complicates any systematic investigation of concentration dependencies. The view of TST on molecular propagation helps in rationalizing why the diffusivities of the individual molecules under multicomponent adsorption (as considered in the MAS PFG NMR measurements) and single-component adsorption (microimaging) do essentially coincide since the jump rate from cage to cage can be expected to be only marginally affected by the composition of the cage population.

A totally different microdynamic situation is reflected by the self-diffusivity data shown in Figure 3 for a series of *n*-alkanes and *n*-alkenes in zeolite silicalite-1 [33]. Pore diameters in silicalite-1 are between 0.51 nm and 0.56 nm and thus notably exceed the critical diameters of the guest molecules considered. The slight difference in the critical diameters of alkenes (ethene) and alkanes (ethane), which has given rise to a substantial difference of the diffusivities in narrow-pore ZIF-8, is now of essentially no influence anymore. Diffusivities of *n*-alkanes and *n*-alkenes in silicalite-1 are found to essentially coincide. For a given number of guest molecules we note, correspondingly, that the diffusivity remains essentially the same under the conditions of single-component and mixture adsorption.

It is shown in Figure 3 that the diffusivities decrease with increasing chain length, also following the pattern well known from previous PFG NMR studies [47–49]. Increase in chain length by one CH_2 element is found to cause a decrease in the diffusivity by a factor of about 0.4.

The measurements reported in [33] did also agree with previous PFG NMR studies [47–49] in the finding (not shown in Figure 3) that the diffusivity monotonically decreases with increasing loading. This behavior also notably deviates from the patterns observed with ZIF-8 where the diffusivities could be found to both decrease and increase with loading [46]. The monotonic decay with increasing loading is easily attributed to the increase in friction between the guest molecules in the more open pore structure. One may observe a remarkably uniform dependence following a relation $d \lg D / d L = -0.55 \pm 0.10$ [33] where L is a measure of the loading, normalized with the understanding that $L = 1$ refers to one molecule per channel intersection (or to four molecules per unit cell).

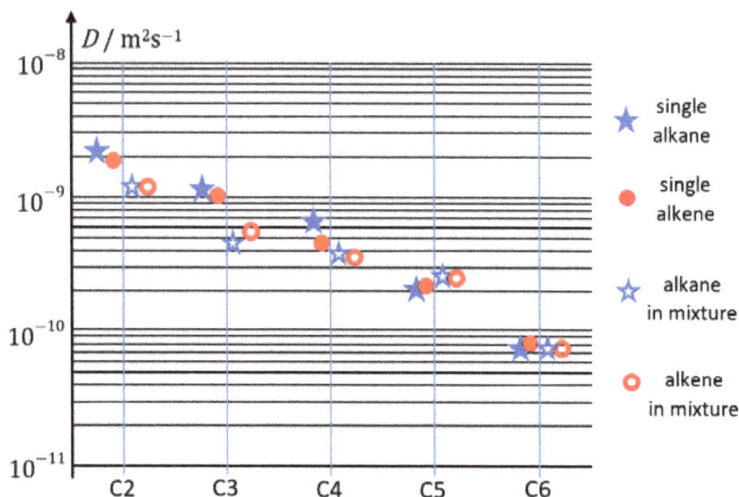

Figure 3. Self-diffusion coefficients of *n*-alkanes, alkenes, and one-to-one alkane/alkene mixtures in silicalite-1 at the temperature of 313 K for a total loading of about one molecule per crossing or less [33].

3.2. Complex Formation in Acetone–n-Alkane Mixtures Revealed via MAS PFG NMR

The properties of *n*-alkanes are known to exhibit minor deviations from a strict monotonic variation with increasing chain lengths. These deviations appear, e.g., in an oscillation in the melting points [50], attributed to an oscillation in the intermolecular forces, with stronger ones for the even-numbered *n*-alkanes [51,52]. This oscillation in interaction has been expected to be as well observable in complex formation with other compounds, notably including ketones [53]. Figure 4 illustrates the favorable conditions offered by MAS PFG NMR for selective diffusion studies with acetone–*n*-alkane mixtures.

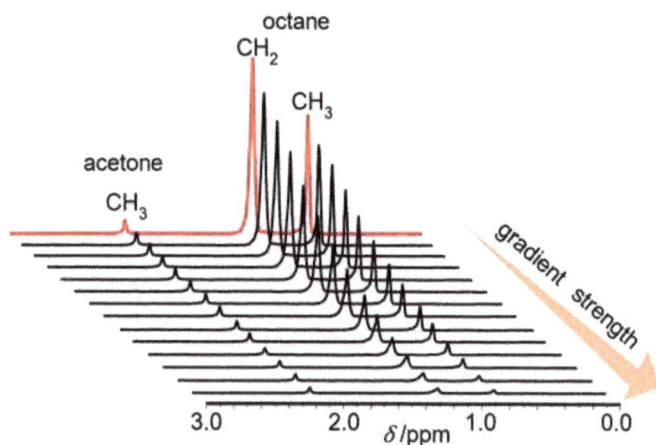

Figure 4. Stack plot of ^{1}H MAS PFG NMR spectra of a 1:10 acetone–*n*-octane mixture adsorbed in porous glass [54].

Figure 5 provides a comparison of the diffusivities of acetone and of various *n*-alkanes (from *n*-hexane up to *n*-nonane) in their binary mixture within a narrow-pore silica gel. In Figure 5a, the mean pore size is about 4 nm and the acetone/*n*-alkane molar ratio is 1:10. Figure 5b shows the results within a larger-pore silica gel with a mean pore size of about 10 nm and an acetone/*n*-alkane molar ratio of 1:20.

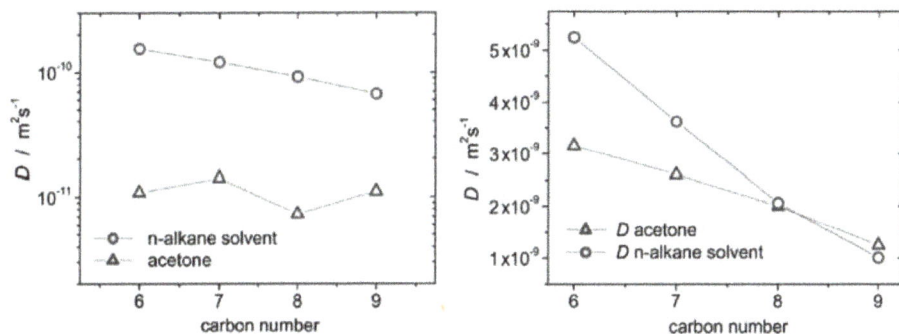

Figure 5. Diffusivities of the two components of the acetone–*n*-alkane mixtures in the narrow-pore silica gel on the left and in the large-pore silica gel on the right in dependence on the *n*-alkane chain length [54].

As a main difference between the two representations, acetone diffusivities in narrow-pore silica gel are found to exhibit a clear oscillation (beyond the uncertainty in measurement which is of the order of the size of the symbols), while there is no similar effect in the large-pore silica gel. We may imply that similarity in pore sizes and in the extension of the complexes formed by the interacting acetone and *n*-alkane molecules amplifies the effect of interaction oscillation. Thus, in complete agreement with the tendency observed with the oscillation of the melting points, even-numbered alkanes are found to give rise to a perceptibly smaller diffusivity, corresponding with larger complex sizes and, thus, stronger mutual interaction. That diffusion in the narrow-pore silica gel is found to be slowed down in comparison with diffusion in the large-pore sample in general, may be easily referred to an associated increase in tortuosity [55–58]. This effect is well known to increase with the size of the diffusing particle [59–61] which explains why the reduction in diffusivity in the narrow-pore silica gel is even more pronounced with acetone than with the *n*-alkanes (since, in contrast to acetone, only a minor part of the *n*-alkanes contributes to the complex formation).

3.3. Diffusion Studies with Nematic Liquid Crystals in Confining Pore Spaces

As considered in the previous section for neat liquids, pore space confinement is expected to also affect the internal dynamics in liquid crystals. As a rule, however, transverse nuclear magnetic relaxation in liquid crystals is known to be dramatically accelerated in comparison with the neat liquid [62,63], excluding the application of conventional PFG NMR for diffusion measurement. As an example of the application of MAS PFG NMR diffusion studies, Figure 6 shows a typical example of the MAS PFG NMR signal attenuation curves (Equation (6)) and an Arrhenius plot of the resulting diffusivities [21].

The measurements were performed with 4′-pentyl-4-cyanobiphenyl (5CB), a nematic liquid crystal commercially available (Merck Ltd., Poole, UK). The transition temperature between solid and the nematic phase is 297.2 K and the isotropization temperature is 308.5 K. Measurements were performed with both the bulk and with the liquid crystal confined within a Bioran porous glass with mean pore diameters of 30 nm and 200 nm. The data shown in Figure 6b are seen to comprise the diffusivities in both the states of nematic crystallinity (low temperatures) and isotropy (high temperatures). Sample heterogeneities and temperature variation over the sample (see Section 2) impeded sample equilibration for the temperatures in between so that in this range no measurements were performed.

While in the isotropic state, diffusivities are seen to decrease with increasing confinement, i.e., with the highest diffusivities in the bulk and with diffusivities in Bioran glasses with pore diameter of 30 nm below those measured for pore diameters of 200 nm, the effect of confinement is reversed in the nematic phase. Here, the diffusivities in the larger pores are seen to be below those in the smaller ones. This dependency reversal did already appear in Figure 6a where at 299 K, i.e., within the nematic phase, the decay of the attenuation curve (being proportional to the diffusivity—see Equation (6)) is seen to decrease with increasing pore size, while it increases at 334 K, i.e., in the isotropic state. A significant difference in the microdynamics of the two phases does as well appear in the activation energies of diffusion where, with (38 ± 6) kJ mol^{-1}, the value determined for the nematic phase notably exceeds that of (27 ± 5) kJ mol^{-1} determined for the isotropic state. Simultaneously, the pre-exponential factor of the diffusivity in the nematic phase is seen to notably exceed that in the isotropic phase, following the correlation pattern of the compensation effect. Further enhancement in measurement accuracy is needed for a clear distinction between these two influences, promoting the elaboration of expedient model conceptions for explaining the observed diffusion anomalies.

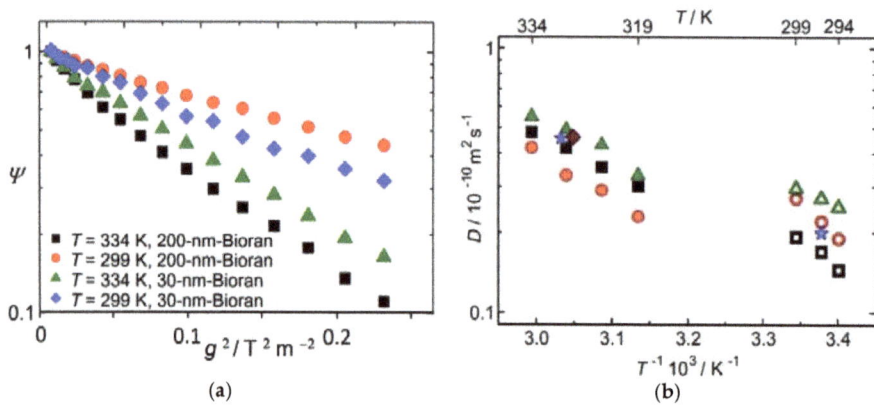

Figure 6. (a) ^{1}H MAS PFG NMR spin echo attenuation, ψ, of the 4'-pentyl-4-cyanobiphenyl (5CB) confined in Bioran glasses with pore diameters of 30 and 200 nm. (b) Temperature dependence of the diffusion coefficient D of bulk 5CB (▲) and of 5CB confined in Bioran glasses with pore diameter of 30 nm (●) and 200 nm (■). The diffusivities were measured at 10 kHz rotation frequency, except three values for 200 nm Bioran glass, which were measured at 5 kHz (★) and without rotation (◆). Empty symbols correspond to the nematic isotropic phase and full symbols correspond to the isotropic phase [21].

3.4. Water Diffusion in Lithium-Exchanged Low-Silica X-Type Zeolites

While in Sections 3.1 and 3.2, resolution enhancement by MAS PFG NMR was exploited for selective diffusion measurement of different components, it may also, equally importantly, be applied for studying the transport patterns of one and the same molecular species in different surroundings. Similarly as in micro-mesoporous hierarchical host materials [64–70] where guest molecules are known to propagate with different diffusion rates depending on their current position (i.e., within micro- or transport pores), overall mass transfer in complex systems is often found to occur in a sequence of subsequent displacements covered with different diffusion rates. First-order simulations of mass transfer in such systems are often based on the two-region approach of PFG NMR [71] where the diffusants are implied to diffuse at two different rates $D_{1,2}$, with the respective probabilities $p_{1,2}$ and the mean life times $\tau_{1,2}$ in either of these states [72–77]. Data analysis is significantly facilitated if the two states of mobility give rise to two different signals [14,78].

Figure 7 introduces a system where such a possibility could be exploited. It shows, on the right-hand side, the electron-microscopic picture of a low-silicon zeolite of type X (LSX) [79,80]. The individual particles are seen to be polycrystalline agglomerates. Diffusion studies by conventional PFG NMR have shown that the boundaries between the individual crystallites act as transport resistances for the water molecules, just as for the lithium cations whose diffusivities could, within these materials, be directly measured via PFG NMR for the very first time [81,82]. The schematics in the bottom of the left side show the zeolite pore structure, with the faujasite cage as the main storage place for the water molecules (accommodating about 30 water molecules) and the sodalite cage (with about 4 water molecules). Exchange between the various faujasite cages appears at an extremely fast rate (with mean lifetimes of the order of nanoseconds) while water in the sodalite cages remains kept over tens of milliseconds [81]. As an effect of its small exchange rates and, moreover, an extremely short transverse magnetic relaxation time, sodalite water does not contribute to the measurement of water diffusion by conventional PFG NMR measurements. Figure 8 illustrates that the situation becomes totally different under the conditions of MAS PFG NMR [83].

We do now note a clearly visible signal stemming also from the water kept within the sodalite units. It is well separated from a notably larger line which is caused by the water molecules in the large faujasite cages. These water molecules are, via the gas phase, in fast exchange within the whole bed of host particles. At a temperature of 313 K, water diffusion within the particles and through the bed is seen to give rise to a signal attenuation. Following the general relationship as provided by Equation (6), signal attenuation increases with increasing observation time. There is no signal attenuation visible for the sodalite water, in complete agreement with our understanding that the water molecules are kept caught within the individual sodalite cages.

At 373 K and for an observation time of 100 ms, signal attenuation is also observed with the water molecules contained in the sodalite cages. This indicates that now, at the increased temperature and for the largest observation time, at least a part of the water molecules within the sodalite cages have been replaced by molecules from the surroundings which, on their way through the sodalite cages and the intercrystalline space, have covered long-enough diffusion pathways giving rise to the observed attenuation.

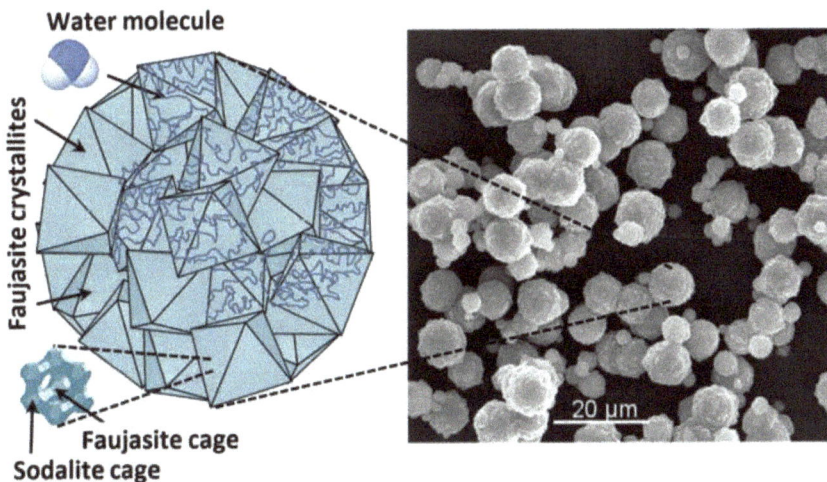

Figure 7. Electron microscopic picture of the LSX zeolite material under study (**right**) and schematics of the structure of the individual zeolite particles, jointly with a model of the pore elements, the "faujasite cage" surrounded by eight "sodalite cages", with the latter ones acting as "traps" on the diffusion path of the water molecules (**left**). Figure 1 from Ref. [83] with permission.

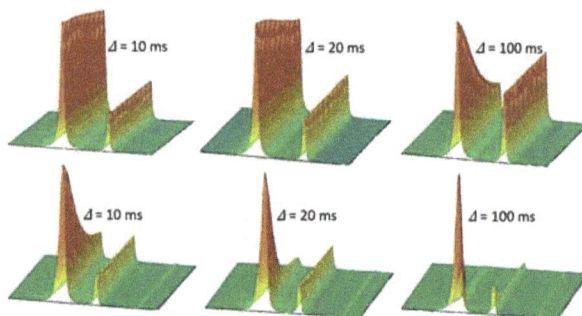

Figure 8. Stack plots of the ^{1}H MAS PFG NMR signal attenuation for water in 100Li-LSX at 313 K (**top**) and 373 K (**bottom**) for the indicated observation times. Chemical shifts increase from 7 ppm to 2 ppm (from left to right), and field gradient amplitudes from 0.011 T m^{-1} to 0.486 T m^{-1}. Figure 4 from Ref. [83] with permission.

Data analysis on the basis of the two-region model of PFG NMR diffusion measurement [14,71,78] yields, as a best fit to the experimentally determined attenuations shown in Figure 8, the dependencies of Figure 9 [83]. We note that this approach yields particularly satisfactory agreement for sodalite water whose mean lifetime within the sodalite cages at 373 K may thus be estimated to amount to 60 ms. Significant differences are observed between measurement and model approaches when considering the water molecules out of the sodalite cages. This may, however, be easily understood as an immediate consequence of the simplifications inherent to the model: while mass transfer outside of the sodalite cages is quite a complex phenomenon including molecular displacements both within the individual host particles and through interparticle space, additionally subject to resistances at the interface between the individual crystallites [82], in the two-region model of PFG NMR, all these influences are lumped together, being represented by one ("effective") diffusivity.

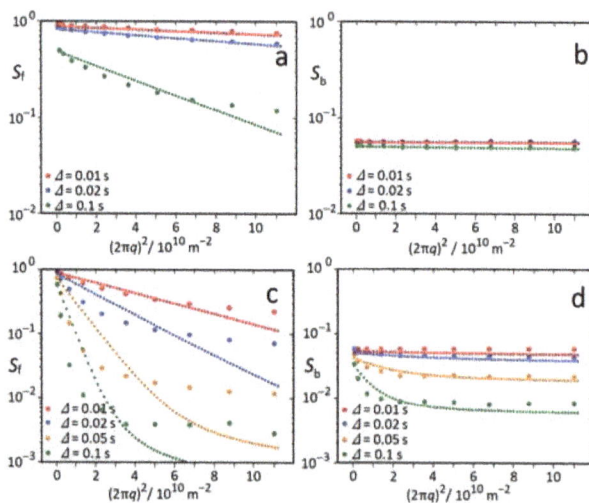

Figure 9. Intensity of the ^{1}H signal (balls) in MAS PFG NMR signal attenuation experiments for water molecules outside (left: (**a,c**)) and inside (right: (**b,d**)) the sodalite cages as a function of the (squared) field gradient pulse intensity in zeolite 100Li-LSX at 313 K (**a,b**) and 373 K (**c,d**) for different observation times Δ and comparison with the best fit (lines) as resulting from application of the two-region model. Figure 6 from Ref. [83] with permission.

3.5. Proton Mobility in Functionalized Mesoporous Materials

Increasing environmental pollution and ebbing away of conventional sources have given rise to steadily increasing efforts searching for alternative energy sources and for higher efficiencies in energy use. These activities include, in particular, the development and exploitation of novel fuel cell technologies. Performance improvements and quests for multifunctionality in application have led to increasingly complex structures, such as to be typically encountered with Polymer Electrolyte Membrane Fuel Cells [84–86]. Variety in charge carriers and the diffusion pathways covered by them make MAS PFG NMR a measuring technique of choice for an in-depth exploration of the elementary steps of charge carrier mass transfer in such systems.

For exemplifying the thus-attainable information, we refer to Figure 10. It shows the (highly resolved) proton spectra and their attenuation with increasing pulsed field gradient intensity for two functionalized siliceous materials. They are candidates for a composite fuel cell membrane consisting of a water-storing solid material and the common polymer material [87,88]. Both materials have been synthesized by co-condensation using tetraethylorthosilicate (TEOS) and mercaptopropyl-trimethoxysilane (MPMS) [87,88] and bear coinciding functional groups ($\equiv$Si–CH_2–CH_2–CH_2–SO_3H). They are linked to the siliceous host framework and possess an acidic hydrogen, representing the crucial "ingredient" for the use of such materials in fuel cells. The host particles of the specimens under study do, however, notably differ in size, which turns out to make, in the present context, the most relevant difference.

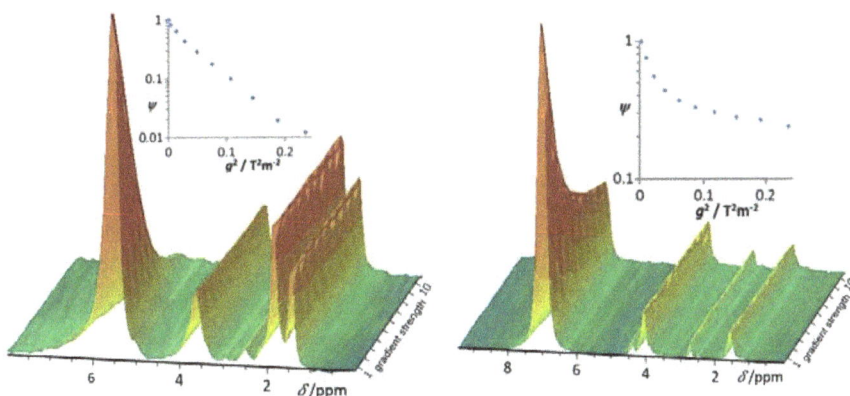

Figure 10. Two-dimensional presentations of the signal decay with linearly increasing strength of the gradient pulses and insets showing the semilogarithmic plot of the signal decay as a function of the squared strength of the gradient pulses. The ^{1}H MAS NMR experiments were performed at 100 °C. Results for sample MCM-48 and sample KIT-6 are presented on the left and right, respectively [88].

From the given material description, we easily recognize, in the various lines between 1 ppm and 4 ppm shown in Figure 10, the contribution of the protons in the CH_2 groups which remain unaffected by the field gradient pulses. The line at about 6 ppm, however, is clearly seen to be subject to signal attenuation by diffusion. It may be attributed, therefore (and also in total agreement with our knowledge about the chemical shift with protons in such systems [25,81,82,87,88]), to the acidic protons and to the protons of the water molecules within the host material, being in mutual fast exchange.

Quantitative analysis yields, for the signal attenuation (inserts in Figure 10) in the case of functionalized MCM-48, purely exponential decay as required by Equation (6), yielding a value of 1.9×10^{-8} m^2 s^{-2}. Since the mean particle size of the MCM-48 material under study (<0.1 µm) is significantly exceeded by the mean diffusion path length considered in this study (about 34 µm for an observation time of 10 ms), the thus-attained diffusivities no longer reflect any features of intracrystalline diffusion. They are rather given by the product of the relative number of molecules outside of the individual particles and their diffusivities. These "long-range" diffusivities are indeed known to be able to attain extremely large values, possibly even exceeding those in the liquid [89,90]. Estimates on the basis of the Nernst-Einstein equation by impedance spectroscopy (i.e., by conductivity measurements) for the identical system did correspondingly yield, with a value of 1.65×10^{-11} m^2 s^{-2}, notably smaller intracrystalline diffusivities [88].

Measurements with KIT-6 could be performed with notably larger host particles, with particle sizes up to 50 μm. Signal attenuation for the mobile protons is now found to be nicely approached by the superposition of two exponentials of the type of Equation (6). The fast decay is, once again, attributed to those water molecules which, during the observation time, have left their particles and could thus "benefit" from the high mobility in intercrystalline space. Their "effective" diffusivity amounts to 4.5×10^{-9} m^2 s^{-2} which is, not unexpectedly, of a similar order of magnitude as already observed for long-range diffusion in MCM-48. Now, however, a substantial amount of the protons (water molecules) is seen to remain exclusively within the intraparticle pore space, giving rise to an intraparticle diffusivity of 2.3×10^{-10} m^2 s^{-2}. This value is indeed of the same order as the result of impedance spectroscopy which yielded a diffusivity of 3.04×10^{-10} m^2 s^{-2} [88], in complete agreement with our understanding.

4. Conclusions

Experimental techniques are, by their very nature, only applicable to certain systems and under certain conditions. In the course of time, these limitations may become increasingly cumbersome—notably with powerful techniques, in view of their rich potential. As a consequence, further research work aiming at a release of these limitations is stimulated. It is often accompanied by an improvement of the performance and an enlargement of the applicability of these techniques, which have become possible based on these developments.

All these features may be recognized in following the introduction and further development of the pulsed field gradient (PFG) technique of NMR. Its application to studying molecular diffusion in complex media, notably in nanoporous host-guest systems, has provided us with a number of surprises, cumulating in a novel view on mass transfer in a number of nanoporous materials like zeolites in comparison with our understanding a couple of decades ago. In addition, this technique is subject to strict limitations, notably concerning the displacements of the molecules under study. Sensitivity towards displacements is—via transverse nuclear magnetic relaxation—particularly limited for systems of low mobility. Difficulties (with the measurement of displacements) are thus found to arise preferably in such cases where they are particularly harmful, namely with systems of particularly low mobility, with particularly low molecular shifts. In this mutual affectedness, one may recognize one of the versions of Murphy's law.

A way out of this conflict is provided by combining magic-angle spinning (MAS) with PFG NMR diffusion measurement. The present review introduces the fundamentals of this approach and illustrates the novel potential with a number of cases. They include diffusion measurements with systems of extremely low mobility just as with multicomponent systems where, for the first time, diffusion coefficients of the various components have become accessible. MAS PFG NMR has thus contributed to a notable broadening of the scope so far considered in PFG NMR diffusion studies, including the investigation of fuel cells and of oscillations in the diffusion patterns of *n*-alkane–ketone mixtures under confinement. A most decisive breakthrough has been achieved in MAS PFG NMR diffusion studies with water in low-silicon lithium-exchanged zeolite X, where even water molecules confined to the sodalite cages have been made accessible to direct diffusion measurement. This type of measurement has all potential for becoming applicable to materials with hierarchically organized pore spaces in general, notably including mesoporous zeolites [56,91,92]. In-depth studies of molecular exchange between the various subspaces are among the great challenges of future investigations aiming at quality enhancement and exploration of the elementary steps behind the overall performance of such materials.

Acknowledgments: We are grateful to all co-authors of our MAS PFG NMR studies which were presented in this review. Financial support by the German Science Foundation (HA 1893-16) and Fonds der Chemischen Industrie is gratefully acknowledged.

Conflicts of Interest: The authors declare no conflict of interest.

References

1. Bunde, A.; Caro, J.; Kärger, J.; Vogl, G. (Eds.) *Diffusive Spreading in Nature, Technology and Society*; Springer International Publishing: Cham, Switzerland, 2018.
2. Ruthven, D.M.; Farooq, S.; Knaebel, K.S. *Pressure Swing Adsorption*; VCH: New York, NY, USA, 1994.
3. Schüth, F.; Sing, K.S.W.; Weitkamp, J. (Eds.) *Handbook of Porous Solids*; Wiley-VCH: Weinheim, Germany, 2002.
4. Ertl, G.; Knözinger, H.; Schüth, F.; Weitkamp, J. (Eds.) *Handbook of Heterogeneous Catalysis*, 2nd ed.; Wiley-VCH: Weinheim, Germany, 2008.
5. Van de Voorde, M.H.; Sels, B. (Eds.) *Nanotechnology in Catalysis: Applications in the Chemical Industry, Energy Development, and Environment Protection*; Wiley-VCH: Weinheim, Germany, 2017.
6. Chen, N.Y.; Degnan, T.F.; Smith, C.M. *Molecular Transport and Reaction in Zeolites*; VCH: New York, NY, USA, 1994.
7. Kärger, J.; Freude, D. Mass transfer in micro- and mesoporous materials. *Chem. Eng. Technol.* **2002**, *25*, 769–778. [CrossRef]
8. Kärger, J.; Ruthven, D.M.; Theodorou, D.N. *Diffusion in Nanoporous Materials*; Wiley-VCH: Weinheim, Germany, 2012.
9. Kärger, J. Diffusionsuntersuchung von Wasser an 13X- sowie 4A- und 5A-Zeolithen mit Hilfe der Methode der gepulsten Feldgradienten. *Z. Phys. Chem.* **1971**, *248*, 27–41. [CrossRef]
10. Kärger, J.; Vasenkov, S. Quantitation of diffusion in zeolite catalysts. *Microporous Mesoporous Mater.* **2005**, *85*, 195–206. [CrossRef]
11. Valiullin, R. *Diffusion NMR of Confined Systems*; Royal Society of Chemistry: Cambridge, UK, 2016.
12. Kärger, J.; Heink, W. The Propagator Representation of Molecular Transport in Microporous Crystallites. *J. Magn. Reson.* **1983**, *51*, 1–7. [CrossRef]
13. Callaghan, P.T. *Principles of NMR Microscopy*; Clarendon Press: Oxford, UK, 1991.
14. Price, W.S. *NMR Studies of Translational Motion*; Cambridge University Press: Cambridge, UK, 2009.
15. Kärger, J. (Ed.) *Leipzig, Einstein, Diffusion*, 3rd ed.; Leipziger Universitätsverlag: Leipzig, Germany, 2014.
16. Zhdanov, S.P.; Khvostchov, S.S.; Feoktistova, N.N. *Synthetic Zeolites*; Gordon and Breach: New York, NY, USA, 1990.
17. Kärger, J. Measurement of diffusion in zeolites—A never ending challenge? *Adsorption* **2003**, *9*, 29–35. [CrossRef]
18. Gaede, H.C.; Gawrisch, K. Multi-dimensional pulsed field gradient magic angle spinning NMR experiments on membranes. *Magn. Reson. Chem.* **2004**, *42*, 115–122. [CrossRef] [PubMed]
19. Pampel, A.; Zick, K.; Glauner, H.; Engelke, F. Studying lateral diffusion in lipid bilayers by combining a magic angle spinning NMR probe with a microimaging gradient system. *J. Am. Chem. Soc.* **2004**, *126*, 9534–9535. [CrossRef] [PubMed]
20. Fernandez, M.; Kärger, J.; Freude, D.; Pampel, A.; van Baten, J.M.; Krishna, R. Mixture diffusion in zeolites studied by MAS PFG NMR and molecular simulation. *Microporous Mesoporous Mater.* **2007**, *105*, 124–131. [CrossRef]
21. Romanova, E.E.; Grinberg, F.; Pampel, A.; Karger, J.; Freude, D. Diffusion studies in confined nematic liquid crystals by MAS PFG NMR. *J. Magn. Reson.* **2009**, *196*, 110–114. [CrossRef] [PubMed]
22. Gratz, M.; Hertel, S.; Wehring, M.; Stallmach, F.; Galvosas, P. Mixture diffusion of adsorbed organic compounds in metal-organic frameworks as studied by magic-angle spinning pulsed-field gradient nuclear magnetic resonance. *New J. Phys.* **2011**, *13*, 045016. [CrossRef]
23. Kimmich, R. *NMR Tomography, Diffusometry, Relaxometry*; Springer: Berlin, Germany, 1997.

24. Valiullin, R.; Kärger, J. Confined Fluids: NMR Perspectives on Confinements and on Fluid Dynamics. In *Diffusion NMR of Confined Systems*; Valiullin, R., Ed.; Royal Society of Chemistry: Cambridge, UK, 2016; pp. 390–434.
25. Harris, R.K.; Becker, E.D.; De Menezes, S.M.C.; Goodfellow, R.; Granger, P. NMR Nomenclature. Nuclear Spin Properties and Conventions for Chemical Shifts–(IUPAC Recommendations 2001). *Pure Appl. Chem.* **2001**, *73*, 1795–1818. [CrossRef]
26. Callaghan, P.T.; Coy, A.; MacGowan, D.; Packer, K.J.; Zelaya, F.O. Diffraction-like effects in NMR diffusion studies of fluids in porous solids. *Nature* **1991**, *351*, 467–469. [CrossRef]
27. Cussler, E.L. *Diffusion: Mass Transfer in Fluid Systems*, 3rd ed.; Cambridge University Press: Cambridge, UK, 2009.
28. Rousselot-Pailley, P.; Maux, D.; Wieruszeski, J.M.; Aubagnac, J.L.; Martinez, J.; Lippens, G. Impurity detection in solid-phase organic chemistry: Scope and limits of HR MAS NMR. *Tetrahedron* **2000**, *56*, 5163–5167. [CrossRef]
29. Schröder, H. High resolution magic angle spinning NMR for analyzing small molecules attached to solid support. *J. Comb. Chem.* **2003**, *6*, 741–753. [CrossRef]
30. Pampel, A.; Fernandez, M.; Freude, D.; Kärger, J. New options for measuring molecular diffusion in zeolites by MAS PFG NMR. *Chem. Phys. Lett.* **2005**, *407*, 53–57. [CrossRef]
31. Schlayer, S.; Stallmach, F.; Horch, C.; Splith, T.; Pusch, A.-K.; Pielenz, F.; Peksa, M. Konstruktion und Test eines Gradientensystems für NMR-Diffusionsuntersuchungen in Grenzflächensystemen. *Chem. Ing. Tech.* **2013**, *85*, 1755–1760. [CrossRef]
32. Romanova, E.E.; Scheffler, F.; Freude, D. Crystallization of zeolite MFI under supergravity, studied in situ by 11B MAS NMR spectroscopy. *Microporous Mesoporous Mater.* **2009**, *126*, 268–271. [CrossRef]
33. Dvoyashkina, N.; Freude, D.; Stepanov, A.G.; Böhlmann, W.; Krishna, R.; Kärger, J.; Haase, J. Alkane/alkene mixture diffusion in silicalite-1 studied by MAS PFG NMR. *Microporous Mesoporous Mater.* **2018**, *257*, 128–134. [CrossRef]
34. Mildner, T.; Ernst, H.; Freude, D. 207Pb NMR detection of spinning-induced temperature gradients in MAS rotors. *Solid State Nucl. Magn. Reson.* **1995**, *5*, 269–271. [CrossRef]
35. Ferguson, D.B.; Haw, J.F. Transient methods for in-situ NMR of reactions on solid catalysts using temperature jumps. *Anal. Chem.* **1995**, *67*, 3342–3348. [CrossRef]
36. Neue, G.; Dybowski, C. Determining temperature in a magic-angle spinning probe using the temperature dependence of the isotropic chemical shift of lead nitrate. *Solid State Nucl. Magn. Reson.* **1997**, *7*, 333–336. [CrossRef]
37. Chmelik, C.; Freude, D.; Bux, H.; Haase, J. Ethane/ethene mixture diffusion in the MOF sieve ZIF-8 studied by MAS PFG NMR diffusometry. *Microporous Mesoporous Mater.* **2012**, *147*, 135–141. [CrossRef]
38. Chmelik, C. Characteristic features of molecular transport in MOF ZIF-8 as revealed by IR microimaging. *Microporous Mesoporous Mater.* **2015**, *216*, 138–145. [CrossRef]
39. Park, K.S.; Ni, Z.; Cote, A.P.; Choi, J.Y.; Huang, R.D.; Uribe-Romo, F.J.; Chae, H.K.; O'Keeffe, M.; Yaghi, O.M. Exceptional chemical and thermal stability of zeolitic imidazolate frameworks. *Proc. Natl. Acad. Sci. USA* **2006**, *103*, 10186–10191. [CrossRef] [PubMed]
40. Bux, H.; Liang, F.; Li, Y.; Cravillon, J.; Wiebcke, M.; Caro, J. Zeolitic imidazolate framework membrane with molecular sieving properties by microwave-assisted solvothermal synthesis. *J. Am. Chem. Soc.* **2009**, *131*, 16000–16001. [CrossRef] [PubMed]
41. Lu, G.; Hupp, J.T. Metal-organic frameworks as sensors: A ZIF-8 based Fabry-Pérot device as a selective sensor for chemical vapors and gases. *J. Am. Chem. Soc.* **2010**, *132*, 7832–7833. [CrossRef] [PubMed]
42. Kärger, J.; Binder, T.; Chmelik, C.; Hibbe, F.; Krautscheid, H.; Krishna, R.; Weitkamp, J. Microimaging of transient guest profiles to monitor mass transfer in nanoporous materials. *Nat. Mater.* **2014**, *13*, 333–343. [CrossRef] [PubMed]
43. Gladstone, S.; Laidler, K.J.; Eyring, H. *The Theory of Rate Processes*; McGraw-Hill: New York, NY, USA, 1941.
44. Ruthven, D.M.; Derrah, R.I. Transition state theory of zeolitic diffusion. *J. Chem. Soc. Faraday Trans. I* **1972**, *68*, 2332–2343. [CrossRef]
45. Kärger, J.; Pfeifer, H.; Haberlandt, R. Application of absolute rate theory to intracrystalline diffusion in zeolites. *J. Chem. Soc. Faraday Trans. I* **1980**, *76*, 1569–1575. [CrossRef]

46. Chmelik, C.; Kärger, J. The predictive power of classical transition state theory revealed in diffusion studies with MOF ZIF-8. *Microporous Mesoporous Mater.* **2016**, *225*, 128–132. [CrossRef]

47. Caro, J.; Bülow, M.; Schirmer, W.; Kärger, J.; Heink, W.; Pfeifer, H.; Zhdanov, S.P. Microdynamics of methane, ethane and propane in ZSM-5 type zeolites. *J. Chem. Soc. Faraday Trans. I* **1985**, *81*, 2541–2550. [CrossRef]

48. Heink, W.; Kärger, J.; Pfeifer, H.; Salverda, P.; Datema, K.P.; Nowak, A.K. High-temperature pulsed field gradient nuclear magnetic resonance self-diffusion measurements of n-Alkanes in MFI-type zeolites. *J. Chem. Soc. Faraday Trans.* **1992**, *88*, 3505–3509. [CrossRef]

49. Jobic, H.; Schmidt, W.; Krause, C.; Kärger, J. PFG NMR and QENS diffusion studies of n-alkane homologues in MFI-type zeolites. *Microporous Mesoporous Mater.* **2006**, *90*, 299–306. [CrossRef]

50. Tsai, H.L.; Sato, S.; Takahashi, R.; Sodesawa, T.; Takenaka, S. Liquid-phase hydrogenation of ketones in the mesopores of nickel catalysts. *Phys. Chem. Chem. Phys.* **2002**, *4*, 3537–3542. [CrossRef]

51. Nguyen, N.Q.; McGann, M.R.; Lacks, D.J. Elastic stability limits of polyethylene and n-alkane crystals from molecular simulation. *J. Phys. Chem. B* **1999**, *103*, 10679–10683. [CrossRef]

52. Kishore, K.; Bharat, S.; Kannan, S. Correlation of Kauzman temperature with odd-even effect in n-alkanes. *J. Chem. Phys.* **1996**, *105*, 11364–11365. [CrossRef]

53. Takahashi, R.; Sato, S.; Sodesawa, T.; Ikeda, T. Diffusion coefficient of ketones in liquid media within mesopores. *Phys. Chem. Chem. Phys.* **2003**, *5*, 2476–2480. [CrossRef]

54. Fernandez, M.; Pampel, A.; Takahashi, R.; Sato, S.; Freude, D.; Kärger, J. Revealing complex-formation in acetone-n-alkane mixtures by MAS PFG NMR diffusion measurement in nanoporous hosts. *Phys. Chem. Chem. Phys.* **2008**, *10*, 4165–4171. [CrossRef] [PubMed]

55. Rincon Bonilla, M.; Bhatia, S.K. Diffusion in Pore Networks: Effective self-diffusivity and the concept of tortuosity. *J. Phys. Chem. C* **2013**, *117*, 3343–3357. [CrossRef]

56. Rincon Bonilla, M.; Titze, T.; Schmidt, F.; Mehlhorn, D.; Chmelik, C.; Valiullin, R.; Bhatia, S.K.; Kaskel, S.; Ryoo, R.; Kärger, J. Diffusion study by IR micro-imaging of molecular uptake and release on mesoporous zeolites of structure type CHA and LTA. *Materials* **2013**, *6*, 2662–2688. [CrossRef] [PubMed]

57. Kärger, J.; Valiullin, R. Mass transfer in mesoporous materials: The benefit of microscopic diffusion measurement. *Chem. Soc. Rev.* **2013**, *42*, 4172–4197. [CrossRef] [PubMed]

58. Mehlhorn, D.; Kondrashova, D.; Küster, C.; Enke, D.; Emmerich, T.; Bunde, A.; Valiullin, R.; Kärger, J. Diffusion in complementary pore spaces. *Adsorption* **2016**, *22*, 879–890. [CrossRef]

59. Stoeckel, D.; Kübel, C.; Hormann, K.; Höltzel, A.; Smarsly, B.M.; Tallarek, U. Morphological analysis of disordered macroporous-mesoporous solids based on physical reconstruction by nanoscale tomography. *Langmuir* **2014**, *30*, 9022–9027. [CrossRef] [PubMed]

60. Kondrashova, D.; Lauerer, A.; Mehlhorn, D.; Jobic, H.; Feldhoff, A.; Thommes, M.; Chakraborty, D.; Gommes, C.; Zecevic, J.; de Jongh, P.; et al. Scale-dependent diffusion anisotropy in nanoporous silicon. *Sci. Rep.* **2017**, *7*, 40207. [CrossRef] [PubMed]

61. Reich, S.-J.; Svidrytski, A.; Höltzel, A.; Florek, J.; Kleitz, F.; Wang, W.; Kübel, C.; Hlushkou, D.; Tallarek, U. Hindered diffusion in ordered mesoporous silicas: Insights from pore-scale simulations in physical reconstructions of SBA-15 and KIT-6 silica. *J. Phys. Chem. C* **2018**, *122*, 12350–12361. [CrossRef]

62. Crawford, G.P.; Vilfan, M.; Doane, J.W.; Vilfan, I. Escaped-radial nematic configuration in submicrometer size cylindrical cavities: Deuterium nuclear-magnetic-resonance study. *Phys. Rev. A* **1991**, *43*, 835–842. [CrossRef] [PubMed]

63. Cramer, C.; Cramer, T.; Arndt, M.; Kremer, F.; Naji, L.; Stannarius, R. NMR and dielectric studies of nano-confined nematic liquid crystals. *Mol. Cryst. Liq. Cryst. Sci. Technol. Sect. A Mol. Cryst. Liq. Cryst.* **2006**, *303*, 209–217. [CrossRef]

64. Gueudré, L.; Milina, M.; Mitchell, S.; Pérez-Ramírez, J. Superior mass transfer properties of technical zeolite bodies with hierarchical porosity. *Adv. Funct. Mater.* **2014**, *24*, 209–219. [CrossRef]

65. Mitchell, S.; Pinar, A.B.; Kenvin, J.; Crivelli, P.; Kärger, J.; Pérez-Ramírez, J. Structural analysis of hierarchically organized zeolites. *Nat. Commun.* **2015**, *6*, 8633. [CrossRef] [PubMed]

66. Hartmann, M.; Schwieger, W. Hierarchically-structured porous materials: From basic understanding to applications. *Chem. Soc. Rev.* **2016**, *45*, 3311–3312. [CrossRef] [PubMed]

67. Coasne, B. Multiscale adsorption and transport in hierarchical porous materials. *New J. Chem.* **2016**, *40*, 4078–4094. [CrossRef]

68. Galarneau, A.; Guenneau, F.; Gedeon, A.; Mereib, D.; Rodriguez, J.; Fajula, F.; Coasne, B. Probing interconnectivity in hierarchical microporous/mesoporous materials using adsorption and nuclear magnetic resonance diffusion. *J. Phys. Chem. C* **2016**, *120*, 1562–1569. [CrossRef]

69. Trogadas, P.; Nigra, M.M.; Coppens, M.-O. Nature-inspired optimization of hierarchical porous media for catalytic and separation processes. *New J. Chem.* **2016**, *40*, 4016–4026. [CrossRef]

70. Coppens, M.O.; Ye, G. Nature-inspired optimization of transport in porous media. In *Diffusive Spreading in Nature, Technology and Society*; Bunde, A., Caro, J., Kärger, J., Vogl, G., Eds.; Springer International Publishing: Cham, Switzerland, 2018; pp. 203–232.

71. Kärger, J. NMR Self-diffusion studies in heterogeneous systems. *Adv. Colloid Interface Sci.* **1985**, *23*, 129–148. [CrossRef]

72. Reginald Waldeck, A.; Hossein Nouri-Sorkhabi, M.; Sullivan, D.R.; Kuchel, P.W. Effects of cholesterol on transmembrane water diffusion in human erythrocytes measured using pulsed field gradient NMR. *Biophys. Chem.* **1995**, *55*, 197–208. [CrossRef]

73. Meier, C.; Dreher, W.; Leibfritz, D. Diffusion in compartmental systems. II. Diffusion weighted measurements of rat brain tissue in vivo and postmortem at very large *b*-values. *Magn. Reson. Med.* **2003**, *50*, 510–514. [CrossRef] [PubMed]

74. Nilsson, M.; Alerstam, E.; Wirestam, R.; Stahlberg, F.; Brockstedt, S.; Lätt, J. Evaluating the accuracy and precision of a two-compartment Kärger model using Monte Carlo simulations. *J. Magn. Res.* **2010**, *206*, 59–67. [CrossRef] [PubMed]

75. Himmelein, S.; Sporenberg, N.; Schönhoff, M.; Ravoo, B.J. Size-selective permeation of water-soluble polymers through the bilayer membrane of cyclodextrin vesicles investigated by PFG-NMR. *Langmuir* **2014**, *30*, 3988–3995. [CrossRef] [PubMed]

76. Melchior, J.-P.; Majer, G.; Kreuer, K.-D. Why do proton conducting polybenzimidazole phosphoric acid membranes perform well in high-temperature PEM fuel cells? *Phys. Chem. Chem. Phys.* **2016**, *19*, 601–612. [CrossRef] [PubMed]

77. Ferreira, A.S.D.; Barreiros, S.; Cabrita, E.J. Probing sol-gel matrices microenvironments by PGSE HR-MAS NMR. *Magn. Reson. Chem.* **2017**, *55*, 452–463. [CrossRef] [PubMed]

78. Cabrita, E.J.; Berger, S.; Brauer, P.; Kärger, J. High-resolution DOSY NMR with spins in different chemical surroundings: Influence of particle exchange. *J. Magn. Reson.* **2002**, *157*, 124–131. [CrossRef] [PubMed]

79. Kühl, G.H. Crystallization of low-silica faujasite. *Zeolites* **1987**, *7*, 451–457. [CrossRef]

80. Schneider, D.; Toufar, H.; Samoson, A.; Freude, D. ^{17}O DOR and other solid-state NMR studies concerning the basic properties of zeolites LSX. *Solid State Nucl. Magn. Reson.* **2009**, *35*, 87–92. [CrossRef] [PubMed]

81. Freude, D.; Beckert, S.; Stallmach, F.; Kurzhals, R.; Täschner, D.; Toufar, H.; Kärger, J.; Haase, J. Ion and water mobility in hydrated Li-LSX zeolite studied by 1H, 6Li and 7Li NMR spectroscopy and diffusometry. *Microporous Mesoporous Mater.* **2013**, *172*, 174–181. [CrossRef]

82. Beckert, S.; Stallmach, F.; Toufar, H.; Freude, D.; Kärger, J.; Haase, J. Tracing water and cation diffusion in hydrated zeolites of type Li-LSX by pulsed field gradient NMR. *J. Phys. Chem. C* **2013**, *117*, 24866–24872. [CrossRef]

83. Lauerer, A.; Kurzhals, R.; Toufar, H.; Freude, D.; Kärger, J. Tracing compartment exchange by NMR diffusometry: Water in lithium-exchanged low-silica X zeolites. *J. Magn. Reson.* **2018**, *289*, 1–11. [CrossRef] [PubMed]

84. Laberty-Robert, C.; Vallé, K.; Pereira, F.; Sanchez, C. Design and properties of functional hybrid organic-inorganic membranes for fuel cells. *Chem. Soc. Rev.* **2011**, *40*, 961–1005. [CrossRef] [PubMed]

85. Hickner, M.A.; Ghassemi, H.; Kim, Y.S.; Einsla, B.R.; McGrath, J.E. Alternative polymer systems for proton exchange membranes (PEMs). *Chem. Rev.* **2004**, *104*, 4587–4612. [CrossRef] [PubMed]

86. Kreuer, K.-D.; Paddison, S.J.; Spohr, E.; Schuster, M. Transport in proton conductors for fuel-cell applications: simulations, elementary reactions, and phenomenology. *Chem. Rev.* **2004**, *104*, 4637–4678. [CrossRef] [PubMed]

87. Sharifi, M.; Wark, M.; Freude, D.; Haase, J. Highly proton conducting sulfonic acid functionalized mesoporous materials studied by impedance spectroscopy, MAS NMR spectroscopy and MAS PFG NMR diffusometry. *Microporous Mesoporous Mater.* **2012**, *156*, 80–89. [CrossRef]

88. Dvoyashkina, N.; Seidler, C.F.; Wark, M.; Freude, D.; Haase, J. Proton mobility in sulfonic acid functionalized mesoporous materials studied by MAS PFG NMR diffusometry and impedance spectroscopy. *Microporous Mesoporous Mater.* **2018**, *255*, 140–147. [CrossRef]

89. D'Orazio, F.; Bhattacharja, S.; Halperin, W.P.; Gerhardt, R. Enhanced self-diffusion of water in restricted geometry. *Phys. Rev. Lett.* **1989**, *63*, 43–46. [CrossRef] [PubMed]

90. Kärger, J.; Pfeifer, H.; Riedel, E.; Winkler, H. Self-diffusion measurements of water adsorbed in NaY zeolites by means of NMR pulsed field gradient techniques. *J. Colloid Interface Sci.* **1973**, *44*, 187–188. [CrossRef]

91. Inayat, A.; Knoke, I.; Spieker, E.; Schwieger, W. Assemblies of mesoporous FAU-type zeolite nanosheets. *Angew. Chem. Int. Ed.* **2012**, *51*, 1962–1965. [CrossRef] [PubMed]

92. Na, K.; Choi, M.; Ryoo, R. Recent advances in the synthesis of hierarchically nanoporous zeolites. *Microporous Mesoporous Mater.* **2013**, *166*, 3–19. [CrossRef]

processes

MDPI

Review

Computational Molecular Modeling of Transport Processes in Nanoporous Membranes

Kevin R. Hinkle [1,2,†], Xiaoyu Wang [2], Xuehong Gu [3], Cynthia J. Jameson [1] and Sohail Murad [2,*]

[1] Department of Chemical Engineering, University of Illinois at Chicago, Chicago, IL 60607, USA; hinklek1@udayton.edu (K.R.H.); cjjames@uic.edu (C.J.J.)
[2] Department of Chemical and Biological Engineering, Illinois Institute of Technology, Chicago, IL 60616, USA; xwang181@hawk.iit.edu
[3] State Key Laboratory of Materials-Oriented Chemical Engineering, Nanjing Tech University, Nanjing 210009 China; xuehonggu@yahoo.com
* Correspondence: murad@iit.edu
† Current Address: Department of Chemical and Materials Engineering, University of Dayton, Dayton, OH 45469, USA.

Received: 19 July 2018; Accepted: 4 August 2018; Published: 9 August 2018

Abstract: In this report we have discussed the important role of molecular modeling, especially the use of the molecular dynamics method, in investigating transport processes in nanoporous materials such as membranes. With the availability of high performance computers, molecular modeling can now be used to study rather complex systems at a fraction of the cost or time requirements of experimental studies. Molecular modeling techniques have the advantage of being able to access spatial and temporal resolution which are difficult to reach in experimental studies. For example, sub-Angstrom level spatial resolution is very accessible as is sub-femtosecond temporal resolution. Due to these advantages, simulation can play two important roles: Firstly because of the increased spatial and temporal resolution, it can help understand phenomena not well understood. As an example, we discuss the study of reverse osmosis processes. Before simulations were used it was thought the separation of water from salt was purely a coulombic phenomenon. However, by applying molecular simulation techniques, it was clearly demonstrated that the solvation of ions made the separation in effect a steric separation and it was the flux which was strongly affected by the coulombic interactions between water and the membrane surface. Additionally, because of their relatively low cost and quick turnaround (by using multiple processor systems now increasingly available) simulations can be a useful screening tool to identify membranes for a potential application. To this end, we have described our studies in determining the most suitable zeolite membrane for redox flow battery applications. As computing facilities become more widely available and new computational methods are developed, we believe molecular modeling will become a key tool in the study of transport processes in nanoporous materials.

Keywords: molecular simulation; membrane separations; ion-transport

1. Introduction

It is estimated that approximately 55% of all energy consumed in chemical processes is spent on separations, of which about 50% is consumed by distillation, 20% by evaporation and 10% by drying, and the remaining 20% by non-thermal separations, include membrane-based separations [1]. Thus, further development of membrane-based separation processes to enable their use in applications currently employing thermal separations can lead to significant energy savings in e chemical process industries. One obstacle to the development of membranes for these energy intensive separations is that many membrane-based separation processes are not well understood at the fundamental molecular level, thereby resulting in membrane synthesis becoming an art rather than a science [2–5]. Computational molecular modeling tools such as molecular dynamics [6] can play a crucial role in clarifying the molecular forces that result in making a membrane effective for a proposed application. Such molecular level understanding can thus greatly assist in the design of new membranes for a desired separation. In addition, such molecular level tools can also assist in understanding unexpected behavior observed during membrane separation processes by providing access to spatial resolution to a fraction of an Angstrom and temporal resolution to a fraction of a femtosecond. This level of resolution is often difficult to achieve in experimental methods. Selectively permeable membranes perform important roles in a wide range of systems from naturally occurring lipid membranes in biological systems to engineered polymeric membranes in filtration and energy technologies. In order to design technologies that incorporate such membranes, it is crucial to understand the behavior of these systems at the molecular level so that optimal performance and maximum efficiencies can be achieved. Computational molecular modeling tools such as molecular dynamics are ideally suited to provide detail at a level that can aid in the understanding of the transport process.

In this paper we briefly summarize three applications that use molecular dynamics techniques to examine intramembrane transport:

1. Transport of water and ions in reverse osmosis (RO) nanoporous membranes and the role played by ion-solvation in such membranes.
2. Ion exchange in zeolite membranes and our finding that the separations were almost completely enthalpically driven rather than entropically.
3. Separation of gases using zeolite membranes and the role of membrane loading and diffusion in the observed separation factors achieved in the membranes.

These brief overviews are followed by more detailed discussions of two recent investigations by our group:

1. We describe our investigations of multiple zeolite framework types to determine their transport behavior regarding water, protons, and vanadium ions, and investigate at the molecular level the requirements for their suitability in ion exchange membrane (IEM) applications. In addition to investigating different zeolite frameworks, the effect of composition is also examined by introducing different levels of aluminum substitution into the crystalline structure of a specific zeolite framework. By investigating two characteristics, membrane loading and intramembrane diffusion, it was possible for us to predict the overall ion permeability with the goal of optimizing the amount of aluminum substitution for high proton permeability while maintaining selectivity to undesirable ions. These and similar studies can be instrumental in designing more efficient membranes for important applications such as water purification/desalination and in many proposed applications in energy sustainability.
2. The second application focuses on how molecular simulations can help to understand unexpected or non-intuitive results obtained during experiments. A recent experimental study on the dehydration of alcohols using zeolite membranes showed [7] that the membrane was effective in dehydrating alcohol when 5% water was present but became surprisingly ineffective when the water content was 1%. This was an intriguing observation that molecular simulation was able to explain because of the fine spatial and temporal resolution accessible in such investigations that may not be possible experimentally.

2. Molecular Simulations Applied to Various Membrane Applications

Here we demonstrate the wide range of applications that can be addressed using molecular simulations.

2.1. Transport in Reverse Osmosis Membranes

Reverse osmosis (RO) separations of aqueous electrolyte solutions (such as the desalination process), is a rather challenging problem, because it is not an obvious case of separations enabled by differences in molecular sizes (steric separation). In the case of an aqueous NaCl solution, for example, the size of Na^+ ion (<0.2 nm) is considerably smaller than that of a water molecule (~0.3 nm). In spite of this difference in sizes, it is well known that desalination membranes, such as those described earlier, are very effective in removing salt from water. Since this permeation could not be explained due to size differences, it was generally believed that this was due to surface interactions between the membrane surface and the solute/solvent molecules. Molecular simulations could therefore provide a useful tool for understanding these molecular forces [8]. This problem was examined in considerable detail using the molecular dynamics method and has led to a considerable improvement in our understanding of the intermolecular forces that play a significant role in the RO-based separation of electrolyte solutions as described below.

The simulations used previously developed intermolecular potential models for both the water and the ions. For water, the simple point charge (SPC) model was used [9], while for the ions the primitive model was used [10]. These models provide a realistic representation of these systems for a wide range of properties and state conditions [8,9]. The reaction field method [11] was used to include long-range forces. The membrane was represented by a thin ZK4 zeolite layer with pores of 0.42 nm diameter (see Figure 1). The atoms forming the membrane were not charged so that essentially only steric (size based) separation would be possible in these membranes.

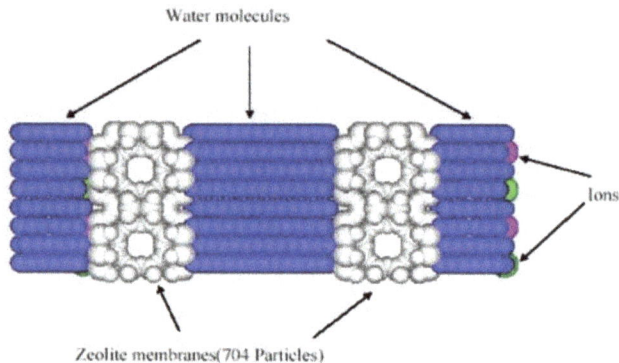

Figure 1. Set up for reverse osmosis separation of brine with zeolite membranes.

One of the most significant observation from the molecular simulation studies of aqueous electrolyte solutions was that no additional intermolecular forces were needed to prevent the much smaller ions from permeating the membrane, while permitting the larger water molecules to readily permeate the membrane. This appeared to be due to the large solvated ionic clusters formed when ions are dissolved in water. The ions were surrounded by the solvating water molecules, thus increasing their effective size to almost 1 nm. A typical cluster observed due to the interaction between the ions and water is shown in Figure 2. These clusters were quite stable, with a rather high energy of de-solvation [10]. In addition to the clusters shown in Figure 2a, larger clusters which involved more than one ion were also presented. Figure 2b shows one such example when two ions separated by an intervening layer of water molecules forming one larger cluster. By increasing the temperature of the solution, it was possible to make ions permeate the membrane. The increased temperature effectively made the larger ionic clusters less stable, and the ions were able to break away from the clusters, and the almost "bare" ions could then permeate the membrane. This was also found to be the case when the simulation system was subjected to an external electric field [12]. The electric field similarly weakened the ionic clusters and again allowed ions to then permeate the membrane. These simulations have shown that the solvation of ions is at least in part responsible for these separations, and these forces must be taken into account as part of the design in RO-based separations of aqueous electrolyte solutions.

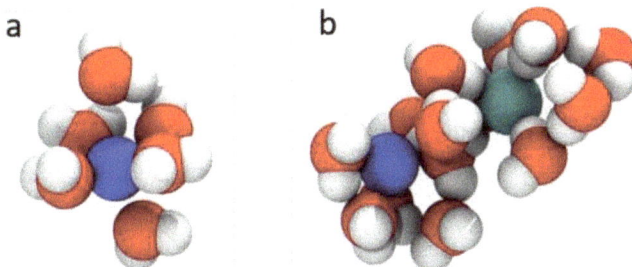

Figure 2. (a) Na^+ ion (blue) solvated with water (red/white) to effectively increase its size; (b) A hydrated ion pair of Na^+ (blue) and Cl^- ions (cyan).

The ionic clusters observed are not limited to just aqueous electrolyte solutions. In fact, similar clusters results were observed in methanolic solutions as well [12]. This shows that such large and stable ionic clusters are a fairly common occurrence whenever ions are dissolved in polar solvents. These clusters are therefore an essential factor in the facilitation of reverse osmosis purification. Since many industrially important solutions include ions in polar solvents, it is important to account for them in separations involving such solvents. The transport of water and ions inside polymide membranes has also been studied using molecular simulation [13].

2.2. Transport in Ion-Exchange Membranes

Another example of the use of molecular simulation to examine transport of water and ions in nanoporous membranes is the ion exchange between an aqueous solution and NaA zeolite. Our group carried out such a simulation to understand the molecular basis of such exchanges [14]. The schematic for such a simulation is shown in Figure 3.

Molecular simulations were used to study both the dynamics and energetics of ion exchanges between monovalent and bivalent cations in aqueous solutions (both supercritical and subcritical/liquid. In this study simulations of up to a nanosecond or more were carried out in which Li^+ and Ca^{++} in aqueous solutions of LiCl and $CaCl_2$ came in contact with an ion exchange membrane (NaA zeolite). NaA zeolites are widely used in many commercially used on exchange processes including detergents. Our results showed that with appropriate driving forces (in this case pressure driven), such ion exchange processes can be clearly observed and investigated using molecular simulations at the time scales accessible in simulations. We were also able to understand the phenomenon of ion-exchange itself at the molecular level. Our simulations have shown that the ion-exchange process is primarily energetically driven and entropic forces do not appear to play a significant role in the ion exchanges observed. For supercritical LiCl solutions, we found rather small differences between the energy of the Li^+ ion inside and outside the membrane. In contrast, for Na^+ there was a significant energetic advantage in being outside the membrane, making the overall exchange process more energetically favorable. In subcritical (liquid) LiCl solutions we found exchange to be more favorable energetically than in supercritical solutions. For Ca^{++} similar behavior was observed, except for the observation that the differences in the energies were much larger (compared to the corresponding Li^+ exchanges), making them more energetically efficient, as has also been observed experimentally [15]. These differences are in clearly shown in Figure 4. In addition to clarifying the molecular basis for these ion exchanges, simulations can also potentially be very useful to determine the behavior (e.g., state dependence, etc.) of hydrodynamic parameters commonly used to characterize ion-exchange processes at a fundamental molecular level, and to determine if the continuum hydrodynamic equations used for ion-exchange processes are applicable to nano-systems.

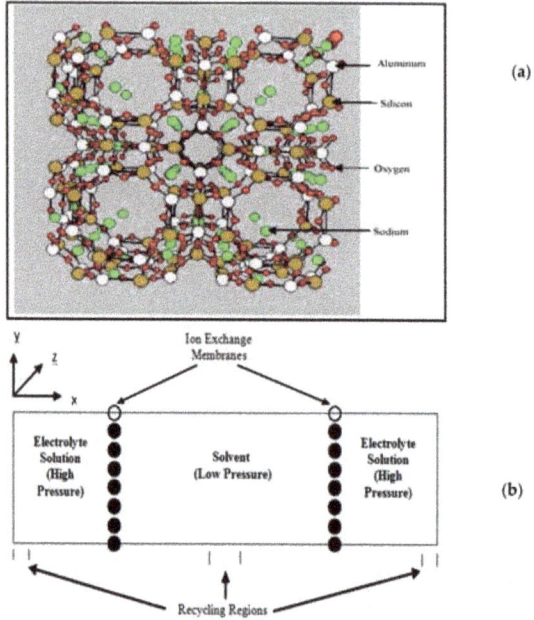

Figure 3. A typical system set up for ion exchange. (**a**) Zeolite membrane; (**b**) Two reservoirs and two membranes in conjunction with periodic boundary conditions are usually used in simulations as shown here.

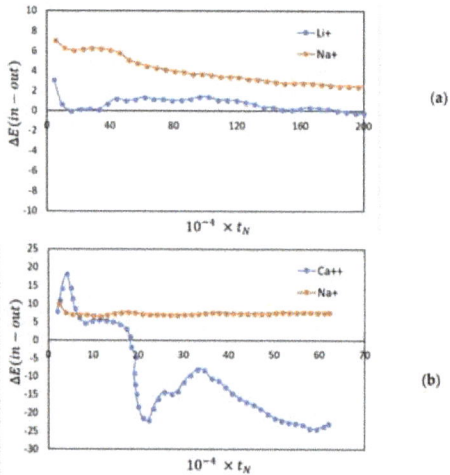

Figure 4. The differences in configurational energy of, Li^+, Ca^{++} and Na^+ ions (kJ/mol) inside and outside the membrane plotted against the number of time steps. A positive number indicates a preference for being outside the membrane. (**a**) Li^+/Na^+; (**b**) Ca^{++}/Na^+.

2.3. Transport in Gas Separations Membranes

Molecular modeling has also been used to study the transport and subsequent separation of gases using nanoporous membranes. A study carried out by our group using a range of zeolite membranes to support gaseous mixtures has shown good agreement with experiments exhibiting the viability of using molecular modeling to study such separations [14]. The simulation setup for such a study is shown in Figure 5.

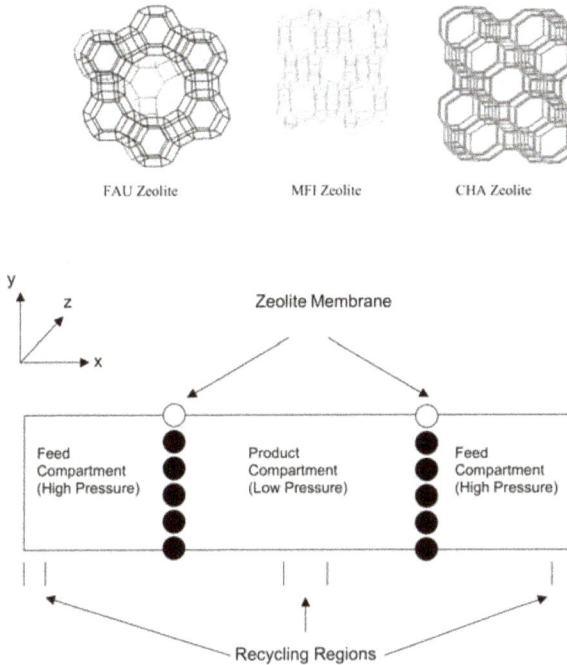

Figure 5. Initial system setup for gas separation simulations and the structure of FAU, MFI and CHA unit cell.

We performed a rather vigorous test of the reliability of the simulation technique to examine the separation factor of CO_2/N_2 and O_2/N_2 mixtures in zeolite membranes. In separations, the ideal separation factor (ISF) is defined as the ratio of the permeabilities of the two gases in their pure state and the usual separation factor (SF) relates to the permeabilities in mixtures (equimolar in our studies). Data obtained from our simulations can be seen in Figure 6 along with experimental results measured on similar systems.

While examining N_2/O_2 mixtures the ISF > SF, in CO_2/N_2 mixtures the opposite trend is observed (SF > ISF). Additional results from our simulations are shown in Figure 7 and confirm this behavior. This phenomenon can be explained as follows, based on observations from our simulations. In the case of N_2/O_2 for pure fluids N_2 has a higher diffusion rate than oxygen which leads to high SF since both have similar loadings. In mixtures this effect is dampened since the narrow pores do not allow N_2 to cross O_2 in mixtures. For pure CO_2 and N_2, CO_2 has a somewhat higher loading than N_2 which leads to CO_2 having a higher selectivity. In mixtures, the loading is exclusively CO_2 because of its high quadrupolar moment, which increases the selectivity in mixtures by almost a factor of 3, correctly predicted by the simulations. In addition, simulations can be a useful tool to determine the type of diffusion in the nanopores. For example, the simulations clearly show surface diffusion of N_2 in the zeolite pores as shown in Figure 7.

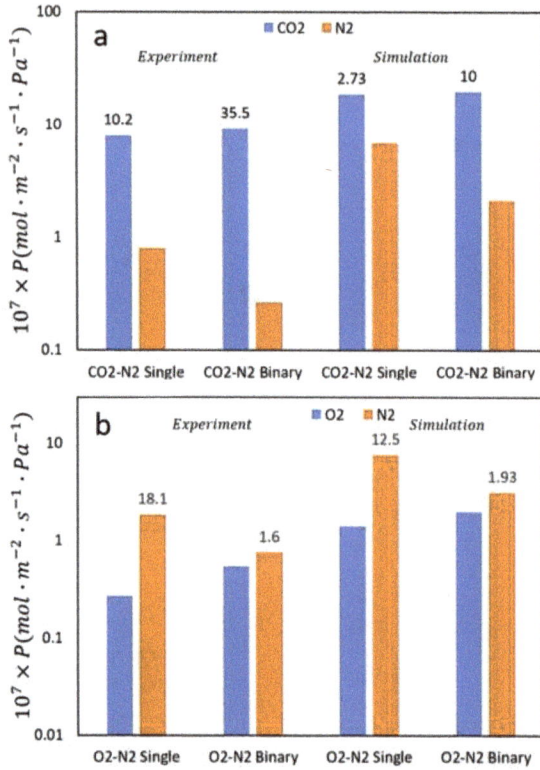

Figure 6. The permeance of (**a**) O_2/N_2 and (**b**) CO_2/N_2 for both pure systems and equimolar binary mixtures compared with experimental results. The first two sets are experimental results while the last two are simulation results as marked. The numbers above the permeances represent the separation factors (SF). Note the contrasting behavior between pure fluids and mixtures for the two systems that the simulation results correctly predict.

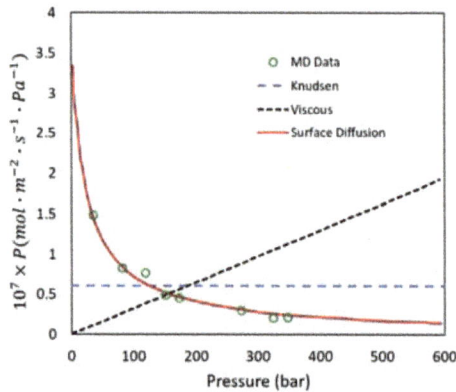

Figure 7. Simulation data on permeance of N_2 at 322 K in an equimolar N_2/O_2 mixture as a function of pressure. Note that except for surface diffusion, the two other mechanisms are clearly qualitatively incorrect.

2.4. Membranes for Redox Flow Batteries

Redox flow batteries (RFBs) show significant potential for energy storage because of their safety, capacity, and small environmental footprint [16,17]. However, this technology is not currently widely available due to problems with the ion-exchange membranes needed in RFBs. For an RFB with high storage capacity and high efficiency: (i) the electrode reactions must be reversible; (ii) both the oxidized and reduced species must have a high solubility in the electrolyte solution; (iii) there must be a large difference between the redox potentials. Several ion pairs satisfy these requirements, among them Fe/Cr [18], Zn/Br [19], and Zn/Ce [20]. Vanadium RFBs [21] constitute a special case in that only a single elemental species is present in the ions on both sides of the ion-exchange membrane (IEM). For this case the two half reactions are shown in Equation (1).

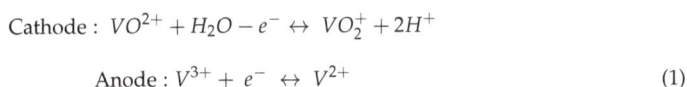

$$Cathode: \ VO^{2+} + H_2O - e^- \leftrightarrow VO_2^+ + 2H^+$$

$$Anode: V^{3+} + e^- \leftrightarrow V^{2+} \tag{1}$$

During charging and discharging process, current is transmitted through an external circuit while protons are transported across an IEM between the two electrode compartments. IEMs therefor play a critical role in the design of a RFB [22]. IEMs while being electrically insulating must also be impermeable to the reactive, vanadium ions species (see reactions above). RFBs currently employ polymeric membranes, more specifically sulfonated fluoropolymer-copolymers (commercially known as Nafion) [23]. While these polymer membranes show acceptable behavior as proton exchange membranes in fuel cells (PEMFC) as well as in direct methanol fuel cells (DMFC) [24], they do work as well in RFBs as they have stability problems in the highly reactive environment, resulting in the crossover of reactive ions and thus the reduction of cell efficiency and lifetime [25]. The highly oxidizing environment in RFBs also tends to degrade the polymer membrane [26,27]. While some improvements have been made in these polymeric membranes to increase selectivity and stability [23], they have not yet overcome all the obstacles for their widespread use in RFBs. New materials must therefore be considered if RFBs are to become an economically viable means of energy storage. We have focused on using zeolite membranes as an alternative to polymeric membranes. Zeolites are aluminosilicate crystals with ordered pore sizes ranging from 0.3 nm to >1.0 nm depending on the framework type. Zeolites with higher Si/Al ratios are electrically insulating and are also extremely stable in both acidic and basic conditions due to their inert chemical nature [28–30]. Previously, thin zeolite membranes have been observed to enable water/ion separation via the size-exclusion mechanism [8,31,32]. We believe that thin film zeolite membranes show considerable potential as IEMs in RFBs. The first experimental study to demonstrate this was carried out by Yang et al. [33].

To test this possibility, we have carried out molecular dynamics studies of six different zeolite frameworks (ERI, LTA, MFI, BEC, CFI, DON) for vanadium RFBs to determine the transport behavior of ions, protons, and solvent in the nanopores of the membranes [34]. The structure of these membranes were obtained from the IZA-SC's Database of Zeolite Structures [35] and constructed to have a thickness of a single unit cell. As discussed previously, the hydration of the vanadium(II) [V^{2+}], vanadium(III) [V^{3+}], oxovanadium(IV) [VO^{2+}], and dioxovanadium(V) [VO$_2^+$] ions plays a key role in ion transport and this effect was examined in detail. We found that a relatively large pore (~7 Å) was necessary for ion transport due to the strongly bound hydration shell that effectively increases the size of the ion. At higher ion concentrations we have observed passive (spontaneous with no external forces) permeation of the pores by the ions. This was observed in our studies when both the ion concentration and the temperature were significantly higher than the normal operating conditions in RFBs (8 mol %, 400 K). We note that even at these extreme conditions, we observed permeation and subsequent transport of only H$_3$O$^+$ ions in ERI, LTA, and MFI membranes. We did not observe any vanadium ion transport across the zeolite (Figure 8b). Therefore, we propose these membranes would be more suited in RFBs because they exhibit the necessary ion selectivity. This agrees with experimental measurements by Xu et al. [36] who have reported similar selective transport using substrate-supported zeolite membranes with ERI frameworks. In contrast, the larger pores in BEC, CFI, and DON frameworks did allow permeation and transport of all ion species V^{2+}, V^{3+}, VO^{2+}, and VO_2^+ in addition to hydronium ions (Figure 8a), which would thus make them unsuitable for vanadium redox flow batteries as they present no proton selectivity.

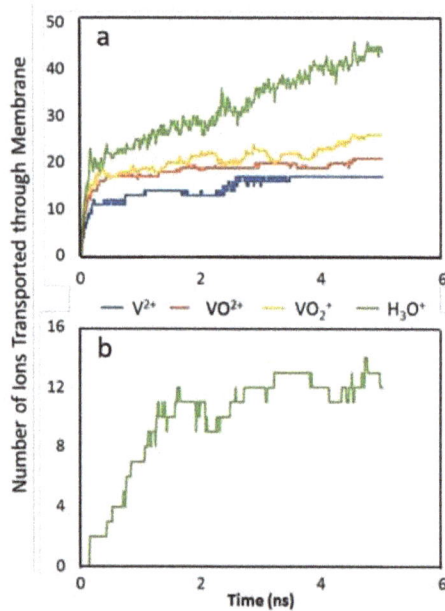

Figure 8. Ion transport events for (**a**) DON framework and (**b**) MFI framework (straight channel) at 8 mol% and 400 K.

This again showed that the size of the hydrated ion complex is a significant factor in zeolite membrane transport. The only ion transported through the membranes with smaller pores (<5 Å) such as ERI, LTA, and MFI zeolite frameworks was the hydronium ion (H_3O^+/proton) which has a kinetic diameter of 0.31 nm compared to 0.58 nm of V^{++}. Therefore, these membranes demonstrate selective transport of hydronium ions over vanadium ions that is an essential requirement for IEMs in vanadium RFBs.

To improve our understanding of the transport mechanism of the hydrated ions through a range of zeolite pores, we placed a single ion on the pore axis and transported it through the membrane at a constant velocity of 0.75 m/s. This steered molecular dynamics (SMD) method differs from the passive diffusion observed previously and permits the calculation of the Kirkwood potential of mean force (PMF) [37] across the membrane. The PMF measures the average force for all configurations along the "reaction path". PMF can be used to determine the energy barriers for membrane permeation [38]. The force needed to keep the hydrated ion on the pore axis (F_c) is assumed to be the opposite of the total pairwise force (F_p) exerted on the ion:

$$F_c = -F_p = \nabla U \tag{2}$$

where U is the total pairwise intermolecular potential. This force can then be integrated along the path maintained by the SMD which, for the one-dimensional path used here, leads to Equation (3):

$$PMF = \int F_c \cdot dr = \int F_{cx} dx \tag{3}$$

The PMFs obtained using this approach are shown in Figure 9.

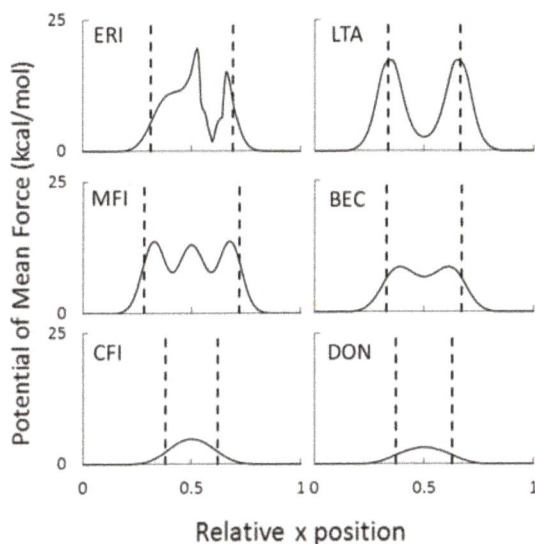

Figure 9. Potentials of mean force for hydrated V^{2+} ion passing through the 6 membrane framework types at 325 K.

The jagged (sharp peaks) behavior of the PMF for the ERI membrane results from changes in the number of water molecules in the hydration shell as it moves across the membrane. These pores are rather small, so in order for the ion to be forced through the pores, some of the hydrating waters must be removed so that the cluster is small enough to enter the pore. Once it leaves the membrane, the ion is hydrated again resulting in the reappearance of the stable cluster. The ERI profile also shows asymmetric behavior. This is because the rather quick movement of the ion in the membrance (0.75 m/s) does not allow for local equilibrium of the changing number of water molecules in the hydration shell as it moves along the pore axis. The local minima shown for LTA, MFI, and BEC membranes also indicates the most favorable location for the hydrated complex within the pore. These locations correspond to the cavity at the intersection of the channels running in 3 directions within the zeolite membrane. For CFI and DON membranes the maxima is observed at the center because their pores are 1-dimensional channels. In Figure 9, it is also clear that the free energy barrier at the entrance decreases with increasing pore size. While results for other vanadium ions all showed similar behavior, the hydronium ion profiles are quite different in magnitude. In Figure 10 we have compared the PMF peak heights for V^{2+} and H_3O^+ ions in the six zeolites membranes shown in Figure 9. The large differences observed in the free energy barriers for the vanadium ions and hydronium ions for zeolites with pores less than 5 Å shows the effectiveness of selective transport, a desirable characteristic of RFB IEMs.

Further examination of the orientation of the hydrated ion complexes during membrane permeation yields additional insights to the relative size of the zeolite pores and the diffusing species. The $[V(H_2O)_6]^{2+/3+}$, $[VO(H_2O)_5]^{2+}$, and $[VO_2(H_2O)_4]^+$ complexes have rather stable octahedral structures, so it is interesting to monitor the changes in orientation of the hydrated ions as they traverse the membrane during our directed simulations. We calculated the angle between: (1) a single $V^{2+/3+}$-water pair and the pore axis for $V^{2+/3+}$; (2) In the case of VO^{2+} the V-O bond and the pore axis, and finally for (3) VO_2^+ one of the V-O bonds and the pore axis. The results obtained are shown in Figure 11 for the BEC membrane. This membrane was chosen as the cluster size closely matches the pore dimensions (~6.08 Å for the cluster and ~6.23 Å for the pore). Unsurprisingly, this analysis shows that outside the membrane, the $[V(H_2O)_6]^{2+}$ complex exhibits random tumbling. However, in the small pore of the BEC membrane the angular motion of the complex is restricted, and it remains in a certain orientation until it leaves the channel.

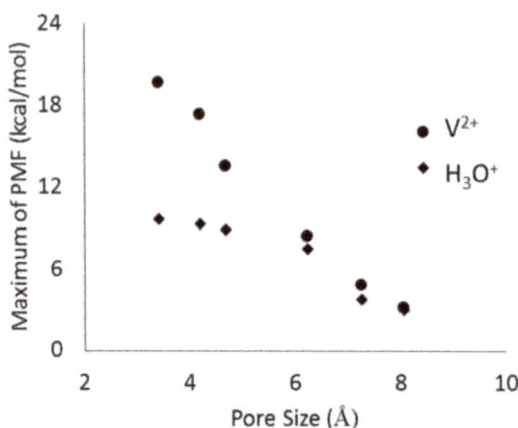

Figure 10. Magnitude of free energy barriers for V^{2+} and H_3O^+ ions as they traverse the pore of different zeolite membranes.

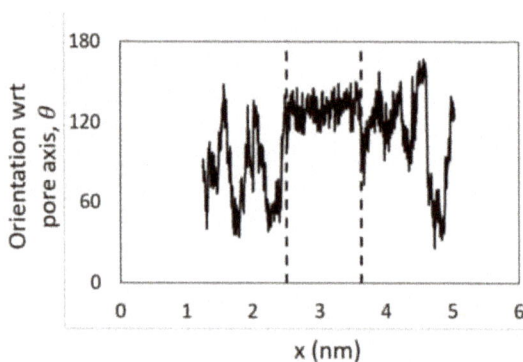

Figure 11. Orientation of the $[V(H_2O)_6]^{2+}$ complex with respect to the pore axis during membrane transport in BEC zeolite membrane. The dashed lines indicate the zeolite boundary. Tumbling, followed by alignment and finally tumbling are observed.

In addition to determining how zeolite pore size affects ion transport, we also investigated how zeolite composition could lead to differing behavior. The membranes described in the above analyses all consisted of solely silicon and oxygen representing a Si/Al ratio of infinity. These membranes were shown to display selectivity of the hydronium ion over the heavy vanadium ions in the vanadium-RFB at pore sizes below a certain threshold. The composition of these membranes was then modified to incorporate various levels of aluminum substitution. Seven membranes of the identical MFI framework, but varying aluminum content were constructed representing Si/Al ratios of infinity, 383, 191, 95, 54, 47, and 31. The aluminum atoms were substituted at sites corresponding to the T7, T10, and T12 sites as described in studies which previously modeled the uptake of gas into MFI zeolites [39,40]. The substitutions also followed Lowenstein's Rule [41,42] which does not allow for the presence of Al-O-Al linkages. In order to ensure that the membrane remained charge neutral, the local negative charges due to the aluminum atoms were spread across the four neighboring oxygens and an extra-framework sodium ion was placed in the channel in the vicinity of the substitution. These cations were allowed to equilibrate and find their lowest energy positions prior to the addition of any water or ions to the system.

The aluminum-substituted MFI membranes were used with the same protocol and system setup to ensure that the local charges on the aluminum atoms did not disrupt the hydrated ion complex or allow vanadium ions to permeate the membrane and alter its selective character. Our results showed that for all levels of Si/Al ratios, we still observed no spontaneous vanadium permeation for the entire simulation time of 10 ns. Once this selective characteristic had again been confirmed, the vanadium cations were replaced with hydronium ions in order to measure the dynamics of proton transport. Since the principle of the RFB mechanism rests on protons being released by the reaction in one half-cell and moving through the membrane to be used in the complimentary reaction in the other half-cell, the membranes not only need to be selective for proton transport, but also must have a high enough permeability that they not limit the reaction. For this reason, it is necessary to observe whether or not the membrane permeability is tunable using the Si/Al ratio in order that the proton flux be maximized. The flux of any species through a membrane is proportional to the concentration difference and the permeability [43] (Equation (4)).

$$j_i = \frac{P_m}{L}(c_i - c_0) \tag{4}$$

Here the permeability, P_m, can be represented as the product of the membrane loading, $\frac{c_m}{c_0}$, and the intramembrane diffusion, D_m (Equation (5)). Each of these quantities were measured independently in order to determine which was more strongly affected by the membrane composition.

$$P_m = D_m \frac{c_m}{c_0} \tag{5}$$

Membrane Loading: In order to observe the ionic loading of the membrane, the system consisting of a single membrane with solution on both sides was studied with the various levels of aluminum substitution. The solution consisted of a 5 mol % HCl solution in which the dissociated proton was modeled as a hydronium ion, or protonated water molecule ($H3O^+$). For each membrane, the system was simulated until the number of hydronium ions absorbed reached an equilibrium value, this process was usually complete after ~2 ns. The number was then averaged over an additional 0.5 ns to obtain the final values presented in Table 1.

Table 1. Hydronium loading in membranes with increasing aluminum substitution levels. Uncertainties represent one standard deviation of the ion loading over 0.5 ns of simulation.

No. of Substitutions Per Unit Cell	Si/Al Ratio	Hydronium Ion Concentration, mol/L
0.00	∞	1.86 ± 0.46
0.25	383	2.02 ± 0.50
0.50	191	2.22 ± 0.57
1.00	95	2.77 ± 0.54
1.75	54	4.36 ± 0.52
2.00	47	5.42 ± 0.63
3.00	31	6.37 ± 0.61

The data clearly shows an increase in the number of framework substitutions results in an increase in the uptake of hydronium ions. This agrees with previous findings regarding water uptake in MFI membranes [44], however the difference is much more pronounced with the charged species in this study. This increase in absorption is due to Coulombic attraction between the positively charged hydronium ions and the local negative charges present on the oxygens where aluminum atoms have been substituted within the zeolite membrane framework.

Intramembrane Diffusion: In molecular dynamics simulations, the calculation of diffusion coefficients in three dimensional systems is commonly performed using Equation (6), where MSD is the mean square displacement of the species of interest.

$$D = \frac{1}{6} \lim_{t \to \infty} \frac{d(MSD)}{dt} \qquad (6)$$

Because we are only interested in the rate of diffusion within the membrane and not in the bulk, care must be taken, when finding the MSD, that one accounts only for the intramembrane ions. This method indicates a slight decrease in the diffusion as more framework substitutions are made (Table 2). This decrease can be explained in a similar manner as the charged framework attracts more hydronium ion and results in an increased loading, this attraction also reduces the mobility of the absorbed ions when compared to those in the bulk solution.

Table 2. Hydronium diffusion in membranes with increasing aluminum substitution levels.

No. of Substitutions Per Unit Cell	Si/Al Ratio	Hydronium Ion Diffusion, $\times 10^5$ cm^2/s
0.00	∞	2.66 ± 0.11
0.25	383	2.56 ± 0.24
0.50	191	2.40 ± 0.11
1.00	95	2.35 ± 0.25
1.75	53.86	2.31 ± 0.23
2.00	47	2.09 ± 0.18
3.00	31	1.98 ± 0.31

However, our model for the proton provides an incomplete picture of what occurs in reality. Protons in aqueous solution are known to diffuse by hopping from one water molecule to another adjacent molecule via a process known as the Grothuss mechanism [45]. Work on this topic has been quite extensive [46–50] and modeling of this phenomenon is usually performed using techniques that account for quantum behavior. Here, the model we are using is solely classical in nature and represents a permanently hydrated water molecule [48]. Due to this quantum hopping, the mobility of H$^+$ ions in water is significantly higher than other monoatomic ions. The diffusion measured above only accounts for what is termed "vehicular diffusion", or the proton riding along with the water molecule. A measure of the "hopping diffusion", which is the quantum effect, would provide a more complete picture of the hydronium diffusion, but is beyond the scope of this study. In addition quantum effects can also play a role in the entry of protons in the zeolite pores and this effect has not been explicitly included in our study, but we believe it is not as significant a contribution as the proton hopping inside the pore.

When the membrane loading and intra-membrane diffusion of the various systems are scaled relative to the pure silica membrane and the product is taken to find the relative permeability, an interesting trend emerges (Figure 12). At high Si/Al ratios (low substitution numbers), the relative permeability increases as the higher ion uptake dominates, but as more framework substitutions are made, the decreasing diffusion rate forces the permeability to stabilize at a constant value.

In this case, we observe that a substitution rate of ~2 per unit cell, corresponding to a Si/Al ratio of ~45 there exists a threshold beyond which we no longer see an increasing ionic flux. In this context, this represents the minimum number of aluminum substitutions per unit cell of zeolite that will maximize the proton transport through the membrane. As the values used here are relative to the pure silica membrane, one should not view these values as absolute, but the trends should be reproducible using experimental methods. Our simulations show that both the resistance to ion entry to the zeolite pores and their subsequent diffusion inside the pore play an important role in the overall permeation rate in the zeolite membrane. For example as shown in Figure 9, other than the LTA membrane the highest resistance to ion transport does not occur at the pore entrance. In these cases the diffusion rate can be the rate determining step for the overall permeation.

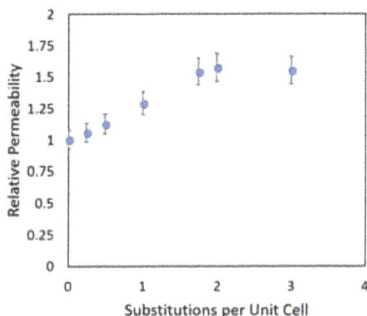

Figure 12. Relative permeability of MFI zeolite membranes at increasing levels of aluminum substitution.

2.5. Dehydration of Alcohol

We discuss finally our recent studies using molecular dynamics to understand unexpected results observed in experiments [7]. Again, we emphasize the high temporal and spatial resolution provided by molecular simulations lead to increased insights into certain non-intuitive phenomena. In this particular case, a NaA zeolite membrane was used to dehydrate alcohol via vapor phase pervaporation. While the membrane was quite effective when the water content was 5% by weight, it became ineffective when the water content dropped to 1%. Such an observation certainly is counter-intuitive because in general membranes should be more effective when the undesirable component is at a lower concentration. The schematic of the system designed for this study is shown in Figure 13. The middle compartment of the simulation system contains the vapor phase mixture being investigated. Two layers of NaA zeolite membranes separate this compartment from the two side compartments which are initially empty (vacuum). This provides the driving force for the vapor to permeate the zeolite membranes. The system size was chosen to ensure that no vapor condensation takes place in the bulk phase of the vapor compartment at the system temperature of 423 K. By removing some atoms that constitute the membrane we are also able to simulate a zeolite membrane with defects of approximately 1 nm—similar to those observed experimentally [7].

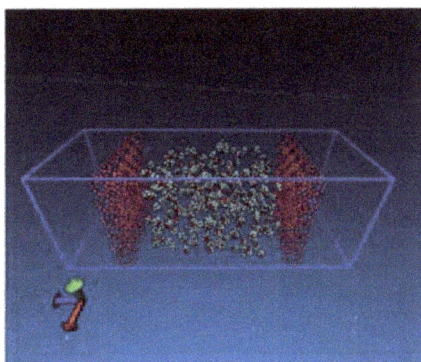

Figure 13. Schematic of the simulation system for vapor permeation through NaA zeolite membranes. Red represents the zeolite framework, water and iso-propyl alcohol (IPA) are between the two zeolite membranes.

Our simulation results, as shown in Figure 14, confirm the non-intuitive trend observed during experiments [7]. In addition, the simulations also provided significant insight into why this unusual and unexpected phenomenon occurred. Simulations were carried out for pure iso-propyl alcohol (IPA), as well as with 5 and 10% by weight water in the IPA. As can be clearly observed, the IPA was able to readily permeate the defect in the absence of water. Once water was included in the mixture the IPA permeation completely stopped in our time frame.

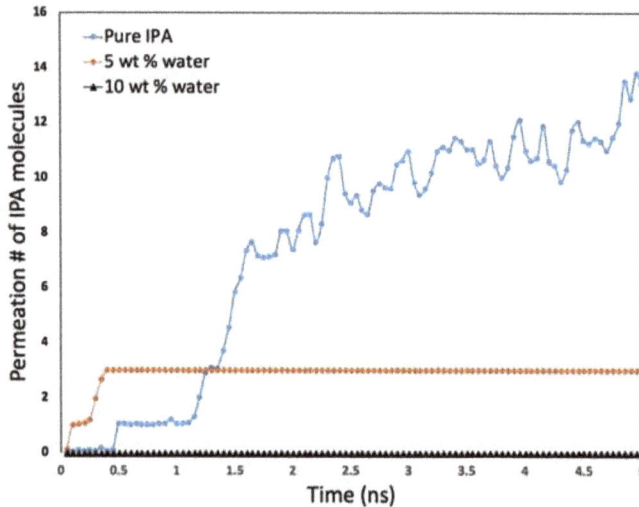

Figure 14. Number of IPA molecules permeating the membrane as a function of time for different compositions.

Upon further investigation, our simulations showed that there were two primary reasons that explain the decrease in IPA permeation when water is present. Firstly, we found that water molecules get adsorbed in zeolite pores as well as the defects, thus effectively reducing the effective size of the defects. When water is present at the defect sites, IPA molecules can also get adsorbed (with high adsorption energy) which further contributes to blocking the defect (Figure 15). In addition, we observed another interesting phenomenon in our simulations. In the bulk vapor phase in the presence of water, the IPA molecules tend to form larger IPA clusters which as a result effectively increases the dynamic diameter of the IPA molecules, making it more difficult for them to permeate the zeolite defects. This can be clearly observed in Figure 16. In addition to this, as can be seen from Figure 16c,d, the presence of water also results in fewer IPA molecules at the surface of the zeolite, which further restricts the permeation of IPA into the zeolite. We also observed that when water was present no IPA permeated the membrane.

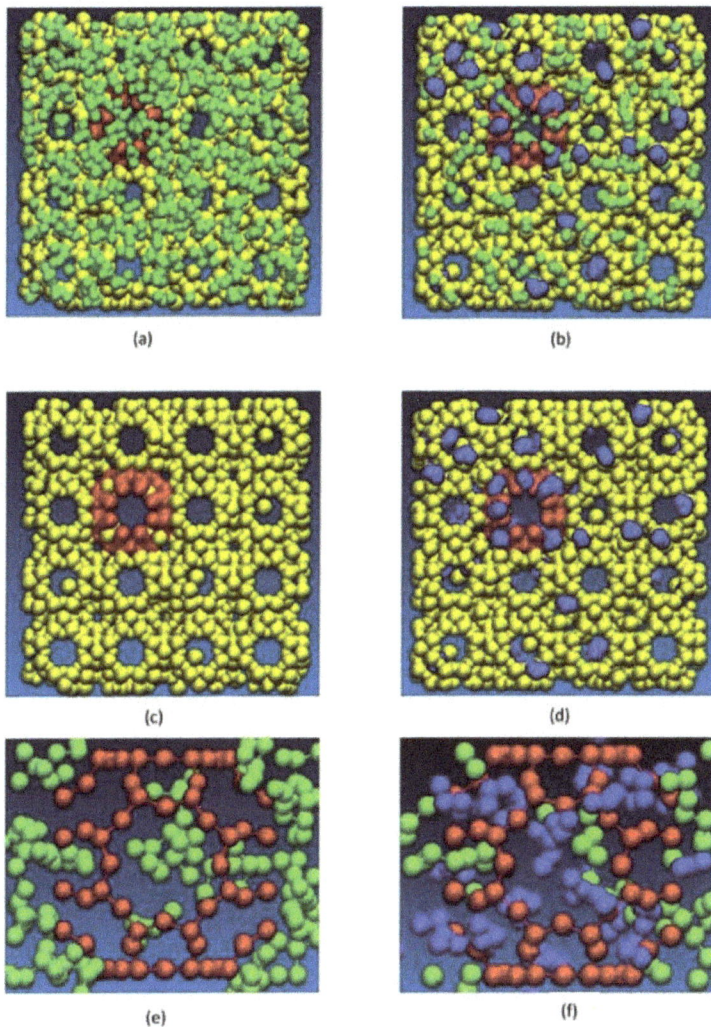

Figure 15. Changes in behavior observed in pure IPA (left side, **a,c,e**) and 5 wt % water (right side, **b,d,f**): axial snapshots of the membrane defect showing (**a**) pure IPA molecules permeating membrane defect and (**b**) water and IPA molecules blocking the membrane defect (water in blue); (**c**) as (**a**) above but IPA molecules removed; (**d**) as (**b**) above but IPA molecules removed; (**e**) cross section views showing pure IPA in cavity and (**f**) cross sectional view of both water and IPA molecules in cavity. The molecular diameters shown in the figure as spheres are not to scale.

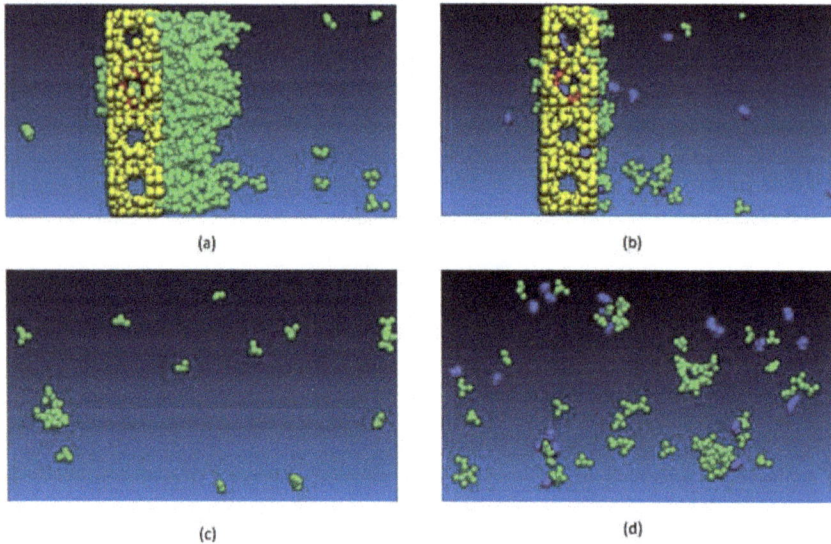

Figure 16. Differences in the observed behavior of the vapor phase when water is present: (**a**) snapshot of the pure IPA system near the zeolite membrane; (**b**) snapshot of system with 5 wt % water in vapor phase near the zeolite membrane. (**c**); snapshot of the pure IPA system away from membrane; (**d**): snapshot with 5 wt % water in vapor phase away from membrane; Key: yellow zeolite membrane; green IPA; and blue water; red: membrane defects.

3. Conclusions

Membrane separations have a number of advantages over other more conventional techniques including minimal maintenance due to the lack of moving parts and the fact that unlike other techniques it does not require a phase change. As an increasing amount of membrane-based separations are developed in fields as diverse as gas separation, desalination, dialysis, batteries, biosensing, and drug delivery, it is crucial that these processes be well understood so that their efficiencies can be maximized. This requires a more complete picture of the molecular level phenomena that occur between the membranes and the species they are separating.

In this paper we have discussed the important role molecular simulations (specifically the technique of molecular dynamics) can play towards the realization of this goal. We have described our previous work which has helped provide an improved understanding of fluid and ionic transport in processes as diverse as reverse osmosis, ion exchange, gas separations, and redox flow batteries. The application of these simulations has shown how they can play two important roles. Foremost, they can allow us to understand phenomena not well understood, e.g., why water can permeate membranes in RO separations and ions cannot; or why N_2/O_2 and CO_2/N_2 show opposite behavior in their separation factors when switching from pure fluids to mixtures; or why zeolite membranes can be effective in dehydrating alcohols via pervaporation when 5% water is present and become ineffective when only 1% water is present. Secondly, they can be effective screening tools for determining which membranes are suitable for a potential application, e.g., simulations clearly identified the zeolite membranes best suited for redox flow batteries while simultaneously revealing details about the ion transport process and how it can be affected by membrane structure and composition. Both of these roles can ultimately help provide a better understanding of the physiochemical behavior of many such membrane-based applications and aid in more efficient process design and optimization.

Author Contributions: All authors contributed equally to all aspects of this publication.

Funding: This research was supported by grants from the National Science Foundation (CBET 1545560) and State Key Laboratory of Materials-Oriented Chemical Engineering (KL13-05).

Conflicts of Interest: The authors declare no conflict of interest.

References

1. Sholl, D.S.; Lively, R.P. Seven chemical separations to change the world. *Nature* **2016**, *532*, 435–437. [CrossRef] [PubMed]
2. Lively, R.P.; Sholl, D.S. From water to organics in membrane separations. *Nat. Mater.* **2017**, *16*, 276–279. [CrossRef] [PubMed]
3. Widodo, S.; Khoiruddin; Ariono, D.; Subagjo; Wenten, I.G. Membrane separation for non-aqueous solution. *IOP Conf. Ser. Mater. Sci. Eng.* **2018**, *285*, 012008. [CrossRef]
4. Yuan, H.; Yu, B.; Cong, H.; Peng, Q.; Yang, R.; Yang, S.; Yang, Z.; Luo, Y.; Xu, T.; Zhang, H.; et al. Modification Progress of Polymer Membranes for Gas Separation. *Rev. Adv. Mater. Sci.* **2016**, *44*, 207–220.
5. Soleimany, A.; Hosseini, S.S.; Gallucci, F. Recent progress in developments of membrane materials and modification techniques for high performance helium separation and recovery: A review. *Chem. Eng. Process. Process Intensif.* **2017**, *122*, 296–318. [CrossRef]
6. Allen, M.P.; Tildesley, D.J. *Computer Simulation of Liquids*, 2nd ed.; Oxford University Press: Oxford, UK, 2017; ISBN 978-0-19-252470-6.
7. Qu, F.; Shi, R.; Peng, L.; Zhang, Y.; Gu, X.; Wang, X.; Murad, S. Understanding the effect of zeolite crystal expansion/contraction on separation performance of NaA zeolite membrane: A combined experimental and molecular simulation study. *J. Membr. Sci.* **2017**, *539*, 14–23. [CrossRef]
8. Lin, J.; Murad, S. A computer simulation study of the separation of aqueous solutions using thin zeolite membranes. *Mol. Phys.* **2001**, *99*, 1175–1181. [CrossRef]
9. Berendsen, H.J.C.; Postma, J.P.M.; van Gunsteren, W.F.; Hermans, J. Interaction Models for Water in Relation to Protein Hydration. In *Intermolecular Forces*; The Jerusalem Symposia on Quantum Chemistry and Biochemistry; Springer: Dordrecht, The Netherlands, 1981; pp. 331–342. ISBN 978-90-481-8368-5.
10. Jorgensen, W.L.; Bigot, B.; Chandrasekhar, J. Quantum and statistical mechanical studies of liquids. 21. The nature of dilute solutions of sodium and methoxide ions in methanol. *J. Am. Chem. Soc.* **1982**, *104*, 4584–4591. [CrossRef]
11. Tironi, I.G.; Sperb, R.; Smith, P.E.; van Gunsteren, W.F. A generalized reaction field method for molecular dynamics simulations. *J. Chem. Phys.* **1995**, *102*, 5451–5459. [CrossRef]
12. Murad, S.; Oder, K.; Lin, J. Molecular simulation of osmosis, reverse osmosis, and electro-osmosis in aqueous and methanolic electrolyte solutions. *Mol. Phys.* **1998**, *95*, 401–408. [CrossRef]
13. Kotelyanskii, M.J.; Wagner, N.J.; Paulaitis, M.E. Atomistic simulation of water and salt transport in the reverse osmosis membrane FT-30. *J. Membr. Sci.* **1998**, *139*, 1–16. [CrossRef]
14. Murad, S.; Jia, W.; Krishnamurthy, M. Ion-exchange of monovalent and bivalent cations with NaA zeolite membranes: A molecular dynamics study. *Mol. Phys.* **2004**, *102*, 2103–2112. [CrossRef]
15. Townsend, R.P.; Harjula, R. Ion Exchange in Molecular Sieves by Conventional Techniques. In *Post-Synthesis Modification I*; Molecular Sieves; Springer: Berlin/Heidelberg, Germany, 2002; pp. 1–42, ISBN 978-3-540-64334-0.
16. Ponce de León, C.; Frías-Ferrer, A.; González-García, J.; Szánto, D.A.; Walsh, F.C. Redox flow cells for energy conversion. *J. Power Sources* **2006**, *160*, 716–732. [CrossRef]
17. Cheng, F.; Liang, J.; Tao, Z.; Chen, J. Functional Materials for Rechargeable Batteries. *Adv. Mater.* **2011**, *23*, 1695–1715. [CrossRef] [PubMed]
18. Codina, G.; Perez, J.R.; Lopez-Atalaya, M.; Vasquez, J.L.; Aldaz, A. Development of a 0.1 kW power accumulation pilot plant based on an Fe/Cr redox flow battery Part I. Considerations on flow-distribution design. *J. Power Sources* **1994**, *48*, 293–302. [CrossRef]
19. Lim, H.S.; Lackner, A.M.; Knechtli, R.C. Zinc-Bromine Secondary Battery. *J. Electrochem. Soc.* **1977**, *124*, 1154–1157. [CrossRef]
20. Clarke, R.L.; Dougherty, B.J.; Harrison, S.; Millington, J.P.; Mohanta, S. Battery with Bifunctional Electrolyte. U.S. Patent 6,986,966, 17 January 2006.

21. Skyllas-Kazacos, M.; Rychcik, M.; Robins, R.G.; Fane, A.G.; Green, M.A. New All-Vanadium Redox Flow Cell. *J. Electrochem. Soc.* **1986**, *133*, 1057–1058. [CrossRef]
22. Díaz-González, F.; Sumper, A.; Gomis-Bellmunt, O.; Villafáfila-Robles, R. A review of energy storage technologies for wind power applications. *Renew. Sustain. Energy Rev.* **2012**, *16*, 2154–2171. [CrossRef]
23. Li, X.; Zhang, H.; Mai, Z.; Zhang, H.; Vankelecom, I. Ion exchange membranes for vanadium redox flow battery (VRB) applications. *Energy Environ. Sci.* **2011**, *4*, 1147–1160. [CrossRef]
24. Li, X. *Principles of Fuel Cells*; Taylor & Francis: Boca Raton, FL, USA, 2005; ISBN 978-1-59169-022-1.
25. Vijayakumar, M.; Bhuvaneswari, M.S.; Nachimuthu, P.; Schwenzer, B.; Kim, S.; Yang, Z.; Liu, J.; Graff, G.L.; Thevuthasan, S.; Hu, J. Spectroscopic investigations of the fouling process on Nafion membranes in vanadium redox flow batteries. *J. Membr. Sci.* **2011**, *366*, 325–334. [CrossRef]
26. Mohammadi, T.; Kazacos, M.S. Evaluation of the chemical stability of some membranes in vanadium solution. *J. Appl. Electrochem.* **1997**, *27*, 153–160. [CrossRef]
27. Dai, H.; Zhang, H.; Zhong, H.; Li, X.; Xiao, S.; Mai, Z. High performance composite membranes with enhanced dimensional stability for use in PEMFC. *Int. J. Hydrog. Energy* **2010**, *35*, 4209–4214. [CrossRef]
28. Lew, C.M.; Cai, R.; Yan, Y. Zeolite Thin Films: From Computer Chips to Space Stations. *Acc. Chem. Res.* **2010**, *43*, 210–219. [CrossRef] [PubMed]
29. Deng, S.G.; Lin, Y.S. Sulfur Dioxide Sorption Properties and Thermal Stability of Hydrophobic Zeolites. *Ind. Eng. Chem. Res.* **1995**, *34*, 4063–4070. [CrossRef]
30. Auerbach, S.M.; Carrado, K.A.; Dutta, P.K. *Handbook of Zeolite Science and Technology*; CRC Press: Boca Raton, FL, USA, 2003; ISBN 978-0-203-91116-7.
31. Li, L.; Dong, J.; Nenoff, T.M.; Lee, R. Desalination by reverse osmosis using MFI zeolite membranes. *J. Membr. Sci.* **2004**, *243*, 401–404. [CrossRef]
32. Lia, L.; Dong, J.; Nenoff, T.M.; Lee, R. Reverse osmosis of ionic aqueous solutions on a MFI zeolite membrane. *Desalination* **2004**, *170*, 309–316. [CrossRef]
33. Yang, R.; Xu, Z.; Yang, S.; Li, L.; Angelopoulos, A.; Dong, J. Nonionic Zeolite Membrane as Potential Ion Separator in Redox-Flow Battery. *J. Membr. Sci.* **2014**, *450*, 12–17. [CrossRef]
34. Hinkle, K.R.; Jameson, C.J.; Murad, S. Transport of Vanadium and Oxovanadium Ions Across Zeolite Membranes: A Molecular Dynamics Study. *J. Phys. Chem. C* **2014**, *118*, 23803–23810. [CrossRef]
35. Baerlocher, C.; McCusker, L.B.; Olson, D.H. *Atlas of Zeolite Framework Types*; Elsevier: New York, NY, USA, 2007; ISBN 978-0-08-055434-1.
36. Xu, Z.; Michos, I.; Wang, X.; Yang, R.; Gu, X.; Dong, J. A zeolite ion exchange membrane for redox flow batteries. *Chem. Commun.* **2014**, *50*, 2416–2419. [CrossRef] [PubMed]
37. Kirkwood, J.G. Statistical Mechanics of Fluid Mixtures. *J. Chem. Phys.* **1935**, *3*, 300–313. [CrossRef]
38. Song, B.; Yuan, H.; Jameson, C.J.; Murad, S. Role of surface ligands in nanoparticle permeation through a model membrane: A coarse-grained molecular dynamics simulations study. *Mol. Phys.* **2012**, *110*, 2181–2195. [CrossRef]
39. Olson, D.H.; Khosrovani, N.; Peters, A.W.; Toby, B.H. Crystal Structure of Dehydrated CsZSM-5 (5.8Al): Evidence for Nonrandom Aluminum Distribution. *J. Phys. Chem. B* **2000**, *104*, 4844–4848. [CrossRef]
40. Sethia, G.; Pillai, R.S.; Dangi, G.P.; Somani, R.S.; Bajaj, H.C.; Jasra, R.V. Sorption of Methane, Nitrogen, Oxygen, and Argon in ZSM-5 with different SiO2/Al2O3 Ratios: Grand Canonical Monte Carlo Simulation and Volumetric Measurements. *Ind. Eng. Chem. Res.* **2010**, *49*, 2353–2362. [CrossRef]
41. Bell, R.G.; Jackson, R.A.; Catlow, C.R.A. Löwenstein's rule in zeolite A: A computational study. *Zeolites* **1992**, *12*, 870–871. [CrossRef]
42. Catlow, C.R.A.; George, A.R.; Freeman, C.M. Ab initio and molecular-mechanics studies of aluminosilicate fragments, and the origin of Lowenstein's rule. *Chem. Commun.* **1996**, *0*, 1311–1312. [CrossRef]
43. Wijmans, J.G.; Baker, R.W. The solution-diffusion model: A review. *J. Membr. Sci.* **1995**, *107*, 1–21. [CrossRef]
44. Xu, Z.; Michos, I.; Cao, Z.; Jing, W.; Gu, X.; Hinkle, K.; Murad, S.; Dong, J. Proton-Selective Ion Transport in ZSM-5 Zeolite Membrane. *J. Phys. Chem. C* **2016**, *120*, 26386–26392. [CrossRef]
45. Agmon, N. The Grotthuss mechanism. *Chem. Phys. Lett.* **1995**, *244*, 456–462. [CrossRef]
46. Lobaugh, J.; Voth, G.A. The quantum dynamics of an excess proton in water. *J. Chem. Phys.* **1996**, *104*, 2056–2069. [CrossRef]
47. Schmitt, U.W.; Voth, G.A. Multistate Empirical Valence Bond Model for Proton Transport in Water. *J. Phys. Chem. B* **1998**, *102*, 5547–5551. [CrossRef]

48. Schmitt, U.W.; Voth, G.A. The computer simulation of proton transport in water. *J. Chem. Phys.* **1999**, *111*, 9361–9381. [CrossRef]
49. Day, T.J.F.; Schmitt, U.W.; Voth, G.A. The Mechanism of Hydrated Proton Transport in Water. *J. Am. Chem. Soc.* **2000**, *122*, 12027–12028. [CrossRef]
50. Kusaka, I.; Wang, Z.-G.; Seinfeld, J.H. Binary nucleation of sulfuric acid-water: Monte Carlo simulation. *J. Chem. Phys.* **1998**, *108*, 6829–6848. [CrossRef]

processes

MDPI

Review

Challenges in Nanofluidics—Beyond Navier–Stokes at the Molecular Scale

Peter J. Daivis [1,*,†] and Billy D. Todd [2,*,†]

1 School of Science and Centre for Molecular and Nanoscale Physics, RMIT University, GPO Box 2476, Melbourne, Victoria 3001, Australia
2 Department of Mathematics, School of Science, Faculty of Science, Engineering and Technology, Swinburne University of Technology, PO Box 218, Hawthorn, Victoria 3122, Australia
* Correspondence: peter.daivis@rmit.edu.au (P.J.D.); btodd@swin.edu.au (B.D.T.); Tel.: +61-3-9925-3393 (P.J.D.)
† Both authors contributed to this work, with P.J.D. taking the leading role in writing the manuscript.

Received: 18 July 2018; Accepted: 21 August 2018; Published: 1 September 2018

Abstract: The fluid dynamics of macroscopic and microscopic systems is well developed and has been extensively validated. Its extraordinary success makes it tempting to apply Navier–Stokes fluid dynamics without modification to systems of ever decreasing dimensions as studies of nanofluidics become more prevalent. However, this can result in serious error. In this paper, we discuss several ways in which nanoconfined fluid flow differs from macroscopic flow. We give particular attention to several topics that have recently received attention in the literature: slip, spin angular momentum coupling, nonlocal stress response and density inhomogeneity. In principle, all of these effects can now be accurately modelled using validated theories. Although the basic principles are now fairly well understood, much work remains to be done in their application.

Keywords: nanofluidics; molecular dynamics; hydrodynamics; slip; spin-coupling; non-local constitutive equations

---

1. Introduction

The classical Navier–Stokes theory describing flow of Newtonian fluids has been remarkably successful, but it is inadequate under certain conditions. If the rate of deformation is high, nonlinear effects such as a shear rate dependent viscosity and normal stress differences may become apparent. If the frequency of oscillatory deformation is high, we may observe viscoelastic effects related to the elastic storage and release of energy. These effects are now quite well understood and can be described using standard treatments of non-Newtonian fluid mechanics [1]. Generally speaking, the theory of fluid flow at macroscopic and microscopic scales is successful and well developed. However, when Navier–Stokes fluid dynamics and its extensions to shear rate dependent and frequency dependent constitutive relations are applied to nano-confined flows, the theory can fail. New physical effects become important and serious errors and inconsistencies can arise if the methods of macroscopic fluid mechanics are used without modification.

The reason for this is that several effects that are negligible or absent in macroscopic flows may become significant or even dominant in nanoflows. Wall slip, spin angular momentum coupling, spatially nonlocal response, and nonlinear, nonlocal coupling of the shear pressure and velocity gradient to a strongly inhomogeneous density field can all modify the flow of a fluid under strong confinement. Recent studies have shown that all of these effects can be successfully modelled. Experimental studies of fluid flow are steadily being extended to smaller and smaller system sizes, but computer simulations have a unique advantage in studies of nanoconfined fluid flow. They are ideally suited to studies of nanoflows because the system sizes and timescales that are accessible in computer simulations match the relevant size and timescales. In addition, computer simulations allow us to study velocity, density and shear pressure profiles at resolutions that would be impossible to achieve experimentally.

In this paper, we review some recent advances in the fluid dynamics of nanoconfined liquids, placing them in context and discussing the conditions under which they become important. In this work, we focus on single component fluids. To study multicomponent systems, we would need to include the composition as an additional variable and consider the many ways that the composition couples to the effects that are already present for pure fluids. Likewise, we have not considered electrical effects, which are so important in microfluidic and lab on a chip systems. These topics have been discussed in detail by other authors [2–5], and we urge interested readers to consult their work. We assume that all flows discussed are time independent, so we restrict our attention to steady states, but it is worth mentioning that nanoscale viscoelasticity remains largely unexplored.

2. Slip

In most macroscopic flow situations, a stick boundary condition, where the fluid velocity at the wall is taken to be equal to the wall velocity, is assumed. At the macroscopic scale, it is only in extreme cases that we must allow for slip, for example when we have plug flow of a paste or polymer melt, or when trapped gas bubbles in superhydrophobic surfaces lead to extreme slip [6]. However, even simple liquids can experience some slip. Strong slip is expected for water flowing near hydrophobic surfaces such as graphene or carbon nanotubes. Given a suitable surface structure, slip can also be observed for relatively hydrophilic surfaces [7]. For Poiseuille flow through a wide channel, the slip velocity, defined as the difference between the wall velocity and the fluid velocity at the wall, is only a very small fraction of the maximum velocity of the fluid in the channel, and so it is safely approximated as zero. On the other hand, for very narrow channels of nanometre scale, the slip velocity can be a significant fraction of the maximum velocity in the channel. Predictions of flow rates based on the assumption of the no-slip boundary condition can therefore be significantly in error, underestimating the true flow rate.

Here, we focus on the slip that occurs at the atomic scale on atomically smooth surfaces, sometimes known as intrinsic slip. Slip that occurs on a larger length scale, involving structured or patterned surfaces, roughness and chemical heterogeneity has been discussed by other authors [8].

The slip velocity depends on the strain rate at the wall and is not a material property. The relevant material property describing slip is the slip friction coefficient, defined below. The ratio of the fluid viscosity to the slip friction coefficient gives us another material property (really a property of the interface between the two materials), called the slip length. Very high spatial resolution measurements of the flow velocity of aqueous solutions near a hydrophobic wall, for example, have shown slip lengths of 80–100 nm [9]. The slip length of water confined by highly hydrophobic graphene and carbon nanotube surfaces has been computed to be around 60 nm, but there is enormous variation in both experimental and simulation results [10], due partly to subtle differences in simulation technique [11] and molecular models, but also error prone data analysis [12] and experimental difficulties. The consensus of careful simulation and experimental studies of flow of water on molecularly smooth hydrophobic surfaces is that slip lengths on these surfaces typically vary from nanometres up to tens of nanometres [10]. Because it has been difficult to reproduce the much larger experimental values sometimes found, it is possible that measurements of some of the larger values could suffer from uncontrolled experimental factors, such as dissolved or trapped gases [13], surface roughness and imperfections [14] or other yet unidentified factors.

One of the difficulties with computation of the slip length and slip friction coefficient by direct evaluation in non-equilibrium molecular dynamics that simulate flow through a channel is that for high slip systems, the velocity profile is almost flat. Extrapolation of such a velocity profile to the position where the velocity is zero is extremely error prone [12]. Methods based on equilibrium correlation functions do not suffer from these problems. Several theoretical discussions of correlation function methods for computation of the slip friction coefficient have been published [15–20] but there is still some doubt concerning the agreement between the different forms. Here, we describe a simple one [18] that has been extensively validated by comparison with both nonequilibrium simulations and experimental results [10,12,21–23].

Navier's slip friction coefficient ζ is defined as the proportionality coefficient relating the shear pressure P_{yx} at the wall to the velocity difference between the wall and the fluid Δv_x, which we call the slip velocity

$$P_{yx} = -\zeta \Delta v_x, \tag{1}$$

where we assume that the flow is in the x direction and the velocity gradient is in the y direction. The flow geometry for flow in the x direction adjacent to a flat solid wall is shown in Figure 1.

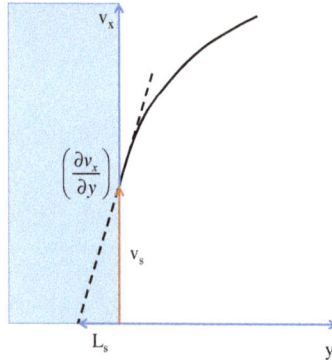

Figure 1. Schematic diagram showing the flow geometry for the definition of the slip length. The magnitude of the velocity gradient in the fluid at the wall is equal to v_s/L_s where L_s is the magnitude of the slip length. Here, we have $\Delta v_x = v_x - 0$ because we assume flow between stationary walls.

The viscous pressure in the fluid at the wall is given by Newton's law of viscosity

$$P_{yx} = -\eta \left(\frac{\partial v_x}{\partial y} \right)_{wall}, \tag{2}$$

where η is the shear viscosity at zero shear rate. Since the shear pressure must be continuous, both values of the shear pressure must be equal. Eliminating the shear pressure, we find

$$\left(\frac{\partial v_x}{\partial y} \right)_{wall} = \frac{\zeta}{\eta} \Delta v_x = \frac{\Delta v_x}{L_s} \tag{3}$$

which defines the slip length as

$$L_s = \frac{\eta}{\zeta}. \tag{4}$$

To derive a correlation function expression for the slip friction coefficient, we must introduce a generalised constitutive equation that describes the fluctuations [18]. It is convenient to formulate this in terms of the wall–fluid shear force, given by $F'(t) = A P_{yx}$

$$F'(t) = -A \int_0^t \zeta(t - \tau) \Delta v_x(\tau) \, d\tau + F'_R \tag{5}$$

where A is the area over which the slip frictional force acts, $\zeta(t - \tau)$ is the friction kernel and F'_R is the random component of the shear pressure. Now, we multiply both sides of the constitutive equation by the slip velocity and ensemble average to form the correlation functions

$$C_{F'v_S}(t) \equiv \langle F'(t) \, v_S(0) \rangle = -\int_0^t \zeta(t-\tau) C_{v_S v_S}(\tau) \, d\tau \tag{6}$$

where

$$C_{v_S v_S} = \langle v_S(t) \, v_S(0) \rangle . \tag{7}$$

When this equation is Laplace transformed, we find

$$\tilde{C}_{F'v_S}(s) = -\tilde{\zeta}(s) \, \tilde{C}_{v_S v_S}(s) . \tag{8}$$

The friction kernel is well approximated by a sum of exponentials [18],

$$\zeta(t) = \sum_i \zeta_i e^{-\lambda_i t} \tag{9}$$

which has the Laplace transform

$$\tilde{\zeta}(s) = \sum_i \frac{\zeta_i}{s + \lambda_i} . \tag{10}$$

In practice, a single exponential is often sufficient [18,21]. The amplitudes and decay rates of the exponentials can then be obtained from fits to the Laplace transformed correlation functions.

From the computational point of view, the wall–fluid shear force is easily evaluated in computer simulations. The slip velocity is more problematic. One way to evaluate it would be to fit the instantaneous velocity profile each time the correlation functions are evaluated. However, this requires the assumption of a fitting function, which could be biassed. Another way is to evaluate the instantaneous velocity of the fluid averaged over a region within a distance Δ of the wall. This has the disadvantage that averaging over a finite region could also result in error, but the calculation of the slip friction coefficient for a given wall–fluid combination can be repeated for different values of Δ, and the most physically meaningful value of the slip friction coefficient chosen. In practice, it is straightforward to choose the most physically meaningful value of the slip friction coefficient because it quickly increases to a broad maximum (plateau) value as Δ is increased, before steadily decreasing thereafter. This maximum value usually occurs when Δ is approximately equal to one molecular diameter. Choosing this value of the slip friction coefficient gives excellent agreement with the results of nonequilibrium molecular dynamics simulations [18,21].

This method has been used to evaluate the slip friction coefficient and the slip length of a simple Lennard–Jones type atomic fluid near a solid planar LJ wall [18] and a graphene wall [21] and more complex molecular fluids such as water against both Lennard–Jones atomic walls and graphene [12]. It has also been adapted to a cylindrical geometry [22] for studies of the slip friction coefficient of water in carbon nanotubes [10,23].

3. Spin Angular Momentum Coupling

It is well known that extended molecules spin in a shear field [24]. What is less well known is that the spin angular velocity is coupled to the translational velocity. This coupling is usually negligible in macroscopic flows, but it can become significant at the nanoscale.

To describe this effect, we begin with the extended Navier–Stokes equations for stable flow in a planar flow geometry (where the convective terms are zero), which are obtained by inserting the relevant linear transport equations into the balance equations for the translational and angular momentum [25,26],

$$\rho \frac{\partial v_x}{\partial t} = (\eta + \eta_r) \frac{\partial^2 v_x}{\partial y^2} + 2\eta_r \frac{\partial \omega_z}{\partial y} + \rho F^e$$

$$\rho \Theta \frac{\partial \omega_z}{\partial t} = (\zeta + \zeta_{rr}) \frac{\partial^2 \omega_z}{\partial y^2} - 2\eta_r \left(\frac{\partial v_x}{\partial y} + 2\omega_z \right) + \rho \Gamma_z^e . \tag{11}$$

The first of these describes the evolution of the translational fluid streaming velocity in the flow direction v_x. A new transport coefficient, the rotational viscosity η_r, which governs the relaxation of spin angular momentum appears in addition to the usual shear viscosity, η. The rotational (or vortex) viscosity η_r describes the rate of conversion between fluid vorticity $\nabla \times \mathbf{v}$ and molecular spin angular momentum ω. If $\eta_r = 0$, then we just have the usual Navier–Stokes equation. In the absence of a pressure gradient (omitted from this equation since it is zero for field driven flows in our flow geometry), flow can be generated by the external body force density ρF^e, but if this is also zero, we see that it is also possible to have a flow driven by the gradient of the angular velocity. We will return to this point later. The second equation, which describes the evolution of the molecular spin angular velocity field includes a diffusive term involving the sum of the spin viscosities $\zeta + \zeta_{rr}$ as well as the rotational viscosity η_r. The spin viscosity ζ describes the diffusive flux of spin angular momentum due to the traceless symmetric part of the spin angular velocity gradient, while ζ_{rr} describes the diffusive flux of spin angular momentum due to the antisymmetric part of the spin angular velocity gradient [25]. These transport coefficients have been evaluated for some molecular fluids, including water [27]. Again, there is a term that accounts for external fields, this time in the form of an external body torque density $\rho \Gamma_z^e$.

When applied to planar Poiseuille flow through a narrow channel driven by an external body force density ρF^e, these equations predict a difference between the velocity field calculated from the Navier–Stokes equation alone compared to the results of the extended Navier–Stokes equations including the spin coupling. Simulations of a molecular fluid consisting of extended linear molecules (buta-triene) show that the flow rate difference is small for channels of width greater than 7 nm, but it grows to around 10% at a channel width of 1 nm [28]. For very narrow channels, accurate prediction of flow rates and velocity profiles requires consideration of the extended Navier–Stokes equations [26,28,29]. Figure 2 shows the flow rate reduction predicted by taking spin angular momentum coupling into account for nanoconfined flows of a dumbbell fluid, liquid butane and liquid water [26].

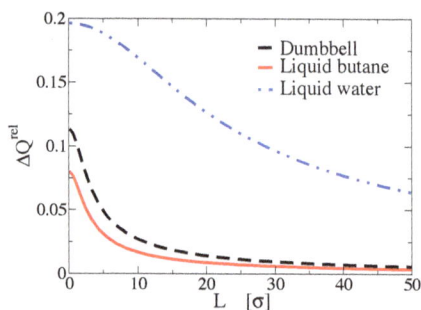

Figure 2. Calculated relative flow rate reduction $\Delta Q^{rel} = (Q_{NS} - Q_{ENS})/Q_{NS}$ between predictions of the Navier–Stokes equations and extended Navier–Stokes equations (including spin angular momentum coupling) for a dumbbell fluid, liquid butane, and liquid water in planar Poiseuille flow. The horizontal axis represents the channel width in units of the Lennard–Jones intermolecular potential parameter σ. $\sigma = 3.92$ Å and 3.17 Å for butane and water, respectively. Reprinted with permission from Hansen, J.S.; Dyre, J.C.; Daivis, P.; Todd, B.D.; Bruus, H., Langmuir **31**, 13275 (2015). Copyright 2015 American Chemical Society [26].

As mentioned above, a translational flow can be generated even in the absence of a translational body force if a body torque is applied instead. With symmetric boundary conditions (for example, with a stick boundary condition on both sides of the channel), equal flow is generated in both directions and no net flow results, but if the boundary conditions are asymmetric, with slip on one wall and stick on the other, a net flow results. This means that an external torque that spins the molecules can be used to pump a fluid. It has been demonstrated that a rotating electric field applied to polar molecules (such as water) under these conditions can generate a net flow in a nanochannel or nanotube, without the need for electrolyte or a pressure gradient [30,31].

4. Nonlocal Response

In macroscopic flows, the most common causes of non-Newtonian behaviour are nonlinear and viscoelastic deviations from Navier–Stokes behaviour. The Pipkin diagram [32] shown in Figure 3 schematically illustrates the different regions of fluid behaviour for typical macroscopic fluids. Here, we are interested in steady (zero frequency) flows so the Deborah number $\omega\tau_v$ is zero.

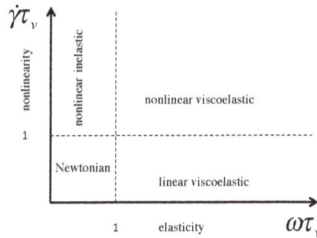

Figure 3. Diagram showing rheological flow regimes for typical macroscopic fluids. The horizontal axis represents the degree of elasticity exhibited by the flow, which is controlled by the Deborah number, the product of the characteristic frequency of the flow ω and the viscoelastic relaxation time of the fluid τ_v. The vertical axis represents the degree of nonlinearity of the flow, controlled by the Weissenberg number, the product of the characteristic strain rate $\dot{\gamma}$ and the viscoelastic relaxation time τ_v.

In most macroscopic situations, spatial nonlocality is unimportant, except possibly when the system is near a glass transition [33] or the viscous correlation length is extraordinarily large for some other reason as it is for example in suspensions of long fibres that exhibit shear banding [34,35]. By contrast, in nanofluidic flows, spatial nonlocality can be a dominant effect that must be taken into account, even for ordinary molecular or atomic fluids. Deformations at sufficiently high wavenumber may differ strongly from homogeneous deformation, particularly in glassy liquids where dynamic heterogeneity with nanoscale dimensions is observed [33]. This means we must consider the concept of a spatially nonlocal response, where the stress at a point becomes a linear functional of the local deformation rate [36]. The stress at a point then depends not only on the strain rate at that point but also on the strain rate at nearby points. An alternative interpretation is that the stress depends not only on the velocity gradient but also on its derivatives, and so there is a contribution to the stress resulting from the spatial derivatives of the strain rate.

To describe spatially inhomogeneous steady state flows where the strain rate is independent of time but it varies strongly in space, we can introduce an analogue of the Pipkin diagram to characterise nonlocality, as shown in Figure 4. The viscous correlation length of the fluid, ξ_v is measured by the width of the nonlocal viscosity kernel, $\eta(y-y')$ defined in Equation (12). For situations where strong density inhomogeneity is also present (discussed below in this section), third and fourth axes could be added, representing the amplitude and spatial frequency of the density inhomogeneity. This diagram does not include these dimensions, so it represents nonlocal response for a uniform density fluid.

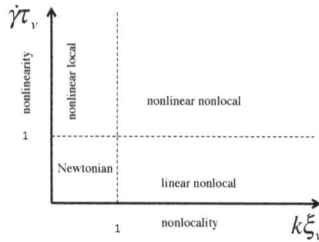

Figure 4. Diagram showing the rheological flow regimes for fluids exhibiting nonlocal shear pressure response under steady flow conditions. The horizontal axis represents the degree of nonlocality exhibited by the flow, which is controlled by the product of the characteristic spatial frequency (or wavenumber) of the flow k and the viscous correlation length of the fluid ξ_v. The vertical axis represents the degree of nonlinearity of the flow, controlled by the product of the characteristic strain rate $\dot{\gamma}$ and the viscoelastic relaxation time τ_v.

To account for the nonlocal response of the shear pressure to a steady velocity gradient of high spatial wavenumber, we apply a spatial convolution equation analogous to the linear viscoelastic constitutive equation

$$P_{yx}(y) = -\int_{-\infty}^{\infty} \eta\left(y - y'\right) \dot{\gamma}\left(y'\right) dy'. \tag{12}$$

This is the most general linear relationship that we can postulate between the shear pressure and the velocity gradient or shear rate $\dot{\gamma}$ for a fluid with spatially homogeneous density where the strain rate varies only in the y-direction. The viscosity kernel $\eta\left(y - y'\right)$ weights the contribution of the strain rate at different distances from the point at which the shear pressure is evaluated. When $y - y' = 0$, we expect the contribution of the shear rate to be greatest, and at larger values of $y - y'$, we expect the effect of the shear rate to be least, so η should be a decreasing function of $y - y'$. At macroscopic length scales, it is reasonable to approximate $\eta\left(y - y'\right)$ as a constant multiplied by a Dirac delta function, and then the convolution integral simply reduces to Newton's law of viscosity with a purely local response to the strain rate field.

A stringent test of this relationship can be made by applying a spatially sinusoidal transverse force to generate a sinusoidal velocity field. Since the velocity is sinusoidal, if the amplitude of the field is sufficiently small, the response is linear and the shear pressure response will also consist of a single sinusoidal component. When the wavelength of the sinusoidal driving force is sufficiently long (greater than a few molecular diameters) the shear pressure is just given by the Newtonian constitutive equation applied locally. In other words, the shear pressure at a given point depends only on the value of the strain rate at that point. However, when the wavelength of the strain rate oscillations is reduced to the order of a few molecular diameters, this procedure fails and the shear pressure is poorly reproduced. Using the nonlocal integral constitutive equation, it is possible to correctly predict the shear pressure, even when the velocity profile varies rapidly in space as displayed in Figure 5 [36].

In nanofluidic systems, it is possible for the strain rate to vary rapidly, and for a nonlocal response of the shear pressure to the rapidly varying strain rate to be observed. However, the rapid variation of the strain rate is also usually associated with rapid variation of the density of the fluid, due to strong molecular packing effects at the wall–fluid interface that occur even at equilibrium [37]. Therefore, we must consider the effect of spatial density variations on the velocity profile and the shear pressure profile.

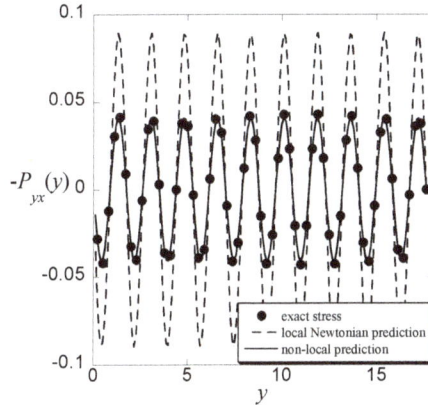

Figure 5. Shear stress $(-P_{yx})$ predictions using the simple Newtonian constitutive equation (dashed line), and the nonlocal constitutive equation (solid line) compared with the exact stress (filled circles) for a simple liquid [36] where the y-position is expressed in units of the Lennard–Jones potential distance parameter σ.

5. Density Inhomogeneity

When a macroscopic fluid system has long wavelength density inhomogeneities, the assumption of constant transport coefficients in the Navier–Stokes equation is clearly inadequate. We can allow for this by making the linear transport coefficients position dependent through the position dependence of the density as well as, if necessary, the temperature and composition. In the simple case of density variation for an isothermal single component fluid, we would write Newton's law of viscosity as

$$P_{yx} = -\eta\left(\rho\left(y\right)\right)\dot{\gamma}. \tag{13}$$

In a nanofluidic system, the fluid is almost always in contact with a solid wall. This induces strong oscillatory density variations near the wall, with local maxima that may exceed typical solid densities. Under these circumstances, it is obviously not viable to use the viscosity at the local density values. Bitsanis and coworkers [38,39] made a significant improvement on the local density model by proposing that the viscosity at a locally averaged value of the density could be used in the Newtonian viscosity equation. Despite the strong density variations near the wall, averaging over a sphere or planar layer of one to two molecular diameters gives an average density that is reasonably close to the liquid value, but still accounts for slow variations in the density. The constitutive equation for the local average density model in a planar geometry is then

$$P_{yx} = -\eta\left(\bar{\rho}\left(y\right)\right)\dot{\gamma} \tag{14}$$

with

$$\bar{\rho} = \frac{1}{\sigma}\int_{-\sigma/2}^{\sigma/2}\rho\left(y+s\right)ds. \tag{15}$$

This introduces an element of nonlocality into the constitutive equation for the shear pressure by making it dependent on the density averaged over a region surrounding the point of interest. The local averaged density model was successfully used by Bitsanis [38] to describe the velocity profile of a highly confined simple fluid. It has been extended by Hoang and Galliero [40–42], who investigated the effectiveness of different averaging kernels in the evaluation of the averaged density, and combined with a model for slip to provide a tractable and efficient hydrodynamic model for nanoflows by Bhadauria et al. [43,44]. The local average density model is, however, only a partial solution to the problem of describing the velocity profile for flow of a nanoconfined fluid. A major deficiency of this model is that it cannot account for the zeroes and velocity gradient reversals that can be seen in the velocity profiles of strongly confined fluids at high densities [45]. Strong molecular packing near the solid-fluid interface results in oscillations in the velocity profile that cannot possibly be described by the local average density model. Any constitutive equation that follows the same functional form as the simple Newtonian one would need to have infinite values of the viscosity at the points where the strain rate goes through zero, which is clearly unphysical. Therefore, we are again led to consider more general, nonlocal constitutive equations.

The fluid density inhomogeneity produced by wall–fluid interactions is uncontrolled. It is not possible to easily and independently control the amplitude and spatial frequency of the density oscillations due to the presence of a confining wall. To study nonlocal constitutive equations for the shear pressure, we need a more flexible framework. By applying a small amplitude sinusoidal longitudinal force (SLF) in a periodic simulation cell, it is possible to induce sinusoidal density variations [46] in a system with periodic boundaries and no explicit walls. When the amplitude of the applied force is increased, a nonlinear density response is generated. By adding together several sinusoidal forces of different frequency, it is possible to Fourier synthesise a spatially periodic density profile consistent with the periodic boundary conditions of the simulation cell that closely resembles the density profile seen in a nanoconfined fluid [47]. A sinusoidal transverse force (STF) can also be applied that generates flow. When both the longitudinal and transverse fields are applied together, we can generate flow in the presence of density inhomogeneity that can be controlled and manipulated at will [48,49].

Considering a system where the external fields vary in the y direction, the combined longitudinal and transverse force can be represented as

$$\mathbf{F}(y) = F^x(y)\mathbf{i} + F^y(y)\mathbf{j}. \tag{16}$$

Figure 6 schematically shows the two applied body forces.

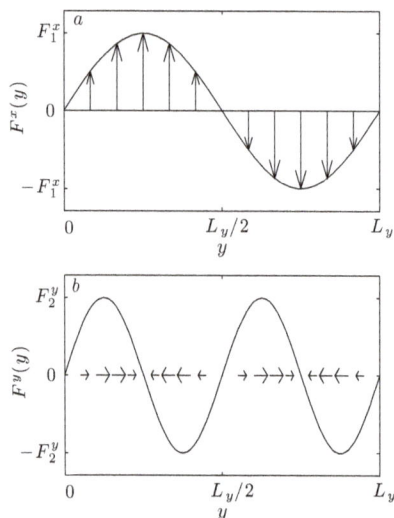

Figure 6. Schematic representation of: the sinusoidal transverse force (STF) (**a**); and the sinusoidal longitudinal force (SLF) (**b**) fields. The arrows show the direction of the forces. The length of the arrows, as well as the sinusoidal line, indicate the strength of the force. The STF is shown for $F^x(y) = F_1^x \sin(k_1 y)$ and the SLF is shown for $F^y(y) = F_2^y \sin(k_2 y)$. Reprinted figure with permission from Glavatskiy, K.S.; Dalton, B.A.; Daivis, P.J.; Todd, B.D., Phys. Rev. E **91**, 062132 (2015). Copyright (2015) by the American Physical Society [48].

The density response consists of contributions due to both the longitudinal and transverse forces. This is because the transverse force results in flow with a strain rate that varies in the y-direction. Heat is produced by the viscous dissipation associated with the velocity gradient, resulting in thermal expansion, which changes the density, in addition to the direct effect of the longitudinal force on the density. In general, the density response is given by a functional expansion that depends on both the longitudinal and transverse forces,

$$
\begin{aligned}
\rho(y) = \rho_0 + \sum_{\alpha_1} \int \chi_{\alpha_1}^{(1)}(y, y')\, F^{\alpha_1}(y')\, dy' \\
+ \frac{1}{2} \sum_{\alpha_1, \alpha_2} \int \chi_{\alpha_1 \alpha_2}^{(2)}(y, y', y'')\, F^{\alpha_1}(y') F^{\alpha_2}(y'')\, dy' dy'' + \cdots,
\end{aligned}
\tag{17}
$$

where α_1 and α_2 can be either x (transverse force) or y (longitudinal force) and the response functions are the functional derivatives

$$
\chi_{\alpha_1 \cdots \alpha_n}^{(n)}(y, y', \cdots, y^{n\prime}) = \left. \frac{\delta^n \rho[F^x(y); F^y(y)]}{\delta F^{\alpha_1}(y') \cdots \delta F^{\alpha_n}(y^{n\prime})} \right|_{F^x, F^y = 0}.
\tag{18}
$$

The response functions are evaluated at equilibrium. Due to the spatial symmetry of the equilibrium system, the response functions must depend only on even powers of the transverse force, because changing its direction cannot change the sign of its contribution to the density. We assume that truncation of the density response at second order in the forces gives a reasonable approximation. It accounts for the density response due to the longitudinal force with terms that are first and second order in the longitudinal field and also the heating and normal pressure effects of the shearing force, which occur to lowest order as a quadratic function of the transverse field,

$$
\begin{aligned}
\rho(y) = \rho_0 + \int \chi_y^{(1)}(y - y') F^y(y') dy' \\
+ \frac{1}{2} \int \chi_{yy}^{(2)}(y - y', y - y'') F^y(y') F^y(y'') dy' dy'' \\
+ \frac{1}{2} \int \chi_{xx}^{(2)}(y - y', y - y'') F^x(y') F^x(y'') dy' dy''.
\end{aligned}
\tag{19}
$$

By expressing both sides of this equation in terms of their Fourier series representations, it is possible to determine the Fourier coefficients of the response functions.

Applying similar arguments to the truncated functional expansion of the strain rate, we find

$$
\begin{aligned}
\dot{\gamma}(y) = \int \xi_x^{(1)}(y - y') F^x(y') dy' \\
+ \int \xi_{xy}^{(2)}(y - y', y - y'') F^x(y') F^y(y'') dy' dy'' \\
+ \int \xi_{xyy}^{(3)}(y - y', y - y'', y - y''') F^x(y') F^y(y'') F^y(y''') dy' dy'' dy'''.
\end{aligned}
\tag{20}
$$

This expression has been limited to terms that are linear in the transverse force and at most quadratic in the longitudinal force since we are mainly interested in cases where the shear is weak but the density inhomogeneity is strong. The shear pressure profile can be written in a similar form, as

$$
\begin{aligned}
\Pi(y) = \int \pi_x^{(1)}(y - y') F^x(y') dy' \\
+ \int \pi_{xy}^{(2)}(y - y', y - y'') F^x(y') F^y(y'') dy' dy'' \\
+ \int \pi_{xyy}^{(3)}(y - y', y - y'', y - y''') F^x(y') F^y(y'') F^y(y''') dy' dy'' dy''',
\end{aligned}
\tag{21}
$$

where $\pi_x^{(1)}$, $\pi_{xy}^{(2)}$, and $\pi_{xyy}^{(3)}$ are the corresponding response functions for shear pressure. Since the system in this case is spatially periodic, the density, strain rate and shear pressure profiles are also periodic and so they have Fourier series representations. The response functions can also be written in terms of their Fourier series representations. For particular combinations of transverse and longitudinal forces, we can isolate each specific response function and vary the spatial frequency to obtain its wavenumber dependence. For second and third order response functions, this can become quite complex, as they are functions of two or three wavenumber arguments. This procedure was described in detail by Dalton et al. [48,49]. To the best of our knowledge, this remains the only validated treatment that allows fully for nonlinear coupling between the density inhomogeneity and shear forces. Recent work by Camargo et al. [50] develops a dynamic density functional theoretical formalism for simple confined fluids, but to our knowledge this formalism has not yet been validated against simulation or experimental data.

Validation of the approach contained in References [46–49] has been provided in those papers. For combinations of single sinusoidal longitudinal and transverse forces at various wavenumbers, it was found that the density, strain rate and shear pressure profiles could all be adequately described by truncating the functional expansion for the density at second order, and truncating the strain rate and shear pressure functional expansions at third order. Some contributions to the third order response could be neglected, leading to considerable simplification.

When the longitudinal force responsible for generating density perturbations was extended to a more complicated superposition of sinusoidal components, it was found sufficient to include the longitudinal force components at most coupled in pairs [49]. This is a highly significant result, because it means that considerable simplification is possible. This approach was validated by applying a combination of longitudinal and transverse forces given by

$$\mathbf{F}(y) = F_1^x \sin(k_1 y)\mathbf{i} + \left(F_6^y \sin(k_6 y) + F_8^y \sin(k_8 y) + F_{10}^y \sin(k_{10} y) \right)\mathbf{j} \tag{22}$$

where $k_n = 2n\pi/L_y$ is the wave number of the STF or SLF, n is a positive integer and $\mathbf{i}$ and $\mathbf{j}$ are unit vectors in the x and y directions. By combining these three Fourier components in the SLF, it was possible to construct a non-trivial periodic density profile somewhat resembling the density profiles found in very narrow planar channels, as shown in Figure 7.

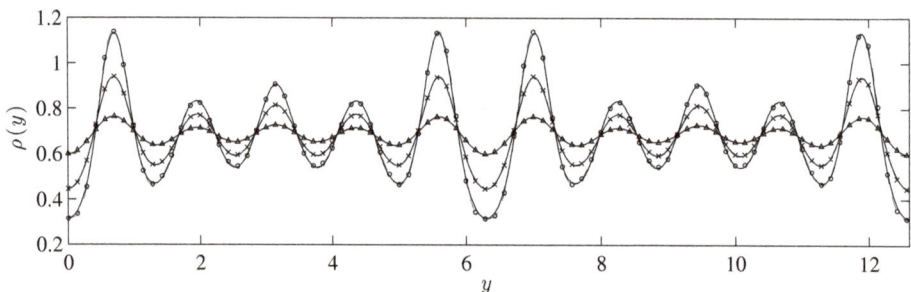

Figure 7. Comparison between density profiles obtained directly from MD simulations and the predictions of the truncated functional expansion using previously computed response functions for a three component SLF with $m_1 = 6$, $m_2 = 8$ and $m_3 = 10$. Bold lines without symbols represent MD simulations. Thin dashed lines indicated with triangles are for $F_6^y = F_8^y = F_{10}^y = 0.5$. Thin dashed lines indicated with crosses are for $F_6^y = F_8^y = F_{10}^y = 1.5$. Thin dashed lines indicated with circles are for $F_6^y = F_8^y = F_{10}^y = 2.5$. Reprinted figure with permission from Dalton, B.A.; Glavatskiy, K.S.; Daivis, P.J.; Todd, B.D. Phys. Rev. E **92**, 012108 (2015). Copyright (2015) by the American Physical Society [49].

With the addition of the transverse force, we can generate flow in a system with strong, and realistically complicated density inhomogeneity. Figure 8 shows the resulting velocity profiles with the predictions obtained from the truncated functional expansion for the velocity gradient. This method is clearly capable of accurately representing the response of a nanoscopic liquid system to a combination of confinement and flow-inducing forces.

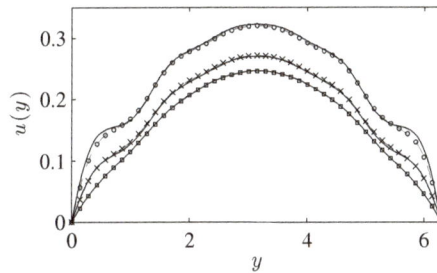

Figure 8. Comparison between velocity profiles obtained directly from MD simulations and predictions of the truncated functional expansion using independently calculated response functions for a single sinusoidal component STF and a three component SLF. Bold lines represent MD simulations, while thin dashed lines with symbols represent predictions. The system parameters labels are the same as those used in Figure 7. Velocity profiles are only shown for half of a wave cycle. Reprinted figure with permission from Dalton, B.A.; Glavatskiy, K.S.; Daivis, P.J.; Todd, B.D. Phys. Rev. E **92**, 012108 (2015). Copyright (2015) by the American Physical Society [49].

6. Conclusions

In this paper, we have provided a brief review of three fundamental phenomena that should be included in an accurate continuum treatment of nanofluidics: slip, spin angular momentum coupling, and non-local response. At the outset, we pointed out that additional complexities would need to be included to account for compositional variation (in binary and multicomponent fluids) and electrostatics, such that a full treatment of ionic solutions and liquids or even polar fluids can be treated. This is clearly some years away. The main complexity is in dealing with the full range of couplings and non-local effects that occur at the nanoscale. A major challenge remains in how to simplify the treatments given in References [46–50] and potentially others not discussed in this brief review such that they can be efficiently applied in predictive nanofluidics. Further advances in the modelling of slip will undoubtedly see the application of both NEMD and EMD methods to complex fluids (e.g., multicomponent mixtures and ionic liquids) and more targeted engineering applications, such as lubrication (see, for example, [51]). The coupling of linear and angular momentum to generate flow poses an intriguing potential application in fluid actuation at the nanoscale. While theoretical and simulation studies exist, there has been no experimental verification to our knowledge.

Author Contributions: P.J.D. and B.D.T. contributed jointly to the conception, supervision and execution of their coauthored work cited in this review. P.J.D. wrote the manuscript, B.D.T. edited and added to it.

Funding: The research undertaken by the authors that has been included in this review was funded by Australian Research Council grant numbers DP0663759 and DP120102976.

Acknowledgments: We would like to acknowledge the contributions made by our collaborators and coauthors to the work reviewed in this article: J.S. Hansen, S.K. Kannam, H. Bruus, S. DeLuca, J.C. Dyre, D. Ostler, F. Frascoli, R. Puscasu, B. Dalton and K. Glavatskiy.

Conflicts of Interest: The authors declare no conflict of interest.

References

1. Huilgol, R.R.; Phan-Thien, N. *Fluid Mechanics of Viscoelasticity*; Elsevier: Amsterdam, The Netherlands, 1997.
2. Qiao, R.; Aluru, N.R. Atomistic simulation of KCl transport in charged silicon nanochannels: Interfacial effects. *Colloids Surf. A Physicochem. Eng. Asp.* **2005**, *267*, 103–109.
3. Joly, L.; Ybert, C.; Trizac, E.; Bocquet, L. Liquid friction on charged surfaces: From hydrodynamic slippage to electrokinetics. *J. Chem. Phys.* **2006**, *125*, 204716.
4. Nilson, R.H.; Griffiths, S.K. Influence of atomistic physics on electro-osmotic flow: An analysis based on density functional theory. *J. Chem. Phys.* **2006**, *125*, 164510.
5. Schoch, R.B.; Han, J.; Renaud, P. Transport phenomena in nanofluidics. *Rev. Mod. Phys.* **2008**, *80*, 839.
6. Scarratt, L.R.J.; Steiner, U.; Neto, C. A review on the mechanical and thermodynamic robustness of superhydrophobic surfaces. *Adv. Colloid Interface Sci.* **2017**, *246*, 133–152.
7. Ho, T.A.; Papavassiliou, D.V.; Lee, L.L.; Striolo, A. Liquid water can slip on a hydrophilic surface. *Proc. Natl. Acad. Sci. USA* **2011**, *108*, 16170.
8. Hendy, S.C.; Lund, N.J. Effective slip boundary conditions for flows over nanoscale chemical heterogeneities. *Phys. Rev. E* **2007**, *76*, 066313.
9. Vinogradova, O.I.; Koynov, K.; Best, A.; Feuillebois, F. Direct Measurements of Hydrophobic Slippage Using Double-Focus Fluorescence Cross-Correlation. *Phys. Rev. Lett.* **2009**, *102*, 118302.
10. Kannam, S.K.; Todd, B.D.; Hansen, J.S.; Daivis, P.J. How fast does water flow in carbon nanotubes? *J. Chem. Phys.* **2013**, *138*, 094701.
11. Bernardi, S.; Todd, B.D.; Searles, D.J. Thermostating highly confined fluids. *J. Chem. Phys.* **2010**, *132*, 244706.
12. Kannam, S.K.; Todd, B.D.; Hansen, J.S.; Daivis, P.J. Slip length of water on graphene: Limitations of non-equilibrium molecular dynamics simulations. *J. Chem. Phys.* **2012**, *136*, 024705.
13. Huang, D.M.; Sendner, C.; Horinek, D.; Netz, R.R.; Bocquet, L. Water Slippage versus Contact Angle: A Quasiuniversal Relationship. *Phys. Rev. Lett.* **2008**, *101*, 226101.
14. Chinappi, M.; Casciola, C.M. Intrinsic slip on hydrophobic self-assembled monolayer coatings. *Phys. Fluids* **2010**, *22*, 042003.
15. Bocquet, L.; Barrat, J.L. Hydrodynamic boundary-conditions, correlation functions, and Kubo relations for confined fluids. *Phys. Rev. E* **1994**, *49*, 3079.
16. Petravic, J.; Harrowell, P. On the equilibrium calculation of the friction coefficient for liquid slip against a wall. *J. Chem. Phys.* **2007**, *127*, 174706.
17. Kobryn, A.E.; Kovalenko, A. Molecular theory of hydrodynamic boundary conditions in nanofluidics. *J. Chem. Phys.* **2008**, *129*, 134701.
18. Hansen, J.S.; Todd, B.D.; Daivis, P.J. Prediction of fluid velocity slip at solid surfaces. *Phys. Rev. E* **2011**, *84*, 016313.
19. Huang, K.; Szlufarska, I. Green-Kubo relation for friction at liquid-solid surfaces. *Phys. Rev. E* **2014**, *89*, 032118.
20. Chen, S.; Wang, H.; Qian, T.; Sheng, P. Determining hydrodynamic boundary conditions from equilibrium fluctuations. *Phys. Rev. E* **2015**, *92*, 043007.
21. Kannam, S.K.; Todd, B.D.; Hansen, J.S.; Daivis, P.J. Slip flow in graphene nanochannels. *J. Chem. Phys.* **2011**, *135*, 144701.
22. Kannam, S.K.; Todd, B.D.; Hansen, J.S.; Daivis, P.J. Interfacial slip friction at a fluid-solid cylindrical boundary. *J. Chem. Phys.* **2012**, *136*, 244704.
23. Kannam, S.K.; Daivis, P.J.; Todd, B.D. Modeling slip and flow enhancement of water in carbon nanotubes. *MRS Bull.* **2017**, *42*, 283–288.
24. Yamakawa, H. *Modern Theory of Polymer Solutions*; Harper and Row: New York, NY, USA, 1971.
25. Todd, B.D.; Daivis, P.J. *Nonequilibrium Molecular Dynamics: Theory, Algorithms and Applications*; Cambridge University Press: New York, NY, USA, 2017.
26. Hansen, J.S.; Dyre, J.C.; Daivis, P.; Todd, B.D.; Bruus, H. Continuum Nanofluidics. *Langmuir* **2015**, *31*, 13275.
27. Hansen, J.S.; Bruus, H.; Todd, B.D.; Daivis, P.J. Rotational and spin viscosities of water: Application to nanofluidics. *J. Chem. Phys.* **2010**, *133*, 144906.
28. Hansen, J.S.; Daivis, P.J.; Todd, B.D. Molecular spin in nano-confined fluidic flows. *Microfluid. Nanfluid.* **2009**, *6*, 785–795.

29. Hansen, J.S.; Dyre, J.C.; Daivis, P.J.; Todd, B.D.; Bruus, H. Nanoflow hydrodynamics. *Phys. Rev. E* **2011**, *84*, 036311.
30. De Luca, S.; Todd, B.D.; Hansen, J.S.; Daivis, P.J. Molecular dynamics study of nanoconfined water flow driven by rotating electric fields under realistic experimental conditions. *Langmuir* **2014**, *30*, 3095–3109.
31. Ostler, D.; Kannam, S.K.; Daivis, P.J.; Frascoli, F.; Todd, B.D. Electropumping of Water in Functionalized Carbon Nanotubes Using Rotating Electric Fields. *J. Phys. Chem. C* **2017**, *121*, 28158–28165.
32. Tanner, R.I. *Engineering Rheology*, 2nd ed.; Oxford Engineering Science Series; Oxford University Press: New York, NY, USA, 2000.
33. Puscasu, R.M.; Todd, B.D.; Daivis, P.J.; Hansen, J.S. Viscosity kernel of molecular fluids: Butane and polymer melts. *Phys. Rev. E* **2010**, *82*, 011801.
34. Dhont, J.K.G.; Kang, K.; Kriegs, H.; Danko, O.; Marakis, J.; Vlassopoulos, D. Nonuniform flow in soft glasses of colloidal rods. *Phys. Rev. Fluids* **2017**, *2*, 043301.
35. Jin, H.; Kang, K.; Ahn, K.H.; Briels, W.J.; Dhont, J.K.G. Non-local stresses in highly non-uniformly flowing suspensions: The shear-curvature viscosity. *J. Chem. Phys.* **2018**, *149*, 014903.
36. Todd, B.D.; Hansen, J.S.; Daivis, P.J. Non-local shear stress for homogeneous fluids. *Phys. Rev. Lett.* **2008**, *100*, 195901.
37. Snook, I.K.; Henderson, D. Monte-Carlo study of a hard-sphere fluid near a hard wall. *J. Chem. Phys.* **1978**, *68*, 2134–2139.
38. Bitsanis, I.; Magda, J.J.; Tirrell, M.; Davis, H.T. Molecular dynamics of flow in micropores. *J. Chem. Phys.* **1987**, *87*, 1733.
39. Bitsanis, I.; Vanderlick, T.K.; Tirrell, M.; Davis, H.T. A tractable molecular theory of flow in strongly inhomogeneous fluids. *J. Chem. Phys.* **1988**, *89*, 3152–3162.
40. Hoang, H.; Galliero, G. Shear viscosity of inhomogeneous fluids. *J. Chem. Phys.* **2012**, *136*, 124902.
41. Hoang, H.; Galliero, G. Local viscosity of a fluid confined in a narrow pore. *Phys. Rev. E* **2012**, *86*, 021202.
42. Hoang, H.; Galliero, G. Local shear viscosity of strongly inhomogeneous dense fluids: From the hard-sphere to the Lennard-Jones fluids. *J. Phys. Condens. Matter* **2013**, *25*, 485001.
43. Bhadauria, R.; Aluru, N.R. A quasi-continuum hydrodynamic model for slit shaped nanochannel flow. *J. Chem. Phys.* **2013**, *139*, 074109.
44. Bhadauria, R.; Sanghi, T.; Aluru, N.R. Interfacial friction based quasi-continuum hydrodynamical model for nanofluidic transport of water. *J. Chem. Phys.* **2015**, *143*, 174702.
45. Travis, K.P.; Gubbins, K.E. Poiseuille flow of Lennard-Jones fluids in narrow slit pores. *J. Chem. Phys.* **2000**, *112*, 1984.
46. Dalton, B.A.; Glavatskiy, K.S.; Daivis, P.J.; Todd, B.D.; Snook, I.K. Linear and nonlinear density response functions for a simple atomic fluid. *J. Chem. Phys.* **2013**, *139*, 044510.
47. Dalton, B.A.; Daivis, P.J.; Hansen, J.S.; Todd, B.D. Effects of nanoscale inhomogeneity on shearing fluids. *Phys. Rev. E* **2013**, *88*, 052143.
48. Glavatskiy, K.S.; Dalton, B.A.; Daivis, P.J.; Todd, B.D. Nonlocal response functions for predicting shear flow of strongly inhomogeneous fluids. I. Sinusoidally driven shear and sinusoidally driven inhomogeneity. *Phys. Rev. E* **2015**, *91*, 062132.
49. Dalton, B.A.; Glavatskiy, K.S.; Daivis, P.J.; Todd, B.D. Nonlocal response functions for predicting shear flow of strongly inhomogeneous fluids. II. Sinusoidally driven shear and multisinusoidal inhomogeneity. *Phys. Rev. E* **2015**, *92*, 012108.
50. Camargo, D.; de la Torre, J.A.; Duque-Zumajo, D.; Espanol, P.; Delgado-Buscalioni, R.; Chejne, F. Nanoscale hydrodynamics near solids. *J. Chem. Phys.* **2018**, *148*, 064107.
51. Ewen, J.P.; Kannam, S.K.; Todd, B.D.; Dini, D. Slip of alkanes confined between surfactant monolayers adsorbed on solid surfaces. *Langmuir* **2018**, *34*, 3864–3873.

processes

MDPI

Review

Modeling Permeation through Mixed-Matrix Membranes: A Review

Gloria M. Monsalve-Bravo and **Suresh K. Bhatia** *

School of Chemical Engineering, The University of Queensland, Brisbane, QLD 4072, Australia;
g.monsalvebravo@uq.edu.au
* Correspondence: s.bhatia@uq.edu.au; Tel.: +61-(07)-3365-4263

Received: 30 August 2018; Accepted: 14 September 2018; Published: 18 September 2018

Abstract: Over the past three decades, mixed-matrix membranes (MMMs), comprising an inorganic filler phase embedded in a polymer matrix, have emerged as a promising alternative to overcome limitations of conventional polymer and inorganic membranes. However, while much effort has been devoted to MMMs in practice, their modeling is largely based on early theories for transport in composites. These theories consider uniform transport properties and driving force, and thus models for the permeability in MMMs often perform unsatisfactorily when compared to experimental permeation data. In this work, we review existing theories for permeation in MMMs and discuss their fundamental assumptions and limitations with the aim of providing future directions permitting new models to consider realistic MMM operating conditions. Furthermore, we compare predictions of popular permeation models against available experimental and simulation-based permeation data, and discuss the suitability of these models for predicting MMM permeability under typical operating conditions.

Keywords: mixed-matrix membrane (MMM); permeation modeling; effective medium approach; simulation of MMM; particle-polymer interface

1. Introduction

In the last few decades, membrane technologies have attracted increasing attention to be used in a variety of industrial applications, which include gas separation [1–3], water desalination [4–6], food processing [7], pervaporation [8,9], membrane contactors [10,11], and membrane reactors [12,13]. In many of these applications, membrane technologies are preferred over conventional separation techniques (e.g., distillation, absorption, and adsorption) due to their superior features such as [14–17]: (i) stable production with high separation efficiency [18]; (ii) low energy consumption with no phase change requirements [15,19]; (iii) simple operation with convenient modular scale-up [20–22]; and (iv) small environmental footprint [18,23]. Nevertheless, implementation of membrane technologies in practical applications has been limited by challenges with the engineering of robust materials able to be effective under a variety of operating conditions and environments [21,24], with only polymer membranes currently available in large-scale applications; yet failing to overcome Robeson's [25,26] trade-off curves between the selectivity and permeability [25–28].

Different alternatives have been explored to enhance polymer membranes to perform beyond Robeson's upper bound, including surface modification [29], facilitated transport [30,31], polymer blends [32], and mixed-matrix membranes (MMMs) [33]. Amongst these alternatives, devoted attention to the synthesis of MMMs has been intensified over the last three-decades [21,34,35], with a myriad of studies focusing on novel materials to increase efficiency of CO_2 capture [18,35,36], natural gas purification [37–39], water purification [40,41], and olefin/paraffin separation [42–44]. Thus, much effort has been devoted to the optimization of MMMs synthesis [34,42,45–48]; with a number of works even reporting fabrication of defect-free MMMs [37,49–51].

Ideally, a mixed-matrix membrane (MMM) consists of a selective inorganic filler phase embedded to continuous polymer matrix [21,35]. In this way, an MMM combines high intrinsic permeability and separation efficiency of advanced molecular sieving materials (e.g., zeolites, carbons, metal-organic frameworks) or nanoscale materials (e.g., carbon nanosheets or nanotubes) with robust processing capabilities and mechanical properties of glassy polymers [23,52]. Consequently, MMMs are expected to have higher efficiency than those based on their polymer counterpart, thus exceeding the trade-off between the permeability and selectivity [15,21,53].

MMMs are commonly prepared either with symmetric or asymmetric structure [17,34,54]. Symmetric MMMs consist of a uniform dense composite film of thickness $20\,\mu m \leq \ell \leq 100\,\mu m$ [41,46] while asymmetric MMMs comprise a thin selective composite skin layer of thickness $2\,\mu m \leq \ell \leq 5\,\mu m$ coated on a highly porous non-selective core layer of thickness $50\,\mu m \leq \ell \leq 300\,\mu m$ [3,55–57]. In this way, thicknesses of MMMs are large enough to disregard effects of interfacial entrance and exit barriers on the transport; these barriers which have been shown to significantly decrease the permeant diffusivity in nanoporous materials only when the overall system thickness is $\ell \leq 0.1\,\mu m$ [58–60]. Such barriers may nevertheless be important at the interfaces of nano-sized fillers and zeolites in nanocomposites, where potential of mean force calculations demonstrate their significance [61]. However, filler size in MMMs is of the order of 0.1–1 μm or larger [34,41], and such barriers are insignificant relative to the internal resistance in the filler particles.

Current models for permeation in MMMs are adaptations of early theories for the transport in heterogeneous media, either following the effective medium approach (EMA) [62–65] or resistance model approach (RMA) [66–69]. In such approaches, the effective permeability is usually based on the permeabilities and volume fractions of the MMM constituent phases [21,34,70–72], with these phase-specific properties largely assumed constant. Thus, applicability of early RMA/EMA models [62–64,66] and later adaptations [65,68,73,74] is often limited to narrow MMM operating conditions (Henry's law region) and ideal polymer-particle morphologies. Furthermore, although experimental studies on MMMs have shown that deviations from Henry's law are common under usual operating pressures (1–4 bar) [15,36,46,75], isotherm nonlinearity is incorporated into the permeation models through the Darken or free volume theories while assuming a uniform field based on the mean permeant concentration [71,76–79], an assumption that needs to be relaxed for further progress.

Over the last decade, increased efforts have been devoted to advance permeation models to integrate effects of filler morphology (e.g., particle size, shape, and agglomeration) [65,80,81] and defects at the particle-polymer interface in the form of a rigidified polymeric [76–79,82–84], and void [69,77,79,85] or pore-blocked [84] regions. However, while considering such non-idealities, these models share the uniform field assumption inherent to the EMA and RMA [69,76,77,86,87]. Thus, effects of the filler morphology together with isotherm nonlinearity are often embedded in a single empirical morphology-related parameter, such as in the Pal [65], Lewis-Nielsen [73], and Higuchi [88] models. Thus, extending the predictive capability of existing models through more rigorous approaches able to be valid over a variety of conditions and systems remains a possibility [15,21,36].

In this work, we review existing approaches for engineering models of permeation through MMMs. To do so, we first introduce the concepts of permeability and selectivity in the context of MMMs and discuss how their mathematical formulation is integrated to existing models for the transport in composite media. Here, we also classify permeation models by approach (i.e., RMA and EMA) and discuss their range of applicability based on their fundamental assumptions. Finally, we compare RMA/EMA models to simulation-based and experimental permeation data for various gases (e.g., CO_2, CH_4, H_2, O_2, N_2) in several MMM systems, and provide future directions on how to progress existing models to undertake realistic MMM operating conditions.

2. Gas Transport through Mixed-Matrix Membranes

Gas separation through membranes can take place by different mechanisms [89,90]. Three main diffusion mechanisms have been well-accepted to describe gas transport through membranes [91,92]: (i) Knudsen diffusion; (ii) molecular sieving (molecular diffusion); and (iii) solution-diffusion (sorption-diffusion), with detailed discussion of these transport mechanisms available elsewhere [91]. In general, the diffusion mechanism is assumed to change from solution-diffusion to Knudsen diffusion with increase of the pore size in the membrane material [18,36]. Based on this consideration, transport through inorganic porous membranes has been largely associated with the Knudsen diffusion [35,91], that in membranes based on nanomaterials such carbon molecular sieves (CMSs) [46], zeolitic imidazolate frameworks (ZIFs) [93,94] and metal organic frameworks (MOFs) [95], has been associated with the molecular diffusion, and that through glassy polymers has been associated with the solution-diffusion [30,89–91,96–98]. While MMMs combine transport principles of both polymer and inorganic membranes, diffusion through them is understood via the solution-diffusion mechanism [21,34]. This mechanism assumes that permeant molecules dissolve (adsorb) on one side of the membrane, diffuse across the membrane and then are released (desorbed) at the other side [14,19,90,99], as depicted in Figure 1 for the CO_2/CH_4 separation.

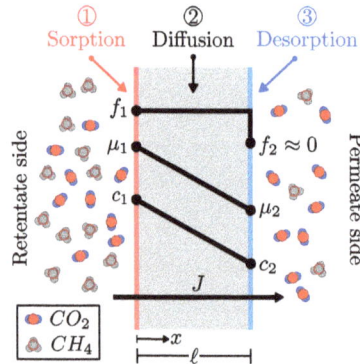

Figure 1. Schematic representation of gas permeation through the solution-diffusion mechanism.

2.1. Permeability and Selectivity in MMMs

In the solution-diffusion mechanism, the permeant transport is driven by the chemical potential gradient ($\nabla\mu$) across the membrane, in which μ is only dependent on the concentration gradient (∇c) while the fugacity (f) is assumed uniform across the membrane (cf. Figure 1) [100]. Under these considerations, the permeant flux (J) can be defined as:

$$J = \overline{DS}(-\Delta f)/\ell = P(-\Delta f)/\ell, \ \Delta f = f_2 - f_1 \tag{1}$$

with detailed derivation of Equation (1) found elsewhere [100,101]. Here, $\overline{D}$ and $\overline{S}$ are the concentration-averaged diffusivity and solubility, respectively; and $\Delta f = f_2 - f_1$ is the fugacity difference between the retentate (f_1) and permeate (f_2) sides of the membrane (cf. Figure 1), respectively [102]. Further, Equation (1) is usually rearranged as [46]:

$$P = J\ell/(-\Delta f), \ \Delta f = f_2 - f_1 \tag{2}$$

as the permeant flux (J), membrane thickness (ℓ), and fugacity difference (Δf) can be measured in practice [42,99,103,104]. In Equation (2), the permeability (P) is calculated in Barrer, with [90,105]:

$$1\ \text{Barrer} = 1 \times 10^{-10} \frac{\text{cm}^3\,(\text{STP})\ \text{cm}}{\text{cm}^2\ \text{s cmHg}} = 3.348 \times 10^{-16} \frac{\text{mol m}}{\text{m}^2\ \text{s Pa}}. \tag{3}$$

For gas mixtures, the permselectivity is used to characterize the MMM separation efficiency, with the permselectivity of species A relative to species B (α^*_{AB}) defined as [91,102,103,106]:

$$\alpha^*_{AB} = P_A/P_B = (D_A S_A)/(D_B S_B) \tag{4}$$

where α^*_{AB} is also known as the ideal selectivity. In Equation (4), the permeability of the slower permeant is usually placed in the denominator, and thus $\alpha^*_{AB} > 1$ [99].

2.2. Diffusion and Sorption in MMMs

Based on the permeant flux definition in Equation (1), the permeability can be defined as the product of a kinetic ($\overline{D}$) and a thermodynamic ($\overline{S}$) contribution, with [91,99]:

$$P = \overline{D}\,\overline{S} \tag{5}$$

where, assuming the retentate side fugacity to be negligibly small (cf. Figure 1), the mean diffusivity ($\overline{D}$) and solubility ($\overline{S}$) can be expressed as [107–109]:

$$\overline{D} = (1/c_1) \int_{c_2 \approx 0}^{c_1} D(c)\,dc \tag{6}$$

$$\overline{S} = (1/f_1) \int_{c_2 \approx 0}^{c_1} dc = c_1/f_1. \tag{7}$$

respectively. In Equations (6) and (7), c is the permeant adsorbed concentration and D is the Fickian diffusivity, well-accepted to follow the Darken relation as $D = D_o(d \ln f / d \ln c)$ [110,111]. Here, D_o is the permeant mobility, assumed concentration-independent [112,113].

At low pressures, Henry's law is often adequate to express the gas concentration, with $c = K_H f$ [89,109]. Thus, by following Equations (6) and (7), both diffusivity and solubility are concentration-independent, with the permeability in the MMM constituent phases given by [109,110]:

$$P_f = D_{of} K_{Hf} \tag{8}$$

$$P_c = D_{oc} K_{Hc} \tag{9}$$

with subscripts f and c denoting the filler and continuous phases, respectively.

At moderate/high pressures, the solubility in the filler phase is commonly assumed to follow the Langmuir isotherm, leading to:

$$\overline{S}_f = K_f c_f^s/(1 + K_f f_1) \tag{10}$$

which often describes well the sorption equilibrium in various porous fillers [42,48,75,114–116]. Here, K_f is the affinity constant and c_f^s the capacity in the filler phase. Then, by following Equation (5) while assuming the diffusivity to be concentration-independent, with $\overline{D} = D$ ($\overline{D} \neq D_o$) based on Equation (6) [117], the permeability in the filler phase is given by [99]:

$$P_f = D_f K_f c_f^s/(1 + K_f f_1) \tag{11}$$

where D_f is the permeant diffusivity in the filler. Similarly, a combination of Henry's law and the Langmuir model is often used to describe the mean polymer solubility, following [75,106]:

$$\overline{S}_c = K_h + K_c c_c^s / (1 + K_c f_1) \tag{12}$$

with K_h being the Henry's law constant, K_c the affinity constant and c_c^s the capacity in the matrix. Thus, by following Equation (5), the permeability in the polymer is given by [117,118]:

$$P_c = D_h K_h + D_c K_c c_c^s / (1 + K_c f_1) \tag{13}$$

where D_h and D_c are the diffusivities in the Henry's law and Langmuir sites, respectively. Equation (13) is known as the dual-mode/partial immobilization model, as the penetrant is assumed fully mobile in the Henry environment and partially mobile in the Langmuir environment [118–121]. In the next section, the permeant permeabilities in the filler (P_f) and continuous (P_c) phases are largely assumed concentration-independent following Equations (8) and (9) [86,122,123] or based on concentration-averaged solubility and diffusivity via Equations (11) and (13) [15,36,99].

3. Models for Gas Permeation in Mixed-Matrix Membranes

While a universal description of the gas transport through MMM is a complex problem [21,71,87,112], the modeling of permeation through MMMs is largely based on early theories for thermal/electrical conduction of heterogeneous media [62,65,124–126]. These theories were extended to MMMs on the basis of the analogy between the thermal/electrical conductivity and the permeability of composite materials [66,81] in the presence of linear flux laws. Based on this analogy, two main approaches have been extensively used to predict the permeability in MMMs: the resistance model approach (RMA) and the effective medium approach (EMA), described in Sections 3.1 and 3.2, respectively. Besides these early approaches, simulation-based rigorous modeling of MMMs has attracted increased attention in recent years, and thus this latter approach is described in Section 3.3. In what follows, subscripts c, f, i, and *eff* refer to the continuous phase (polymer matrix), filler phase (selective phase), polymer-filler interface and MMM as a whole, respectively.

3.1. Resistance Model Approach

The resistance model approach (RMA) relies on analogy between the current flow through a series-parallel array of resistors (Ohm's law) and the permeation rate through a composite membrane (Fick's law) [66,67,125,127]. Under this consideration, the MMM permeability is inversely proportional to the overall transport resistance, with the fundamental equation of the RMA being [87,128–132]:

$$F = (-\Delta f)/R_{eff}, \quad \Delta f = f_2 - f_1 \tag{14}$$

where R_{eff} is the overall transport (permeation) resistance and $F = A_{cross} J$ is the permeant flow rate, with A_{cross} the cross-sectional area in the flow direction. Consequently, the permeant flux (J) through the membrane can also be expressed as:

$$J = (-\Delta f)/(A_{cross} R_{eff}), \quad \Delta f = f_2 - f_1. \tag{15}$$

On comparing flux definitions in Equations (1) and (15), the resistance (R_{eff}) can be equated as [66,68]:

$$R_{eff} = \ell / (P_{eff} A_{cross}). \tag{16}$$

In this way, if an expression for the equivalent resistance (R_{eff}) is known, the MMM permeability (P_{eff}) can be calculated via Equation (16) [125,132].

A number of permeation models have been proposed for MMMs [68,69,72,76,82,87,133], based on Equations (15) and (16), as listed in Table 1, with the most popular models considering platelet or cubic filler particles [72,130,133] and few newly developed models considering tubular particles [69,76,82]. The simplest RMA models idealize the MMM as a two-phase laminated composite comprising multiple sheets of polymer and selective material alternated in series or in parallel to the flow direction [70,134], as depicted in Figure 2a,b, respectively. Figure 2 also depicts electrical circuit analogues used to calculate the overall transport resistance of both composites.

Table 1. Models based on the resistance model approach.

Model	Key Equations	
Series [47]	$P_{eff} = \dfrac{P_c P_f}{P_c \phi_f + P_f (1 - \phi_f)}$	(17)
Parallel [47]	$P_{eff} = P_f \phi_f + P_c (1 - \phi_f)$	(18)
Te Hennepe [130]	$P_{eff} = P_c \left[(1 - \phi_f^{\frac{1}{3}}) + \dfrac{\frac{3}{2}\phi_f^{\frac{1}{3}} P_c}{P_c(1-\phi_f)+\frac{3}{2}P_f\phi_f} \right]^{-1}$	(19)
Cussler [72]	$P_{eff} = P_c \left[(1 - \phi_f) + \left(\dfrac{P_f}{\phi_f P_c} + \dfrac{4(1-\phi_f)}{\lambda_f^2 \phi_f^2} \right)^{-1} \right]^{-1}$	(20)
Ebneyamini [68]	$P_{eff} = \tau \left[(1 - \phi^{2/3}) P_c + \dfrac{P_c P_f \phi^{2/3}}{\phi^{1/3} P_c + (1 - \phi^{1/3}) P_f} \right]$	(21)
KJN [69]	$P_{eff}^{Oriented} = P_c \left[\left(1 - \dfrac{\cos\theta}{\cos\theta + \lambda_f \sin\theta} \phi_f \right) + \dfrac{P_c}{P_f} \left(\dfrac{1}{\cos\theta + \lambda_f \sin\theta} \right) \phi_f \right]^{-1}$	(22)
	$P_{eff}^{Random} = \frac{\pi}{2} P_c \left[\int_0^{\frac{\pi}{2}} \dfrac{P_c}{P_{eff}^{Oriented}(\theta)} d\theta \right]^{-1}$	(23)

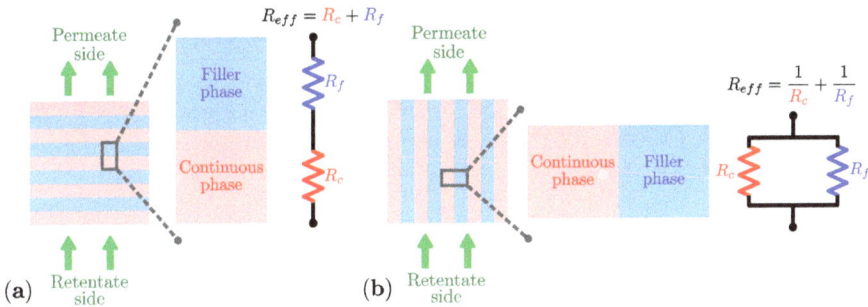

Figure 2. Resistance model approach for a multilayer composite in: (**a**) series and (**b**) parallel.

Following the electrical circuit analog in Figure 2 and the above definition for the permeation resistance in Equation (16), the permeability for the multilayer composite in series yields Equation (17) in Table 1, while that of a multilayer composite in parallel yields Equation (18) [47,70]. The series model in Equation (18) is assumed to provide the lower bound for the permeability of a given penetrant in an ideal MMM [135]. Alternatively, the parallel model in Equation (18) is assumed to provide to the upper bound for the effective permeability of a given penetrant in an ideal MMM [47,70].

In addition to the two-resistance based models in Equations (17) and (18), more complex models, including three-resistances or more, have been proposed following the RMA [68,69,72,82,87,130]. In these models, the additional diffusion resistances are intended to account for tortuosity effects in the permeant diffusion path when there are large differences in permeabilities amongst the filler and polymer phases [68,87] or defects in MMM structure [69,82,87]. Amongst existing RMA models, those of Te Hennepe [130] in Equation (19) and Cussler [72] in Equation (20), based on three-resistance circuit analogs, have widely been applied to zeolite-polymer MMMs [114,128,136].

Te Hennepe et al. [130] considered the one-dimensional transport in zeolite-rubber MMMs. They idealized the MMM as a lamella containing composite layers [131], in which each composite layer comprised two regions. The first region consisted of polymer and the second one of polymer and zeolite particles (mixed-region). In this model, the polymer region was assumed in series with parallel resistances of the second mixed-region [130,131], which led to Equation (19) in Table 1. Alternatively, Cussler [72] considered two-dimensional transport in the MMM. He assumed the resistance of the polymer region in series with that of a second mixed-region, similar to Te Hennepe et al. [130]. However, transport in the mixed-region was assumed to occur in the permeation direction through filler phase and perpendicular to the permeation direction through polymer phase [15,36]. This assumption led to Equation (20) in Table 1, in which $\lambda_f = w_f/\ell_f$ is the aspect ratio of the filler phase with w_f and ℓ_f being the flake width and thickness [21,136].

Recently, Ebneyamini et al. [68] proposed a semi-empirical four-resistance model for ideal MMMs comprising cubical particles. To do so, an empirical correction factor (τ) was introduced to a one-dimensional four-resistance model. The final model is given by Equation (21) in Table 1, and referred here as the Ebneyamini model. In this model, τ was estimated via simulation of the 3D particle-polymer system and adjusted to follow Langmuir-type equations. Thus, τ is assumed to accommodate tortuosity effects arising from large differences amongst the MMM constituent phase permeabilities.

Modeling MMMs with tubular filler has received less attention than those having cubic or platelet fillers, with only few studies [69,87] developing RMA models for nanotube-MMMs. The first of these RMA models was proposed by Kang et al. [69], who accommodated the orientation of tubular fillers in the calculation of the overall transport resistance. The final model was named by authors as the Kang-Jones-Nair (KJN) model, and for uniformly oriented fillers is given by Equation (22) in Table 1. For randomly oriented fillers, the KJN model is rewritten as Equation (23) in Table 1, with $P_{eff}^{Oriented}(\theta)$ in Equation (23) following Equation (22). Here, $\theta \in [0, \pi/2]$ is the orientation of the tubular filler, measured with respect the permeation direction, and $\lambda_f = d_f/\ell_f$ is the aspect ratio of the tubular filler with d_f and ℓ_f being the diameter and length of the cylindrical particle, respectively. The predictions of the KJN model are always lower than those based on the series model, with the KJN model simplifying to the series model when $\theta = 0$. Figure 3 depicts a comparison of the permeability (P_{eff}) profiles based on models of Table 1 with $\alpha_{fc} = P_f/P_c = 10$ in all models and $\lambda_f = 0.25$ in the Cussler and KJN models. Figure 3 also depicts the MMM structure assumed by each model.

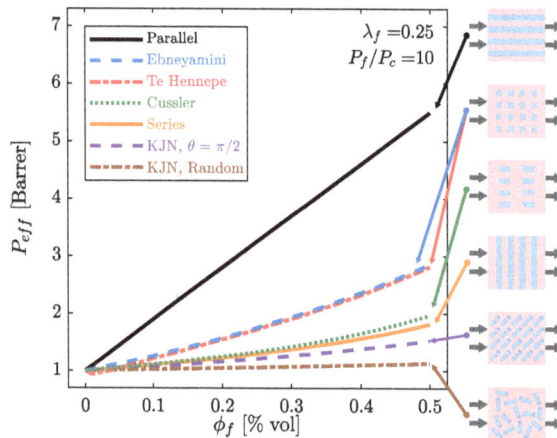

Figure 3. Comparison of permeability profiles based on models of Table 1 with $\alpha_{fc} = P_f/P_c = 10$ $\lambda_f = 0.25$. Right-hand side depicts the composite structure considered by each model.

3.2. *Effective Medium Approach*

The crux of the effective medium approach (EMA) lies in the substitution of a given composite system by an equivalent effective homogeneous one having the properties of the composite [137–139]. The resulting effective composite properties are generally functions of the volume fraction and permeabilities of the composite constituent phases, similar to RMA-based models (cf. Section 3.1). However, EMA differs from RMA in the way the filler phase is considered within the composite. While most RMA models assume regular distributions (e.g., simple cubic or body centered cubic lattices) of platelet and/or cubic particles [47,68,131,140], EMA models consider random distributions of spherical inclusions [64,65,141–143]. For ease of analysis, we here classify EMA models in two main groups. The first group corresponds to EMA models following Maxwell's theory and second to those following Bruggeman's theory, with models associated with each theory described in Sections 3.2.1 and 3.2.2, respectively.

3.2.1. Maxwell Theory

Maxwell [62] calculated the electrical conductivity of infinitely diluted cluster of particles embedded in an infinite matrix [138]. To do so, he assumed that the far field potential of this cluster was equivalent to that of a homogeneous sphere having the original composite volume [65,135,137], as depicted in Figure 4. Based on this assumption, Maxwell defined the composite conductivity as that of the homogeneous sphere [21,141,144,145], which led to Equation (24) in Table 2, with $\beta_{fc} = (\alpha_{fc} - 1)/(\alpha_{fc} + 2)$ and $\alpha_{fc} = P_f/P_c$. Further, because the Maxwell model disregards particle interaction, it is only applicable to dilute suspensions ($\phi_f \leq 0.2$).

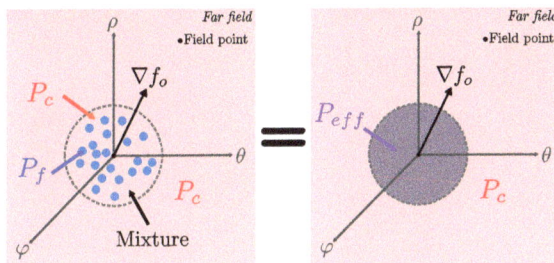

Figure 4. Schematic representation of Maxwell's theory. A composite sphere comprised of spherical particles (phase *f*) in a matrix (phase *c*), immersed in an infinite matrix of phase *c*.

The Maxwell model was later extended to spheroids by Wagner and Sillars [126]. To do so, the conduction problem was reformulated with oriented spheroidal inclusion, with these spheroids oriented along the axis of the potential difference [81,126]. This consideration led to Equation (25) in Table 2, referred as the Maxwell-Wagner-Sillars model [81]. Here, $\lambda_f \in [0,1]$ is the particle shape factor, and for prolate spheroids $\lambda_f \in [0,1/3]$ while for oblate spheroids $\lambda_f \in [1/3,1]$. In the limits, when $\lambda_f = 0$, $\lambda_f = 1/3$, and $\lambda_f = 1$, Equation (25) reduces to the parallel model, series model (cf. Table 1) and Maxwell model, respectively. The Maxwell-Wagner-Sillars model is only applicable to diluted ellipsoid dispersions ($\phi_f \leq 0.2$), similar to the Maxwell model [3,21,41,146].

Several attempts have been made to extend Maxwell's equation to concentrated composites ($\phi_f > 0.2$) [64,65,73,141,142], of which, the one with the most significant results is that of Jeffrey [142,147]. He showed that relative conductivity (P_r) of a dispersion can be expressed in series of ϕ_f following the form $P_r = P_{eff}/P_c = 1 + K_1\phi_f + K_2\phi_f^2 + \cdots$ [141,148], in which each K-term in the series ($K_1, K_2, \ldots, K_n$) accommodates the interaction of successively larger sets of particles [141,149]. Further, Jeffrey [142,147] demonstrated that only the first-order term of the Maxwell model is exact, where Equation (24) yields $P_r = P_{eff}/P_c = 1 + 3\beta_{fc}\phi_f + \mathcal{O}(\phi_f^2)$ with $K_1 = 3\beta_{fc}$ when $\phi \to 0$. This limiting equation was later used by Bruggeman [139] and Pal [65], as described in Section 3.2.2.

Although Jeffrey [142,147] hypothesized that K_2 in the series was dependent on β_{fc} and ϕ_f, his final equation corresponds to the low-density limit of interacting spheres [141]. Thus, Jeffrey's model provides very similar predictions to the Maxwell model [150]. Later, Chiew and Glandt [141] estimated K_2 in the series using pair-correlation functions of hard-sphere fluid simulations, which led to Equation (26) in Table 2. In this work, the resulting values of K_2 were tabulated as function of α_{fc} and ϕ_f [141,149]. Later, Gonzo et al. [145] fitted Chiew and Glandt's results for K_2, which led to Equation (27) in Table 2. Further, the Chiew-Glandt model corresponds to the exact solution to the series truncated after K_2-term [141,149], with the model applicable to moderate/high filler loadings ($\phi_f \leq 0.645$) [141,145,151].

Table 2. Models based on Maxwell's theory.

Model	Key Equations	
Maxwell [62]	$P_{eff} = P_c\left[\frac{1+2\beta_{fc}\phi_f}{1-\beta_{fc}\phi_f}\right] = P_c\left[\frac{P_f+2P_c-2\phi_f(P_c-P_f)}{P_f+2P_c+\phi_f(P_c-P_f)}\right]$	(24)
Maxwell-Wagner-Sillars model [81]	$P_{eff} = P_c\frac{\lambda_f P_f+(1-\lambda_f)P_c-(1-\lambda_f)\phi_f(P_c-P_f)}{\lambda_f P_f+(1-\lambda_f)P_c+\lambda_f\phi_f(P_c-P_f)}$	(25)
Chiew-Glandt [141]	$P_{eff} = P_c\left[\frac{1+2\beta_{fc}\phi_f+(K_2-3\beta_{fc}^2)\phi_f^2}{1-\beta_{fc}\phi_f}\right]$	(26)
	$K_2 = a + b\phi_f^{\frac{3}{2}}$	(27)
	$a = -0.002254 - 0.123112\beta_{fc} + 2.93656\beta_{fc}^2 + 1.6904\beta_{fc}^3$	(28)
	$b = 0.0039298 - 0.803494\beta_{fc} - 2.16207\beta_{fc}^2 + 6.48296\beta_{fc}^3 + 5.27196\beta_{fc}^4$	(29)
MB-B model [152]	$\frac{1}{R^{i-1}}\frac{d}{dR}\left[R^{i-1}P_m(R)\frac{dC_m}{dR}\right] = 0$	(30)
	$\overline{\phi}_f(R) = \frac{3}{2r_o^3}\int_0^{r_o}\int_0^{\pi}\phi_f(R'(R,r,\theta))r^2\sin\theta d\theta dr$	(31)
Lewis-Nielsen [73]	$P_r = \frac{1+2\beta_{fc}\phi_f}{1-\beta\phi_{fc}\psi_m}$	(32)
	$\psi_m = 1 + [(1-\phi_m)/\phi_m^2]\phi_f$	(33)
Higuchi [88]	$P_{eff} = P_c\left[1 + \frac{3\phi_f\beta_{fc}}{1-\phi_f\beta_{fc}-K_H(1-\phi_f)\beta_{fc}^2}\right]$	(34)
Felske model [77]	$P_{eff} = P_c\left[\frac{2(1-\phi_{fi})+(1+2\phi_i)(\eta/\gamma)}{(2+\phi_{fi})+(1-\phi_i)(\eta/\gamma)}\right]$	(35)
	$\eta = (2+\delta^3)\alpha_{fc} - 2(1-\delta^3)\alpha_{ic}$	(36)
	$\gamma = (1+\delta^3) - (1-\delta^3)\alpha_{fi}$	(37)
	$\phi_{fi} = \frac{\phi_f^N}{\phi_f^N+[(1-\phi_f^N)/\delta^3]}$	(38)
Pseudo-two-phase Maxwell [85]	$P_{eff} = P_c\left[\frac{P_{eff}^f+2P_c-2\phi_{fi}(P_c-P_{eff}^f)}{P_{eff}^f+2P_c+\phi_{fi}(P_c-P_{eff}^f)}\right]$	(39)
	$P_{eff}^f = P_i\left[\frac{P_f+2P_i-2\phi_s(P_i-P_f)}{P_f+2P_i+\phi_s(P_i-P_f)}\right]$	(40)
	$\phi_s = 1/(1+\ell_i/r_o)^3$	(41)

Recently, Monsalve-Bravo and Bhatia [151–153] used the Chiew-Glandt model in conjunction with the one-dimensional transport equation for the permeant to describe the permeability for various gases in flat and hollow fiber MMMs. The semi-analytical model is given by Equation (30) in Table 2, referred to as the MB-B model. Here, $i = 1$ for a flat MMM and $i = 2$ for a hollow fiber MMM while C_m is the position-dependent pseudo-bulk concentration in the MMM, with boundary conditions $C_m = C_{m1} = f_1/R_g T_g$ at $R = R_1$ and $C_m = C_{m2} = f_2/R_g T_g$ at $R = R_1$. In Equation (30), $P_m(R)$ is the effective local MMM permeability, estimated using the Chiew-Glandt model in Equation (26), in which the constituent phase permeabilities (P_f and P_c) are concentration-dependent via the Darken model [154,155]. Further, the locally averaged filler volume fraction ($\overline{\phi}_f$) in Equation (31) averages the filler volume fraction (ϕ_f) over the particle volume at a given membrane location, with r_o being the filler particle radius and $R'(R,r,\theta)$ the location in the particle relative to the position R in the MMM. Thus, the MB-B model incorporates effects of particle size [151], membrane geometry [152] and isotherm nonlinearity [153] in the calculation of the MMM permeability. The MB-B model reduces to the Chiew-Glandt model when $r_o/\ell \to 0$.

As an alternative to the exact second order solution of Chiew and Glandt [141], several empirical modifications of the Maxwell model [80,156,157] have been proposed for concentrated composites. These models empirically embed packing-related effects [158] and variation in filler properties, such as particle agglomeration, size, and shape [71,80,86], within a single parameter in the model [80,86,153], with the Lewis-Nielsen model in Equation (32) being one of most popular of these models. In this model, packing-related effects are accommodated via the maximum filler volume fraction (ϕ_m) [124,159], with the Lewis-Nielsen model simplifying to that of Maxwell when $\phi_m \to 1$. Alternatively, Higuchi [88] introduced empirical parameter, K_H, to the Maxwell model to account for particle-particle interactions and asphericity effects arising from particle shape variation, with the Higuchi model given by Equation (34) [70,85] and $0 \leq K_H \leq 0.78$ for spherical particles [160]. Thus, Equation (34) reduces to the Maxwell model when $K_H = 0$ [74]. Figure 5 compares the permeability (P_{eff}) profiles based on ideal models in Table 2, with $\alpha_{fc} = P_f/P_c = 10$ and specific model parameters values listed in the legend.

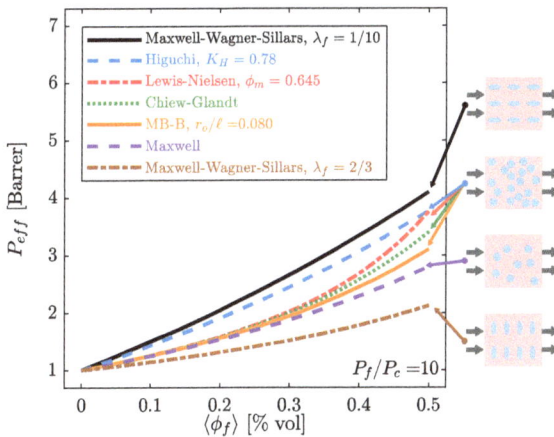

Figure 5. Comparison of permeability profiles based on the ideal models of Table 2 with $\alpha_{fc} = P_f/P_c = 10$, depicting the composite structure considered in each model.

Based on Maxwell's theory, several models have been proposed to account for non-ideal polymer-particle morphologies [77,78,84,86,160], either by: (i) solving the transport problem in an MMM comprising spherical core-shell inclusions [77,86] or (ii) assuming the particle-interface system as a pseudo-phase dispersed in the polymer matrix [71,85]. In the first group of non-ideal models are found that of Felske [77], and later Pal's adaptation [86]. In the second group are found the pseudo-two-phase Maxwell model [85], and its adaptations [84,161]. The Felske model in Equation (35) [77] is the exact first order solution to the transport problem through a dispersed composite comprising non-interacting core-shell particles [160]. In this model, $\phi_{fi} = \phi_f + \phi_i$ is the volume fraction of the total dispersed phase following Equation (38), with $\delta = (\ell_i + r_o)/r_o$ and ϕ_f^N the nominal filler volume fraction [70,86,160]. Further, $\alpha_{fc} = P_f/P_c$, $\alpha_{ic} = P_i/P_c$, and $\alpha_{fi} = P_f/P_i$ in Equations (36) and (37). The Felske model is only applicable to dilute suspensions ($\phi_{fi} \leq 0.2$) [71,160,162].

In the pseudo-two-phase Maxwell model in Equation (39) [85], it is assumed that the three-phase composite can be idealized as pseudo two-phase composite [163,164], with the polymer matrix being one phase and the combined filler-interface system being the other phase (i.e., pseudo dispersed filler phase) [21,71]. In Equation (39), ϕ_{fi} is the volume fraction of total dispersed phase, given by Equation (38) [70]. Further, P_{eff}^f is the permeability of the combined filler-interface system, given by Equation (40). Here, ϕ_s is the volume fraction of the filler in the filler-interface composite, and following Equation (41) [70,85,163]. Further, P_i is assumed well described by Knudsen diffusion mechanism when interfacial voidage is considered [70,163]. Alternatively, when polymer rigidification is considered at the filler surface $P_i = P_c/\sigma$, with $\sigma \in [3,4]$ being the chain immobilization factor; an empirical parameter used to differentiate the rigid polymer permeability from that in the bulk polymer [21,163,164]. Later, Li et al. [84] modified the pseudo-two-phase Maxwell model to accommodate both partial pore blockage and polymer rigidification. In this later work, the Maxwell model is accordingly applied three times to include both effects.

3.2.2. Bruggeman's Theory

The Bruggeman's theory proceeds from the premise that the field of neighboring particles can be taken into account by randomly adding the dispersed phase incrementally while considering the surrounding medium as the existing composite with effective transport properties at each stage [64,65,139]. This concept of incremental homogenization is illustrated in Figure 6, in which P_{eff}^1, P_{eff}^2 and P_{eff} are the MMM permeabilities at different stages of the homogenization process.

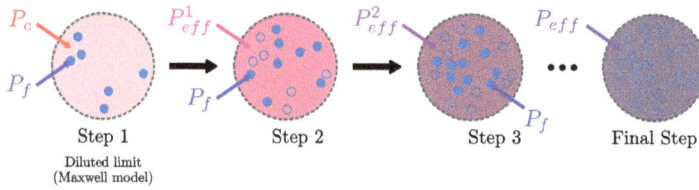

Figure 6. Schematic representation of the Bruggeman's theory.

Bruggeman [64] based his expression for the dielectric permeability of dispersed composites on the assumption that the Maxwell model described well the composite permeability in the diluted limit [65,139], as depicted in Figure 6. To do so, He considered that the differential increment in the dielectric permeability (dP) resulting from addition of new particles was well described by $P_r = P_{eff}/P_c = 1 + 3\beta_{fc}\phi_f$, that upon substitution of $P_c \rightarrow P$, $P_m \rightarrow P + dP$, and $\phi_f \rightarrow d\phi_f/(1 - \phi_f)$ leads to Equation (42) in Table 3 after integration [138,165]. The Bruggeman model is assumed to be applicable to moderate filler loadings $(0 \leq \phi_f \leq 0.35)$ [135,139,146].

Table 3. Models based on Bruggeman's theory.

Model	Key Equations	
Bruggeman [64]	$\left[\frac{P_{eff}}{P_c}\right]^{\frac{1}{3}}\left[\frac{\alpha_{fc}-1}{\alpha_{fc}-(P_{eff}/P_c)}\right] = \left[1 - \phi_f\right]^{-1}$	(42)
Pal [65]	$\left[\frac{P_{eff}}{P_c}\right]^{\frac{1}{3}}\left[\frac{\alpha_f-1}{\alpha_f-(P_{eff}/P_c)}\right] = \left[1 - \frac{\phi_f}{\phi_m}\right]^{-\phi_m}$	(43)
Pseudo-two phase Bruggeman model [78]	$\left[\frac{P_{eff}}{P_c}\right]^{\frac{1}{3}}\left[\frac{(P_{eff}^f/P_c)-1}{(P_{eff}^f/P_c)-(P_{eff}/P_c)}\right] = \left[1 - \phi_{fi}\right]^{-1}$	(44)
	$\left[P_{eff}^f/P_i\right]^{\frac{1}{3}} = 1 - \phi_s, \ P_f \ll P_i$	(45)
	$\left[\frac{P_{eff}^f}{P_i}\right]^{\frac{1}{3}}\left[\frac{(P_f/P_i)-1}{(P_f/P_i)-(P_{eff}^f/P_i)}\right] = [1 - \phi_s]^{-1}, \ P_f \geq P_i$	(46)
Pseudo-two phase Pal model [71]	$\left[\frac{P_{eff}}{P_c}\right]^{\frac{1}{3}}\left[\frac{(P_{eff}^f/P_c)-1}{(P_{eff}^f/P_c)-(P_{eff}/P_c)}\right] = \left[1 - \frac{\phi_{fi}}{\phi_m}\right]^{-\phi_m}$	(47)
	$\left[\frac{P_{eff}^f}{P_i}\right]^{\frac{1}{3}}\left[\frac{(P_f/P_i)-1}{(P_f/P_i)-(P_{eff}^f/P_i)}\right] = \left[1 - \frac{\phi_s}{\phi_m}\right]^{-\phi_m}$	(48)

Similar to Bruggeman [64], Pal [65] assumed that the increment dP resulting from the addition of new particles was well-described by $P_r = P_{eff}/P_c = 1 + 3\beta_{fc}\phi_f$, with $P_c \to P$ and $P_m \to P + dP$; however, $\phi_f \to d\phi_f/(1 - \phi_f/\phi_m)$ was used instead of $\phi_f \to d\phi_f/(1 - \phi_f)$, which leads to Equation (43) in Table 3 after integration [86,145]. In this model, ϕ_m has the same connotation as in the Lewis-Nielsen model (cf. Section 3.2.1) [80,86,151,153]. The Pal model reduces to Bruggeman's result when $\phi_m = 1$. Further, the Pal model always predicts higher values for the permeability (P_{eff}) than that of Bruggeman for a given system.

Analogous to the pseudo-two-phase Maxwell [85], both Bruggeman and Pal models have been extended to describe non-ideal MMMs. The former of these adaptations is known as the pseudo-two phase Bruggeman model in Equation (44) [78], with the model describing the effect on the MMM permeability of a voided or rigidified interfacial region at the filler particle surface [79]. In this model, ϕ_{fi} is given by Equation (38) [70] and ϕ_s given by Equation (41) [70,85,163]. Further, P_{eff}^f is given by Equation (45) for voided interphase($P_f \ll P_i$) [81] and by Equation (46) for rigidified interface [166]. Similarly, the pseudo-two phase Pal model in Equation (48) [71] accounts for the effect of polymer rigidification on the MMM permeability [79,85,160,164]. In this way, ϕ_{fi} and ϕ_s are given by Equations (38) and (41), respectively [70]. Here, P_{eff}^f and P_i have the same connotation as in the pseudo-two-phase Maxwell model (cf. Section 3.2.1) [79]. Recently, Idris et al. [79] modified the pseudo-two-phase Bruggeman model to account for both effects of polymer rigidification and presence of voids at the particle surface. In this later work, the Bruggeman model is accordingly applied three times to include both effects.

Finally, Bruggeman also developed a symmetric theory for the transport in composites [126,137]. However, models such as those of Landauer [63] and Böttcher [167], associated with this symmetric theory, are not discussed here. This is because the symmetric Bruggeman's theory considers the composite to be a random assembly of spherical particles of different materials [138,168], in which all components in the composite are continuous in the medium [141]. This condition is not met in MMMs, where one phase is preferentially assumed dispersed in the other.

3.3. Simulation-Based Rigorous Modeling Approach

The simulation-based rigorous modeling approach (SMA) is based on numerical solution of coupled partial differential equations (PDEs) describing the transport through the MMM via the finite-element method (FEM) [20,151,169], or any other suitable method (e.g., finite volume method or boundary element method), to discretize the 3D computational system [123]. Thus, the SMA is based on the assumption that the steady-state transport through the MMM is well-described by the continuity equation [151,170], as:

$$\nabla \cdot J = 0 \tag{49}$$

where J is the permeant steady-state flux, with the Fick's law describing the flux in both dispersed (J_f) and continuous (J_c) phases, as [153]:

$$J_f = D_f(C_f)(-\nabla C_f) \tag{50}$$

$$J_c = D_c(C_c)(-\nabla C_c). \tag{51}$$

respectively. In Equations (50) and (51), ∇C and D are the concentration gradient and diffusivity in a given phase, respectively. In this way, Equations (49)–(51) are solved in the 3D MMM [151], in which the resulting steady-state flux is used to calculate the MMM permeability via Equation (2) [152].

The SMA offers several benefits in comparison to earlier RMA and EMA approaches, amongst which the main advantage is that this approach can easily incorporate the permeability dependence on the concentration field [153] and finite-system size effects [151], largely disregarded in the former approaches. In this way, the SMA can accommodate effects on the permeability of intrinsic system properties such filler particle size [151] and shape [171,172], isotherm nonlinearity [153], and membrane geometry [152,169]. These effects are empirically treated in most RMA/EMA models [20], such in the Ebneyamini [68], Higuchi [88], and Pal [65] models.

The effect of filler size on the MMM permeability has been the focus of several studies [20,151,169], with Singh et al. [20] being the first to investigate such an effect for spherical particles and MMMs operating in the Henry's law region. In this work, the relative permeability (P_{eff}/P_c) was found independent of the particle size in the range of sizes investigated. However, later studies [151,169] suggested that the MMM permeability decreased with increase of particle size, and associated this depletion in the permeability with decrease of the specific polymer-filler interfacial area with increase of particle size at fixed volume fraction [151,173]. In these latter studies, the discrepancy amongst studies on the effect of particle size on the MMM permeability was associated with poor mesh quality in the FEM implementation in the former study [169], where the permeability was also found sensitive to changes of the product ratio $K_f D_f/K_c D_c$; inconsistent with definition in Equation (5) (cf. Section 2.2) [174]. Further, for MMMs operating at low pressure, other investigations have been focused on tubular [172] and layered fillers [171], with these investigations indicating sensitivity of the MMM permeability to filler shape, orientation, and packing.

Recently, the effect of isotherm nonlinearity was evaluated in MMMs with spherical fillers [153]. In this study, the MMM permeability was found more sensitive to isotherm nonlinearity in the filler phase than in the continuous phase. Similar to earlier studies in the Henry's law region [151,169], increase of filler particle size was found to decrease the MMM permeability. Further, comparison of the simulation predictions to EMA models suggested that the Chiew-Glandt model in Equation (26) describes well the effective permeability of ideal MMMs when the relative particle size is negligible ($r_o/\ell \to 0$), as in such case the effects of isotherm nonlinearity and finite-system size effects are negligible in the system. Figure 7 depicts a comparison of several EMA models to simulation results [151] for small filler particle size, and showing the Chiew-Glandt model to match well the simulation results, with the smallest percentage deviation (4.7%).

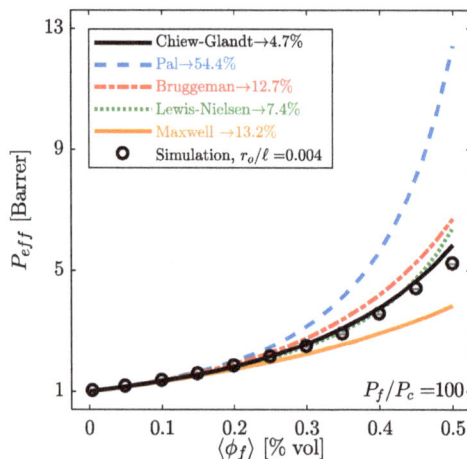

Figure 7. Comparison of the permeability profiles based on various effective medium approach (EMA) models to simulation results for $r_o/\ell = 0.004$ from [151], with $\alpha_{fc} = P_f/P_c = 100$.

Following the SMA, the effect of the membrane geometry has been evaluated [153,169]. Yang et al. [169] were the first to evaluate such an effect, by comparing the permeability of full-scale hollow fiber and flat-dense MMMs. By considering randomly distributed spherical inclusion and constant filler and polymer diffusivities, they found the hollow-fiber MMMs to have higher permeabilities than flat-dense MMMs. The increased efficiency of hollow fiber MMMs was associated with decrease in the length of the permeant diffusion pathways in the hollow fiber relative to those in a flat MMM. However, a more recent study found opposite behavior [152], by comparing MMMs having the same filler particle size, thickness, and volume, with the permeability in the hollow fiber found lower than in the flat MMM. Further, this tendency was associated with asymmetric filler phase distribution at the MMM ends, and arising from curvature change in the hollow fiber. The discrepancy amongst the different studies was associated with inaccuracies in the FEM implementation, which also led to the incorrect finding that the effective permeability was sensitive to variation of the ratio K_f/K_c and D_{of}/D_{oc} at constant $K_f D_{of}/K_c D_{oc}$ [149].

4. Predicting the Effective Permeability in Mixed-Matrix Membranes

Although there are a myriad of models for permeation in MMMs [62,68,80,85,147,151,175], with the popular ones described in Section 3, only a few of these match experimental permeation data [120,145,153,176]. While ideal models are often used as reference in practice [45,93,99], they are commonly shown to poorly describe experimental MMM permeabilities [71,128,166]. Alternatively, models incorporating interfacial non-idealities have been largely shown to match experimental data [70,71,78,79,161]. However, properties of the interfacial layer are always empirically fitted in these models [160,166,177]. When compared to experimental data, RMA and EMA models are used to either estimate the permeability in: (i) the MMM or (ii) any of its constituent phases by fitting MMM permeability data. In the first case, filler and continuous phase permeabilities are experimentally measured, and EMA/RMA are used to estimate the permeability in the MMM [99,107,110]. In the second case, the MMM and continuous phase permeabilities are experimentally measured, and EMA/RMA models are used to estimate the filler phase permeability [42,81], by fitting MMM permeation data. Here, we discuss the use of EMA/RMA models to predict filler permeabilities in Section 4.1 and MMM permeabilities in Section 4.2.

4.1. Estimation of the Filler Phase Permeability through EMA and RMA Models

Amongst the studies estimating the filler permeability through EMA/RMA models [42,81,128], that of Bouma et al. [81] estimated the relative permeability of O_2 in the filler phase ($\alpha_{fc} = P_f/P_c$) in an MMM comprising co-polyvinylidene fluoride-hexafluoropropene (co-PVDF) as continuous phase and a nematic liquid crystalline mixture (E7) as dispersed phase. In this work, Bouma et al. [81] found the Maxwell model to lead to $\alpha_{fc} = P_f/P_c = 15.0 \pm 8.1$ while that of Bruggeman to $\alpha_{fc} = P_f/P_c = 9.2 \pm 3.8$. Based on the relative error from each model fit, they concluded that the Bruggeman model better fitted the MMM permeabilities at higher filler concentrations.

Later, Gonzo et al. [145] found that the Chiew-Glandt model better fitted Bouma's experimental permeation data [81] as well as in other gases (e.g., CO_2, He, H_2, O_2, N_2) for various MMMs upon comparison of the Maxwell, Bruggeman, Chiew-Glandt, Higuchi and Landauer models (cf. Section 3.2). Furthermore, they found the Chiew-Glandt model to describe well the MMM permeability at both low and moderate filler volume fractions ($0 \leq \phi_f \leq 0.4$) [145]. Gonzo et al. [145] associated the overall success of the Chiew-Glandt model with the way particle-particle interactions are accommodated in this model [141,149], as the asymmetric integration technique used by Bruggeman [64] considers that newly added spheres are too dilute to interact among themselves, and interact only with the previously homogenized system [135,139,158]. This assumption may be inaccurate at high filler loadings.

In a similar fashion to Bouma et al. [81], Erdem-şenatalar et al. [128] estimated the filler phase permeability (P_f) by fitting experimental permeation data of various gases (e.g., CO_2, O_2, N_2, CH_4, $n - C_4H_{10}$, and $i - C_4H_{10}$) in Zeolite/Polymer MMMs (e.g., Zeolite 4A, 5A, 13X, and Silicalite-1) to the Series, Parallel, Te Hennepe, Maxwell, and Landauer models. They found large differences in P_f estimations amongst models for all gases and associated poor P_f predictions with the limited capability of RMA/EMA models to account for interfacial defects. However, here we associate these differences in P_f with the use of RMA/EMA models beyond their range of applicability, as most of these models are only suitable at low filler loadings.

Figure 8 depicts a comparison of the experimental effective permeability and that predicted by RMA/EMA models for one of the experimental data sets used by Erdem-şenatalar et al. [128]. However, we here calculated P_f via nonlinear regression of each model with the experimental permeation data in the range $0 \leq \phi_f \leq 0.2$, as only at low volume fractions RMA/EMA models provide comparable predictions of P_{eff} for a given $\alpha_{fc} = P_f/P_c$ (cf. Figures 3 and 5). In Figure 8, P_f predictions and percentage deviations between RMA/EMA models and experimental permeabilities are shown in the legend. Here, the experimental permeation data were originally reported by Duval et al. [178] for CO_2 in silicalite-1/EPDM MMMs, in which size of zeolite particles ranged between $1 - 5 \, \mu m$, membrane thicknesses between $50 - 200 \, \mu m$, and no morphological defects were reported in the MMMs structure.

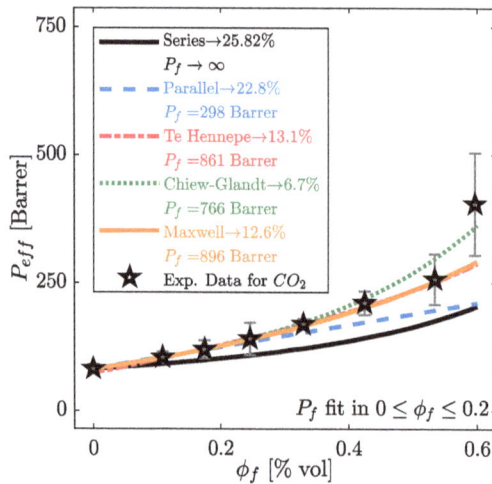

Figure 8. Comparison of resistance model approach (RMA)/effective medium approach (EMA) predictions for the filler phase permeability (P_f), obtained by fitting P_{eff} to experimental permeation data of CO_2 in Zeolite-13X/EPDM MMMs from [178], for filler fraction in the range $0 \leq \phi_f \leq 0.2$.

In Figure 8, we used the Chiew-Glandt model in place of the Landauer model, as the Landauer model has been shown not suitable for MMMs [81,141]. Here, the series model ($P_f \rightarrow \infty$) under-predicts the experimental permeabilities, with deviations of about 25%. This tendency is expected, as the series model corresponds to the lower bound for the permeability in ideal MMMs (cf. Section 3.1). Further, the filler phase permeability varies from $298 < P_f < 896$ Barrer, with the Chiew-Glandt model providing the best match to the experimental MMM permeabilities (deviation of 6.7%). This suggests that the Chiew-Glandt models describes well the MMM permeability when membrane structure has no associated defects, and particle size is negligible in comparison to the membrane thickness.

4.2. Prediction of the MMM Permeability through EMA and RMA Models

Among the studies comparing simulation-based [20,151,152,169,171,172] or experimental permeabilities of MMMs to EMA/RMA models [45,120,145,153,166,179], that of Monsalve-Bravo and Bhatia [151] compared simulation-based MMM permeabilities to various EMA models. In this work, they showed the Chiew-Glandt model to describe well the simulation-based MMM permeabilities when the ratio of the particle size (radius) to the membrane thickness is small (cf. Figure 7), and that the MB-B model (cf. Section 3.2.1) matched well the simulation-based MMM permeabilities when relative filler particles sizes were in the range $0.004 \leq r_0/\ell \leq 0.16$.

Figure 9 depicts a comparison of the simulations-based MMM permeabilities reported in [151] and the MB-B, Pal, Bruggeman, Lewis-Nielsen, Chiew-Glandt, and Maxwell models, with $\alpha_{fc} = P_f/P_c = 100$ in all models, $r_o/\ell = 0.040$ in the MB-B model, and $\phi_m = 0.645$ in the Pal and Lewis-Nielsen models. Further, percentage deviations between EMA models and simulation results are shown in the legend. In Figure 9, the MB-B model matches the simulation-based permeabilities, with a percentage deviation of 2.0%. Further, Monsalve-Bravo and Bhatia [151] showed that increase of filler particle size decreased the effective permeability of MMMs and associated this behavior with both depletion of the filler volume fraction at the membrane ends (due to the finite character of the system) and decrease of total available filler phase surface area, at a given volume fraction, with increase of the particle size.

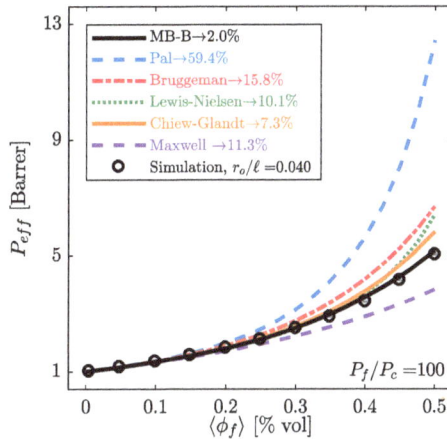

Figure 9. Comparison of the permeability profiles based on various EMA models to simulation results for $r_o/\ell = 0.040$ from [151], with $\alpha_{fc} = P_f/P_c = 100$.

Monsalve-Bravo and Bhatia [153] later extended their EMA model to accommodate isotherm nonlinearity using a self-consistent approach to calculate the filler phase permeability. In this work, they compared their model predictions to experimental permeation data of Vu et al. [46], who fabricated CMS-based MMMs for separation of CO_2/CH_4 and O_2/N_2 using Matrimid 5218 and Ultem 1000 as matrices. For the CMS/Ultem 1000 system, Monsalve-Bravo and Bhatia [153] showed that their modified effective medium theory (EMT) better predicted the permeability of all gases (O_2, N_2, CO_2, and CH_4) at different MMM feeding pressures and concluded that isotherm nonlinearity in the filler phase has a strong effect on the MMM permeability, with this effect becoming less significant with decrease of the particle size in the membrane system.

Figure 10 depicts a comparison of the experimental permeabilities reported in [46] for O_2/N_2 and the MB-B, Pal, Bruggeman, Lewis-Nielsen, Chiew-Glandt, and Maxwell models, with $r_o/\ell = 0.040$ in MB-B model and $\phi_m = 0.645$ in the Pal and Lewis-Nielsen models. Here, percentage deviations between the experimental permeabilities and EMA models are shown in the legend of each plot. In both Figure 10a,b, the MB-B models is in good agreement with the experimental permeabilities, having the smallest deviation amongst considered EMA models.

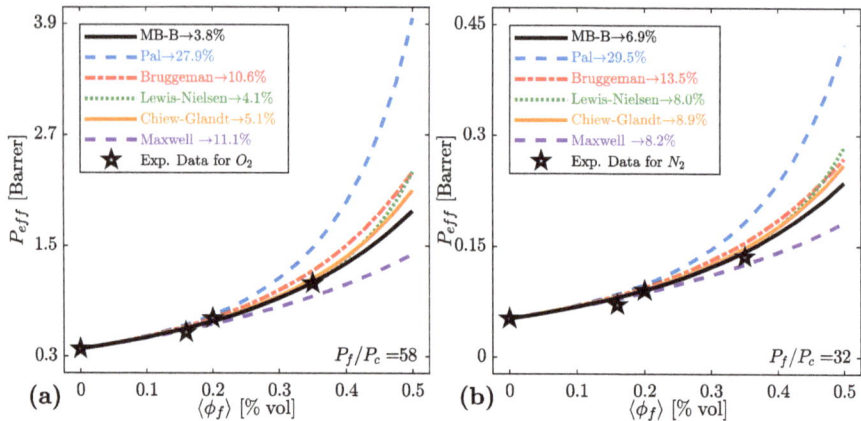

Figure 10. Comparison of the effective permeability profiles based on various EMA models to experimental permeation data in CMS/Ultem 1000 MMMs from [46]: (**a**) O_2 and (**b**) N_2.

While Monsalve-Bravo and Bhatia [153] studied the effect of isotherm nonlinearity for one of the membrane systems (CMS/Ultem 1000) reported by Vu et al. [46], the experimental permeabilities of O_2, N_2, CO_2, and CH_4 in the other system (CMS/Matrimid 5218) have been fitted to several non-ideal models [71,120,166]. This is because Vu et al. [46] hypothesized that polymer rigidification at the particle surface was the cause of the decreased permeabilities of all gases in the CMS/Matrimid 5218 MMMs. Among these works, Vu et al. [120] fitted the experimental permeability of CO_2 and CH_4 of [46] to the pseudo-two phase Maxwell model (cf. Section 3.2.1). They estimated the thickness of the rigidified interfacial layer as $\ell_i = 0.075\,\mu\text{m}$ by assuming filler particle diameter equal to $d_o = 1\,\mu\text{m}$ and interfacial permeability to $P_i = P_c/\sigma$, with $\sigma = 3$. In this work, different values of σ ranging between 1 and 4 were also used to adjust the interfacial thickness at fixed particle size ($d_o = 1\,\mu\text{m}$), which led to the conclusion that increase of σ yields a decrease in the interfacial thickness. A tendency later corroborated by Moore et al. [164] for zeolite-based MMMs.

Figure 11a depicts a comparison of the experimental permeabilities for from [46] and that predicted by Felske, pseudo-two-phase Maxwell and pseudo-two-phase Bruggeman models (cf. Section 3.2). Similar to Vu et al. [120], we here assumed $\sigma = 3$ and $d_o = 1\,\mu\text{m}$ to calculate ℓ_i in all models. Further, ℓ_i predictions and percentage deviation of each model from the experimental permeation data are shown in the legend of Figure 11a. Here, non-ideal models (overlapped) are in fair agreement with the experimental permeabilities, having percentage deviations of about 20%.

Alternatively, Figure 11b depicts a comparison the predicted interfacial thickness with increase of the chain immobilization factor (σ) while assuming the filler phase particle size $d_o = 1\,\mu\text{m}$. Here, an increase in the chain immobilization factor (σ), decreases the thickness of the interfacial layer (ℓ_i), similar to qualitative findings of Vu et al. [120] based on the pseudo-two phase Maxwell model. Further, the Felske and pseudo-two phase Maxwell model provide similar predictions of ℓ_i while the pseudo-two phase Bruggeman model predicts larger ℓ_i values. This behavior is expected because for a given MMM system the original Bruggeman model predicts higher permeabilities than the Maxwell model (cf. Figures 7 and 10), and thus this effect is transferred to the interfacial thickness at fixed P_f, P_i, and P_c.

In practice, the chain immobilization factor is commonly assumed in the range $3 \leq \sigma \leq 4$ [120,161,163,164]. This assumption often leads to interfacial layer thicknesses between the $0.05\,\mu\text{m} \leq \ell_i \leq 0.1\,\mu\text{m}$ [120], as depicted in Figure 11b. However, Dutta and Bhatia [180], via equilibrium molecular dynamics simulations, recently found that thickness of the rigidified layer to be $\ell_i \approx 0.001\,\mu\text{m}$ for polyimide near MFI (mordenite framework inverted) zeolite. While we recognize that thickness of the interfacial layer may vary between MMM systems, Dutta and Bhatia's findings [180] suggests that interfacial permeability (P_i) is much lower than that in the bulk polymer (P_c), also evident in Figure 11b. Thus, further theoretical developments are needed to accurately assess the permeant properties in the interface, as current models rely on empirically fitting these interfacial properties.

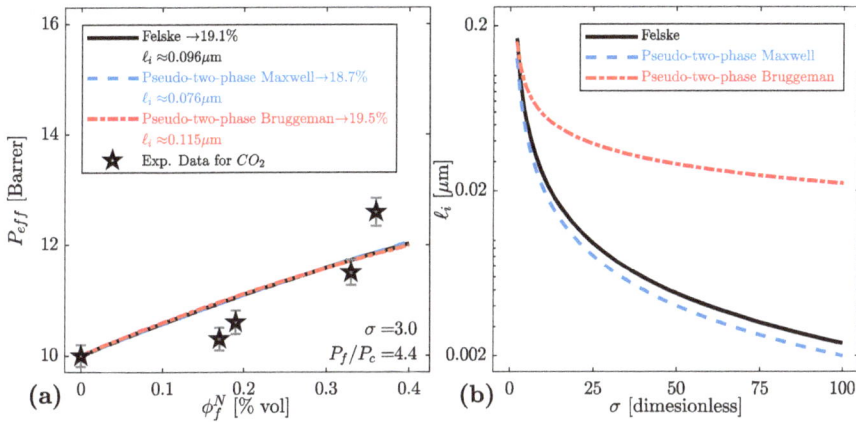

Figure 11. Comparison of the effective permeability and interfacial thickness profiles based on various non-ideal EMA models by fitting experimental permeation data for CO_2 in CMS/Matrimid 5218 MMMs from [46], with $\alpha_{fc} = P_f/P_c = 4.4$: (**a**) effective permeability vs. nominal filler volume fraction and (**b**) thickness vs. chain immobilization factor.

Finally, while RMA is much less popular for describing permeation in MMMs. With the introduction of nano-flake [181–183] and nanotube [184–186] based MMMs, models such as those of Cussler [51] and KJN [48] have become more popular as reference to compare with experimental [76,114,187] or simulation-based permeation data [171,172]. Amongst these works, Wang et al. [171,172] performed simulations of 3D MMMs comprising ideal platelet and tubular particles, and compared simulation predictions to the Cussler and KJN models (cf. Section 3.1). For platelet fillers, they found the Cussler model to be in good agreement with their simulations only when isotropic diffusion was considered in the simulations [171]. For tubular particles, they found the KJN model to under-predict the simulation-based permeabilities [172]. While these simulation studies considered concentration-independent phase permeabilities (P_f and P_c), difference amongst simulation results and RMA models suggest that particle shape has a strong effect on the MMM permeability. Thus, new theoretical developments are required that overcome the limitation of current EMA/RMA models on embedding filler-related effects within empirical fitting parameters, such as in the Ebneyamini [68], Lewis-Nielsen [73], Higuchi [88], and Pal [65] models.

5. Conclusions

Existing models for transport through mixed-matrix membranes (MMMs) are adaptations of well-established theories for the thermal/electric conductivity of heterogeneous media, grounded either in the resistance model approach (RMA) or the effective medium approach (EMA). In these approaches, the MMM permeability is based on both volume fractions and permeabilities of the MMM constituent phases while assuming uniformity of the field across the MMM, and negligible filler phase particle sizes compared to membrane thickness. These considerations lead to concentration-independent permeabilities across the MMM, which mask effects of isotherm nonlinearity and finite filler size in the effective permeability calculation. Furthermore, it has been recently shown [153] that deviations to the Henry's law, common under usual operating conditions of MMMs, lead to nonlinear concentration profiles and non-uniform permeabilities across the MMM.

Thus, EMA/RMA models need to progress beyond these limitations to appropriately characterize the transport of species across the membrane.

Among existing permeation models, that of Maxwell and its adaptations (cf. Table 2), corresponding to the low-density limit of non-interacting dispersed spheres, have remain popular as reference in practical applications. However, this model has also been commonly shown to under-predict experimental MMM permeation data [45,51,81,145]. For concentrated composites, a number of analytical expressions have been proposed, including those corresponding to the Chiew-Glandt, Lewis-Nielsen, Higuchi (cf. Table 2), Bruggeman, and Pal models (cf. Table 3). Here, the Chiew-Glandt result is the exact second order solution of the transport problem through a dispersion. This model has been shown to fit simulation-based and experimental permeation data when filler particle size is negligible (cf. Figures 7 and 8). Besides, an extended version of the Chiew-Glandt model (MB-B model in Table 2), accommodating both particle size and isotherm nonlinearity, has been shown to describe well the permeability for various gases in CMS-Ultem 1000 MMMs (cf. Figure 10). Alternatively, the Lewis-Nielsen, Higuchi, Bruggeman, and Pal models correspond to empirical modifications of the Maxwell or Bruggeman models to integrate particle interaction or packing-related effects; however, assuming negligible particle size in the MMM. Thus, while these models have been shown to match experimental permeation data [81], their associated empirical parameters are always fitted against the experimental MMM permeabilities, which mask effects of isotherm nonlinearity and finite filler size. Therefore, new theoretical models need to advance beyond this limitation and distinguish between effects of different mechanism to be able to achieve breakthroughs for the optimization of key separation processes.

Much effort has been devoted to mathematically represent effects of defective interfacial polymer-particle morphologies, in the form of a rigidified polymer [77,86], interfacial voids [163] or a blocked [161,179] layer around the filler particle surface, as well as particle shape (e.g., cubic, spheroidal, tubular) [69,72]. Nevertheless, while incorporating such morphological effects, existing models inherit the uniform field assumption of early RMA/EMA models, and thus disregard effects of isotherm nonlinearity and finite system size. While this drawback is addressed via simulation-based rigorous modeling of MMMs [152,171,172], EMA/RMA models need to progress beyond this limitation to appropriately characterize the transport of species across the membrane and its dependence on inherent matrix-filler properties. Furthermore, when non-ideal MMM morphologies are considered, the fitting of thickness of the interfacial layer based on an assumed permeability in this layer using MMM permeability data and EMA/RMA models is misleading (cf. Figure 11), and advances to independently characterize the interfacial layer thickness and its permeability are required.

Finally, while modeling of MMM through the RMA have received considerably less attention than EMA, with the introduction novel filler nanomaterials (e.g., carbon nanotubes and graphene nanosheets), models such that Te Hennepe, Cussler, and KJN have gained some popularity to be used as reference upon comparison with experimental and/or simulation-based permeation data. These models have been shown to successfully represent simulated MMM with flake-like and tubular particles operating at low pressures [171,172]. However, because they share the uniform concentration field assumption with EMA models, their applicability is limited to the Henry's law region. With the upcoming advances on 3D printing for synthesis of membrane materials [6], these effects of filler particle shape and packing distribution together with isotherm nonlinearity on the MMM permeability will need to be appropriately incorporated in the new generation of models.

Author Contributions: Conceptualization, G.M.M.-B. and S.K.B.; Writing—Original Draft Preparation, G.M.M.-B.; Writing—Review & Editing, G.M.M.-B. and S.K.B.

Funding: This research has been supported by a grant (No. DP150101996) from the Australian Research Council, through the Discovery scheme.

Conflicts of Interest: The authors declare no conflict of interest.

References

1. Sanders, D.F.; Smith, Z.P.; Guo, R.; Robeson, L.M.; McGrath, J.E.; Paul, D.R.; Freeman, B.D. Energy-efficient polymeric gas separation membranes for a sustainable future: A review. *Polymer* **2013**, *54*, 4729–4761. [CrossRef]
2. Kosinov, N.; Gascon, J.; Kapteijn, F.; Hensen, E.J.M. Recent developments in zeolite membranes for gas separation. *J. Membr. Sci.* **2016**, *499*, 65–79. [CrossRef]
3. Aroon, M.A.; Ismail, A.F.; Matsuura, T.; Montazer-Rahmati, M.M. Performance studies of mixed matrix membranes for gas separation: A review. *Sep. Purif. Technol.* **2010**, *75*, 229–242. [CrossRef]
4. Werber, J.R.; Deshmukh, A.; Elimelech, M. The Critical Need for Increased Selectivity, Not Increased Water Permeability, for Desalination Membranes. *Environ. Sci. Technol. Lett.* **2016**, *3*, 112–120. [CrossRef]
5. Han, Y.; Xu, Z.; Gao, C. Ultrathin Graphene Nanofiltration Membrane for Water Purification. *Adv. Funct. Mater.* **2013**, *23*, 3693–3700. [CrossRef]
6. Lee, J.-Y.; Tan, W.S.; An, J.; Chua, C.K.; Tang, C.Y.; Fane, A.G.; Chong, T.H. The potential to enhance membrane module design with 3D printing technology. *J. Membr. Sci.* **2016**, *499*, 480–490. [CrossRef]
7. Mohammad, A.W.; Ng, C.Y.; Lim, Y.P.; Ng, G.H. Ultrafiltration in Food Processing Industry: Review on Application, Membrane Fouling, and Fouling Control. *Food Bioprocess Technol.* **2012**, *5*, 1143–1156. [CrossRef]
8. Ong, Y.K.; Shi, G.M.; Le, N.L.; Tang, Y.P.; Zuo, J.; Nunes, S.P.; Chung, T.-S. Recent membrane development for pervaporation processes. *Prog. Polym. Sci.* **2016**, *57*, 1–31. [CrossRef]
9. Jia, Z.; Wu, G. Metal-organic frameworks based mixed matrix membranes for pervaporation. *Microporous Mesoporous Mater.* **2016**, *235*, 151–159. [CrossRef]
10. Zhang, Z.; Chen, F.; Rezakazemi, M.; Zhang, W.; Lu, C.; Chang, H.; Quan, X. Modeling of a CO_2-piperazine-membrane absorption system. *Chem. Eng. Res. Des.* **2018**, *131*, 375–384. [CrossRef]
11. Asfand, F.; Bourouis, M. A review of membrane contactors applied in absorption refrigeration systems. *Renew. Sustain. Energy Rev.* **2015**, *45*, 173–191. [CrossRef]
12. Zheng, Y.; Zhang, W.; Tang, B.; Ding, J.; Zheng, Y.; Zhang, Z. Membrane fouling mechanism of biofilm-membrane bioreactor (BF-MBR): Pore blocking model and membrane cleaning. *Bioresour. Technol.* **2018**, *250*, 398–405. [CrossRef] [PubMed]
13. Molinari, R.; Lavorato, C.; Argurio, P. Recent progress of photocatalytic membrane reactors in water treatment and in synthesis of organic compounds. A review. *Catal. Today* **2017**, *281*, 144–164. [CrossRef]
14. Liu, C.; Kulprathipanja, S. Mixed-Matrix Membranes. In *Zeolites in Industrial Separation and Catalysis*; Wiley-VCH Verlag GmbH & CO. KGaA: Weinheim, Germany, 2010; pp. 329–353. ISBN 9783527629565.
15. Zornoza, B.; Tellez, C.; Coronas, J.; Gascon, J.; Kapteijn, F. Metal organic framework based mixed matrix membranes: An increasingly important field of researCH with a large application potential. *Microporous Mesoporous Mater.* **2013**, *166*, 67–78. [CrossRef]
16. Hua, D.; Ong, Y.K.; Wang, Y.; Yang, T.; Chung, T.-S. ZIF-90/P84 mixed matrix membranes for pervaporation dehydration of isopropanol. *J. Membr. Sci.* **2014**, *453*, 155–167. [CrossRef]
17. Basu, S.; Cano-Odena, A.; Vankelecom, I.F.J. MOF-containing mixed-matrix membranes for CO_2/CH_4 and CO_2/N_2 binary gas mixture separations. *Sep. Purif. Technol.* **2011**, *81*, 31–40. [CrossRef]
18. Rezakazemi, M.; Ebadi Amooghin, A.; Montazer-Rahmati, M.M.; Ismail, A.F.; Matsuura, T. State-of-the-art membrane based CO_2 separation using mixed matrix membranes (MMMs): An overview on current status and future directions. *Prog. Polym. Sci.* **2014**, *39*, 817–861. [CrossRef]
19. Goh, P.S.; Ismail, A.F.; Sanip, S.M.; Ng, B.C.; Aziz, M. Recent advances of inorganic fillers in mixed matrix membrane for gas separation. *Sep. Purif. Technol.* **2011**, *81*, 243–264. [CrossRef]
20. Singh, T.; Kang, D.-Y.; Nair, S. Rigorous calculations of permeation in mixed-matrix membranes: Evaluation of interfacial equilibrium effects and permeability-based models. *J. Membr. Sci.* **2013**, *448*, 160–169. [CrossRef]
21. Vinh-Thang, H.; Kaliaguine, S. Predictive Models for Mixed-Matrix Membrane Performance: A Review. *Chem. Rev.* **2013**, *113*, 4980–5028. [CrossRef] [PubMed]
22. Vinh-Thang, H.; Kaliaguine, S. A comprehensive computational strategy for fitting experimental permeation data of mixed matrix membranes. *J. Membr. Sci.* **2014**, *452*, 271–276. [CrossRef]
23. Park, H.B.; Kamcev, J.; Robeson, L.M.; Elimelech, M.; Freeman, B.D. Maximizing the right stuff: The trade-off between membrane permeability and selectivity. *Science* **2017**, *356*. [CrossRef] [PubMed]

24. Fan, H.; Shi, Q.; Yan, H.; Ji, S.; Dong, J.; Zhang, G. Simultaneous Spray Self-Assembly of Highly Loaded ZIF-8–PDMS Nanohybrid Membranes Exhibiting Exceptionally High Biobutanol-Permselective Pervaporation. *Angew. Chem. Int. Ed.* **2014**, *53*, 5578–5582. [CrossRef] [PubMed]

25. Robeson, L.M. The upper bound revisited. *J. Membr. Sci.* **2008**, *320*, 390–400. [CrossRef]

26. Robeson, L.M. Correlation of separation factor versus permeability for polymeric membranes. *J. Membr. Sci.* **1991**, *62*, 165–185. [CrossRef]

27. Burns, R.L.; Koros, W.J. Defining the challenges for C3H6/C3H8 separation using polymeric membranes. *J. Membr. Sci.* **2003**, *211*, 299–309. [CrossRef]

28. Swaidan, R.J.; Ma, X.; Litwiller, E.; Pinnau, I. Enhanced propylene/propane separation by thermal annealing of an intrinsically microporous hydroxyl-functionalized polyimide membrane. *J. Membr. Sci.* **2015**, *495*, 235–241. [CrossRef]

29. Jongok, W.; Ho, K.M.; Soo, K.Y.; Chae, P.H.; Young, K.U.; Chang, C.S.; Keun, K.S. Surface modification of polyimide and polysulfone membranes by ion beam for gas separation. *J. Appl. Polym. Sci.* **2000**, *75*, 1554–1560. [CrossRef]

30. Koros, W.J.; Mahajan, R. Pushing the limits on possibilities for large scale gas separation: WhiCH strategies? *J. Membr. Sci.* **2000**, *175*, 181–196. [CrossRef]

31. Zhao, S.; Wang, Z.; Qiao, Z.; Wei, X.; Zhang, C.; Wang, J.; Wang, S. Gas separation membrane with CO2-facilitated transport highway constructed from amino carrier containing nanorods and macromolecules. *J. Mater. Chem. A* **2013**, *1*, 246–249. [CrossRef]

32. Mannan, H.A.; Mukhtar, H.; Murugesan, T.; Nasir, R.; Mohshim, D.F.; Mushtaq, A. Recent Applications of Polymer Blends in Gas Separation Membranes. *Chem. Eng. Technol.* **2013**, *36*, 1838–1846. [CrossRef]

33. Zhang, C. *Zeolitic Imidazolate Framework (Zif)-Based Membranes and Sorbents for Advanced Olefin/Paraffin Separations*; Georgia Institute of Technology: Atlanta, GA, USA, 2014.

34. Chung, T.-S.; Jiang, L.Y.; Li, Y.; Kulprathipanja, S. Mixed matrix membranes (MMMs) comprising organic polymers with dispersed inorganic fillers for gas separation. *Prog. Polym. Sci.* **2007**, *32*, 483–507. [CrossRef]

35. Vinoba, M.; Bhagiyalakshmi, M.; Alqaheem, Y.; Alomair, A.A.; Pérez, A.; Rana, M.S. Recent progress of fillers in mixed matrix membranes for CO2 separation: A review. *Sep. Purif. Technol.* **2017**, *188*, 431–450. [CrossRef]

36. Seoane, B.; Coronas, J.; Gascon, I.; Benavides, M.E.; Karvan, O.; Caro, J.; Kapteijn, F.; Gascon, J. Metal-organic framework based mixed matrix membranes: A solution for highly efficient CO2 capture? *Chem. Soc. Rev.* **2015**, *44*, 2421–2454. [CrossRef] [PubMed]

37. Kim, S.; Marand, E.; Ida, J.; Guliants, V. V Polysulfone and Mesoporous Molecular Sieve MCM-48 Mixed Matrix Membranes for Gas Separation. *Chem. Mater.* **2006**, *18*, 1149–1155. [CrossRef]

38. Askari, M.; Chung, T.-S. Natural gas purification and olefin/paraffin separation using thermal cross-linkable CO-polyimide/ZIF-8 mixed matrix membranes. *J. Membr. Sci.* **2013**, *444*, 173–183. [CrossRef]

39. Amedi, H.R.; Aghajani, M. Gas separation in mixed matrix membranes based on polyurethane containing SiO2, ZSM-5, and ZIF-8 nanoparticles. *J. Nat. Gas Sci. Eng.* **2016**, *35*, 695–702. [CrossRef]

40. Galizia, M.; Chi, W.S.; Smith, Z.P.; Merkel, T.C.; Baker, R.W.; Freeman, B.D. 50th Anniversary Perspective: Polymers and Mixed Matrix Membranes for Gas and Vapor Separation: A Review and Prospective Opportunities. *Macromolecules* **2017**. [CrossRef]

41. Tanh Jeazet, H.B.; Staudt, C.; Janiak, C. Metal-organic frameworks in mixed-matrix membranes for gas separation. *Dalton Trans.* **2012**, *41*, 14003–14027. [CrossRef] [PubMed]

42. Zhang, C.; Dai, Y.; Johnson, J.R.; Karvan, O.; Koros, W.J. High performance ZIF-8/6FDA-DAM mixed matrix membrane for propylene/propane separations. *J. Membr. Sci.* **2012**, *389*, 34–42. [CrossRef]

43. Faiz, R.; Li, K. Olefin/paraffin separation using membrane based facilitated transport/chemical absorption techniques. *Chem. Eng. Sci.* **2012**, *73*, 261–284. [CrossRef]

44. Liu, J.; Bae, T.-H.; Qiu, W.; Husain, S.; Nair, S.; Jones, C.W.; Chance, R.R.; Koros, W.J. Butane isomer transport properties of 6FDA–DAM and MFI–6FDA–DAM mixed matrix membranes. *J. Membr. Sci.* **2009**, *343*, 157–163. [CrossRef]

45. Song, Q.; Nataraj, S.K.; Roussenova, M.V.; Tan, J.C.; Hughes, D.J.; Li, W.; Bourgoin, P.; Alam, M.A.; Cheetham, A.K.; Al-Muhtaseb, S.A.; et al. Zeolitic imidazolate framework (ZIF-8) based polymer nanocomposite membranes for gas separation. *Energy Environ. Sci.* **2012**, *5*, 8359–8369. [CrossRef]

46. Vu, D.Q.; Koros, W.J.; Miller, S.J. Mixed matrix membranes using carbon molecular sieves: I. Preparation and experimental results. *J. Membr. Sci.* **2003**, *211*, 311–334. [CrossRef]

47. Zimmerman, C.M.; Singh, A.; Koros, W.J. Tailoring mixed matrix composite membranes for gas separations. *J. Membr. Sci.* **1997**, *137*, 145–154. [CrossRef]

48. Kiyono, M.; Williams, P.J.; Koros, W.J. Effect of polymer precursors on carbon molecular sieve structure and separation performance properties. *Carbon N. Y.* **2010**, *48*, 4432–4441. [CrossRef]

49. Fernández-Barquín, A.; Casado-Coterillo, C.; Etxeberria-Benavides, M.; Zuñiga, J.; Irabien, A. Comparison of Flat and Hollow-Fiber Mixed-Matrix Composite Membranes for CO_2 Separation with Temperature. *Chem. Eng. Technol.* **2017**, *40*, 997–1007. [CrossRef]

50. Yun, Y.N.; Sohail, M.; Moon, J.; Tae, W.K.; Kyeng, M.P.; Dong, H.C.; Young, C.P.; Cho, C.; Kim, H. Defect-Free Mixed-Matrix Membranes with Hydrophilic Metal-Organic Polyhedra for Efficient Carbon Dioxide Separation. *Chem. Asian J.* **2018**, *13*, 631–635. [CrossRef] [PubMed]

51. Ma, X.; Swaidan, R.J.; Wang, Y.; Hsiung, C.; Han, Y.; Pinnau, I. Highly Compatible Hydroxyl-Functionalized Microporous Polyimide-ZIF-8 Mixed Matrix Membranes for Energy Efficient Propylene/Propane Separation. *ACS Appl. Nano Mater.* **2018**, *1*, 3541–3547. [CrossRef]

52. Bushell, A.F.; Attfield, M.P.; Mason, C.R.; Budd, P.M.; Yampolskii, Y.; Starannikova, L.; Rebrov, A.; Bazzarelli, F.; Bernardo, P.; Carolus Jansen, J.; et al. Gas permeation parameters of mixed matrix membranes based on the polymer of intrinsic microporosity PIM-1 and the zeolitic imidazolate framework ZIF-8. *J. Membr. Sci.* **2013**, *427*, 48–62. [CrossRef]

53. Adams, R.; Carson, C.; Ward, J.; Tannenbaum, R.; Koros, W. Metal organic framework mixed matrix membranes for gas separations. *Microporous Mesoporous Mater.* **2010**, *131*, 13–20. [CrossRef]

54. Zhang, Y.; Feng, X.; Yuan, S.; Zhou, J.; Wang, B. Challenges and recent advances in MOF-polymer composite membranes for gas separation. *Inorg. Chem. Front.* **2016**, *3*, 896–909. [CrossRef]

55. Nordin, N.A.H.M.; Ismail, A.F.; Mustafa, A.; Murali, R.S.; Matsuura, T. The impact of ZIF-8 particle size and heat treatment on CO_2/CH_4 separation using asymmetric mixed matrix membrane. *RSC Adv.* **2014**, *4*, 52530–52541. [CrossRef]

56. Zhang, C.; Zhang, K.; Xu, L.; Labreche, Y.; Kraftschik, B.; Koros, W.J. Highly scalable ZIF-based mixed-matrix hollow fiber membranes for advanced hydrocarbon separations. *AIChE J.* **2014**, *60*, 2625–2635. [CrossRef]

57. Carreon, M.; Dahe, G.; Feng, J.; Venna, S.R. Mixed Matrix Membranes for Gas Separation Applications. In *Membranes for Gas Separations*; World Scientific Series in Membrane Science and Technology: Biological and Biomimetic Applications, Energy and the Environment; World Scientific: Singapore, 2016; Volume 1, pp. 1–57. ISBN 978-981-320-770-7.

58. Liu, L.; Nicholson, D.; Bhatia, S.K. Interfacial Resistance and Length-Dependent Transport Diffusivities in Carbon Nanotubes. *J. Phys. Chem. C* **2016**, *120*, 26363–26373. [CrossRef]

59. Glavatskiy, K.S.; Bhatia, S.K. Effect of pore size on the interfacial resistance of a porous membrane. *J. Membr. Sci.* **2017**, *524*, 738–745. [CrossRef]

60. Glavatskiy, K.S.; Bhatia, S.K. Thermodynamic Resistance to Matter Flow at The Interface of a Porous Membrane. *Langmuir* **2016**, *32*, 3400–3411. [CrossRef] [PubMed]

61. Hinkle, K.R.; Jameson, C.J.; Murad, S. Transport of Vanadium and Oxovanadium Ions Across Zeolite Membranes: A Molecular Dynamics Study. *J. Phys. Chem. C* **2014**, *118*, 23803–23810. [CrossRef]

62. Maxwell, J.C. *A treatise on Electricity and Magnetism*; Clarendon Press: Oxford, UK, 1873.

63. Landauer, R. The Electrical Resistance of Binary Metallic Mixtures. *J. Appl. Phys.* **1952**, *23*, 779–784. [CrossRef]

64. Bruggeman, D.A.G. Berechnung verschiedener physikalischer Konstanten von heterogenen Substanzen. *Ann. Phys.* **1935**, *24*, 636. [CrossRef]

65. Pal, R. New models for thermal conductivity of particulate composites. *J. Reinf. Plast. Compos.* **2007**, *26*, 643–651. [CrossRef]

66. Henis, J.M.S.; Tripodi, M.K. Composite hollow fiber membranes for gas separation: The resistance model approach. *J. Membr. Sci.* **1981**, *8*, 233–246. [CrossRef]

67. Pinnau, I.; Wijmans, J.G.; Blume, I.; Kuroda, T.; Peinemann, K.V. Gas permeation through composite membranes. *J. Membr. Sci.* **1988**, *37*, 81–88. [CrossRef]

68. Ebneyamini, A.; Azimi, H.; Tezel, F.H.; Thibault, J. Mixed matrix membranes applications: Development of a resistance-based model. *J. Membr. Sci.* **2017**, *543*, 351–360. [CrossRef]

69. Kang, D.-Y.; Jones, C.W.; Nair, S. Modeling molecular transport in composite membranes with tubular fillers. *J. Membr. Sci.* **2011**, *381*, 50–63. [CrossRef]

70. Shen, Y.; Lua, A.C. Theoretical and experimental studies on the gas transport properties of mixed matrix membranes based on polyvinylidene fluoride. *AIChE J.* **2013**, *59*, 4715–4726. [CrossRef]

71. Shimekit, B.; Mukhtar, H.; Murugesan, T. Prediction of the relative permeability of gases in mixed matrix membranes. *J. Membr. Sci.* **2011**, *373*, 152–159. [CrossRef]

72. Cussler, E.L. Membranes containing selective flakes. *J. Membr. Sci.* **1990**, *52*, 275–288. [CrossRef]

73. Lewis, T.B.; Nielsen, L.E. Dynamic mechanical properties of particulate-filled Composites. *J. Appl. Polym. Sci.* **1970**, *14*, 1449–1471. [CrossRef]

74. Higuchi, W.I.; Higuchi, T. Theoretical Analysis of Diffusional Movement Through Heterogeneous Barriers. *J. Am. Pharm. Assoc. (Sci. Ed.)* **1960**, *49*, 598–606. [CrossRef]

75. Moore, T.T.; Koros, W.J. Gas sorption in polymers, molecular sieves, and mixed matrix membranes. *J. Appl. Polym. Sci.* **2007**, *104*, 4053–4059. [CrossRef]

76. Chehrazi, E.; Sharif, A.; Omidkhah, M.; Karimi, M. Modeling the Effects of Interfacial Characteristics on Gas Permeation Behavior of Nanotube–Mixed Matrix Membranes. *ACS Appl. Mater. Interfaces* **2017**, *9*, 37321–37331. [CrossRef] [PubMed]

77. Felske, J.D. Effective thermal conductivity of composite spheres in a continuous medium with contact resistance. *Int. J. Heat Mass Transf.* **2004**, *47*, 3453–3461. [CrossRef]

78. Shariati, A.; Omidkhah, M.; Pedram, M.Z. New permeation models for nanocomposite polymeric membranes filled with nonporous particles. *Chem. Eng. Res. Des.* **2012**, *90*, 563–575. [CrossRef]

79. Idris, A.; Man, Z.; Abdulhalim, S.M.; Uddin, F. Modified Bruggeman models for prediction of CO_2 permeance in polycarbonate/silica nanocomposite membranes. *Can. J. Chem. Eng.* **2017**, *95*, 2398–2409. [CrossRef]

80. Pal, R. On the Lewis–Nielsen model for thermal/electrical conductivity of composites. *Compos. Part A Appl. Sci. Manuf.* **2008**, *39*, 718–726. [CrossRef]

81. Bouma, R.H.B.; Checchetti, A.; Chidichimo, G.; Drioli, E. Permeation through a heterogeneous membrane: The effect of the dispersed phase. *J. Membr. Sci.* **1997**, *128*, 141–149. [CrossRef]

82. Funk, C.V.; Lloyd, D.R. Zeolite-filled microporous mixed matrix (ZeoTIPS) membranes: Prediction of gas separation performance. *J. Membr. Sci.* **2008**, *313*, 224–231. [CrossRef]

83. Nan, C.-W.; Liu, G.; Lin, Y.; Li, M. Interface effect on thermal conductivity of carbon nanotube composites. *Appl. Phys. Lett.* **2004**, *85*, 3549–3551. [CrossRef]

84. Li, Y.; Guan, H.-M.; Chung, T.-S.; Kulprathipanja, S. Effects of novel silane modification of zeolite surface on polymer chain rigidification and partial pore blockage in polyethersulfone (PES)–zeolite A mixed matrix membranes. *J. Membr. Sci.* **2006**, *275*, 17–28. [CrossRef]

85. Mahajan, R.; Koros, W.J. Mixed matrix membrane materials with glassy polymers. Part 1. *Polym. Eng. Sci.* **2002**, *42*, 1420–1431. [CrossRef]

86. Pal, R. Permeation models for mixed matrix membranes. *J. Colloid Interface Sci.* **2008**, *317*, 191–198. [CrossRef] [PubMed]

87. Hashemifard, S.A.; Ismail, A.F.; Matsuura, T. A new theoretical gas permeability model using resistance modeling for mixed matrix membrane systems. *J. Membr. Sci.* **2010**, *350*, 259–268. [CrossRef]

88. Higuchi, W.I. A New Relationship for the Dielectric Properties of Two Phase Mixtures. *J. Phys. Chem.* **1958**, *62*, 649–653. [CrossRef]

89. Javaid, A. Membranes for solubility-based gas separation applications. *Chem. Eng. J.* **2005**, *112*, 219–226. [CrossRef]

90. Ismail, A.F.; Khulbe, K.C.; Matsuura, T. *Gas Separation Membranes: Polymeric and Inorganic*; Springer: Berlin/Heidelberg, Germany, 2015.

91. Koros, W.J.; Fleming, G.K. Membrane-based gas separation. *J. Membr. Sci.* **1993**, *83*, 1–80. [CrossRef]

92. Ghosh, A.; Mistri, E.A.; Banerjee, S. 3—Fluorinated Polyimides: Synthesis, Properties, and Applications. In *Handbook of Specialty Fluorinated Polymers*; William Andrew Publishing: Norwich, NY, USA, 2015; pp. 97–185. ISBN 978-0-323-35792-0.

93. Zhang, C.; Lively, R.P.; Zhang, K.; Johnson, J.R.; Karvan, O.; Koros, W.J. Unexpected Molecular Sieving Properties of Zeolitic Imidazolate Framework-8. *J. Phys. Chem. Lett.* **2012**, *3*, 2130–2134. [CrossRef] [PubMed]

94. Yujie, B.; Zhengjie, L.; Yanshuo, L.; Yuan, P.; Hua, J.; Wenmei, J.; Ang, G.; Po, W.; Qingyuan, Y.; Chongli, Z.; et al. Confinement of Ionic Liquids in Nanocages: Tailoring the Molecular Sieving Properties of ZIF-8 for Membrane-Based CO_2 Capture. *Angew. Chem. Int. Ed.* **2015**, *54*, 15483–15487. [CrossRef]

95. Li, W.; Zhang, Y.; Zhang, C.; Meng, Q.; Xu, Z.; Su, P.; Li, Q.; Shen, C.; Fan, Z.; Qin, L.; et al. Transformation of metal-organic frameworks for molecular sieving membranes. *Nat. Commun.* **2016**, *7*, 11315. [CrossRef] [PubMed]

96. Budd, P.M.; Msayib, K.J.; Tattershall, C.E.; Ghanem, B.S.; Reynolds, K.J.; McKeown, N.B.; Fritsch, D. Gas separation membranes from polymers of intrinsic microporosity. *J. Membr. Sci.* **2005**, *251*, 263–269. [CrossRef]

97. Thran, A.; Kroll, G.; Faupel, F. Correlation between fractional free volume and diffusivity of gas molecules in glassy polymers. *J. Polym. Sci. Part B Polym. Phys.* **1999**, *37*, 3344–3358. [CrossRef]

98. Vrentas, J.S.; Vrentas, C.M. Evaluation of the free-volume theory of diffusion. *J. Polym. Sci. Part B Polym. Phys.* **2003**, *41*, 501–507. [CrossRef]

99. Das, M.; Perry, J.D.; Koros, W.J. Gas-Transport-Property Performance of Hybrid Carbon Molecular Sieve−Polymer Materials. *Ind. Eng. Chem. Res.* **2010**, *49*, 9310–9321. [CrossRef]

100. Wijmans, J.G.; Baker, R.W. The solution-diffusion model: A review. *J. Membr. Sci.* **1995**, *107*, 1–21. [CrossRef]

101. Wijmans, J.G.; Baker, R.W. The Solution–Diffusion Model: A Unified Approach to Membrane Permeation. In *Materials Science of Membranes for Gas and Vapor Separation*; Freeman, B., Yampolskii, Y., Pinnau, I., Eds.; John Wiley & Sons, Ltd.: Hoboken, NJ, USA, 2006.

102. Ghosal, K.; Freeman, B.D. Gas separation using polymer membranes: An overview. *Polym. Adv. Technol.* **1994**, *5*, 673–697. [CrossRef]

103. O'Brien, K.C.; Koros, W.J.; Barbari, T.A.; Sanders, E.S. A new technique for the measurement of multicomponent gas transport through polymeric films. *J. Membr. Sci.* **1986**, *29*, 229–238. [CrossRef]

104. Swaidan, R.J.; Ma, X.; Pinnau, I. Spirobisindane-based polyimide as efficient precursor of thermally-rearranged and carbon molecular sieve membranes for enhanced propylene/propane separation. *J. Membr. Sci.* **2016**, *520*, 983–989. [CrossRef]

105. Stern, S.A. The "barrer" permeability unit. *J. Polym. Sci. Part A-2 Polym. Phys.* **1968**, *6*, 1933–1934. [CrossRef]

106. An, H.; Park, S.; Kwon, H.T.; Jeong, H.-K.; Lee, J.S. A new superior competitor for exceptional propylene/propane separations: ZIF-67 containing mixed matrix membranes. *J. Membr. Sci.* **2017**, *526*, 367–376. [CrossRef]

107. Liu, G.; Chernikova, V.; Liu, Y.; Zhang, K.; Belmabkhout, Y.; Shekhah, O.; Zhang, C.; Yi, S.; Eddaoudi, M.; Koros, W.J. Mixed matrix formulations with MOF molecular sieving for key energy-intensive separations. *Nat. Mater.* **2018**, *17*, 283–289. [CrossRef] [PubMed]

108. Kamaruddin, D.H.; Koros, W.J. Some observations about the application of Fick's first law for membrane separation of multicomponent mixtures. *J. Membr. Sci.* **1997**, *135*, 147–159. [CrossRef]

109. Koros, W.J. Simplified analysis of gas/polymer selective solubility behavior. *J. Polym. Sci. Polym. Phys. Ed.* **1985**, *23*, 1611–1628. [CrossRef]

110. Ruthven, D.M. Sorption kinetics for diffusion-controlled systems with a strongly concentration-dependent diffusivity. *Chem. Eng. Sci.* **2004**, *59*, 4531–4545. [CrossRef]

111. Ruthven, D.M. *Principles of Adsorption and Adsorption Processes*; John Wiley & Sons: Hoboken, NJ, USA, 1984; ISBN 0471866067.

112. Hashemifard, S.A.; Ismail, A.F.; Matsuura, T. Prediction of gas permeability in mixed matrix membranes using theoretical models. *J. Membr. Sci.* **2010**, *347*, 53–61. [CrossRef]

113. Paul, D.R.; Koros, W.J. Effect of partially immobilizing sorption on permeability and the diffusion time lag. *J. Polym. Sci. Polym. Phys. Ed.* **1976**, *14*, 675–685. [CrossRef]

114. Sheffel, J.A.; Tsapatsis, M. A semi-empirical approach for predicting the performance of mixed matrix membranes containing selective flakes. *J. Membr. Sci.* **2009**, *326*, 595–607. [CrossRef]

115. Fu, S.; Sanders, E.S.; Kulkarni, S.S.; Koros, W.J. Carbon molecular sieve membrane structure–property relationships for four novel 6FDA based polyimide precursors. *J. Membr. Sci.* **2015**, *487*, 60–73. [CrossRef]

116. Ning, X.; Koros, W.J. Carbon molecular sieve membranes derived from Matrimid®polyimide for nitrogen/methane separation. *Carbon N. Y.* **2014**, *66*, 511–522. [CrossRef]

117. Barrer, R.M. Diffusivities in glassy polymers for the dual mode sorption model. *J. Membr. Sci.* **1984**, *18*, 25–35. [CrossRef]

118. Barbari, T.A.; Koros, W.J.; Paul, D.R. Polymeric membranes based on bisphenol-A for gas separations. *J. Membr. Sci.* **1989**, *42*, 69–86. [CrossRef]

119. Esekhile, O.; Qiu, W.; Koros, W.J. Permeation of butane isomers through 6FDA-DAM dense films. *J. Polym. Sci. Part B Polym. Phys.* **2011**, *49*, 1605–1620. [CrossRef]

120. Vu, D.Q.; Koros, W.J.; Miller, S.J. Mixed matrix membranes using carbon molecular sieves II. Modeling permeation behavior. *J. Membr. Sci.* **2003**, *211*, 335–348. [CrossRef]
121. Najari, S.; Hosseini, S.S.; Omidkhah, M.; Tan, N.R. Phenomenological modeling and analysis of gas transport in polyimide membranes for propylene/propane separation. *RSC Adv.* **2015**, *5*, 47199–47215. [CrossRef]
122. Koros, W.J.; Chern, R.T.; Stannett, V.; Hopfenberg, H.B. A model for permeation of mixed gases and vapors in glassy polymers. *J. Polym. Sci. Polym. Phys. Ed.* **1981**, *19*, 1513–1530. [CrossRef]
123. Sarra, Z.; Matthieu, Z.; Eliane, E. Modeling diffusion mass transport in multiphase polymer systems for gas barrier applications: A review. *J. Polym. Sci. Part B Polym. Phys.* **2018**, *56*, 621–639. [CrossRef]
124. Nielsen, L.E. Thermal conductivity of particulate-filled polymers. *J. Appl. Polym. Sci.* **1973**, *17*, 3819–3820. [CrossRef]
125. Henis, J.M.S.; Tripodi, M.K. A Novel ApproaCH to Gas Separations Using Composite Hollow Fiber Membranes. *Sep. Sci. Technol.* **1980**, *15*, 1059–1068. [CrossRef]
126. Bánhegyi, G. Comparison of electrical mixture rules for Composites. *Colloid Polym. Sci.* **1986**, *264*, 1030–1050. [CrossRef]
127. Karode, S.K.; Patwardhan, V.S.; Kulkarni, S.S. An improved model incorporating constriction resistance in transport through thin film composite membranes. *J. Membr. Sci.* **1996**, *114*, 157–170. [CrossRef]
128. Erdem-Şenatalar, A.; Tatlier, M.; Tantekin-Ersolmaz, Ş.B. Questioning the validity of present models for estimating the performances of zeolite-polymer mixed matrix membranes. *Chem. Eng. Commun.* **2003**, *190*, 677–692. [CrossRef]
129. Lopez, J.L.; Matson, S.L.; Marchese, J.; Quinn, J.A. Diffusion through composite membranes: A two-dimensional analysis. *J. Membr. Sci.* **1986**, *27*, 301–325. [CrossRef]
130. Te Hennepe, H.J.C.; Smolders, C.A.; Bargeman, D.; Mulder, M.H.V. Exclusion and Tortuosity Effects for Alcohol/Water Separation by Zeolite-Filled PDMS Membranes. *Sep. Sci. Technol.* **1991**, *26*, 585–596. [CrossRef]
131. Te Hennepe, H.J.C.; Boswerger, W.B.F.; Bargeman, D.; Mulder, M.H.V.; Smolders, C.A. Zeolite-filled silicone rubber membranes experimental determination of concentration profiles. *J. Membr. Sci.* **1994**, *89*, 185–196. [CrossRef]
132. Fouda, A.; Chen, Y.; Bai, J.; Matsuura, T. Wheatstone bridge model for the laminated polydimethylsiloxane/polyethersulfone membrane for gas separation. *J. Membr. Sci.* **1991**, *64*, 263–271. [CrossRef]
133. Nielsen, L.E. Models for the Permeability of Filled Polymer Systems. *J. Macromol. Sci. Part A Chem.* **1967**, *1*, 929–942. [CrossRef]
134. Petropoulos, J.H. A Comparative study of approaches applied to the permeability of binary composite polymeric materials. *J. Polym. Sci. Polym. Phys. Ed.* **1985**, *23*, 1309–1324. [CrossRef]
135. Carson, J.K.; Lovatt, S.J.; Tanner, D.J.; Cleland, A.C. Thermal conductivity bounds for isotropic, porous materials. *Int. J. Heat Mass Transf.* **2005**, *48*, 2150–2158. [CrossRef]
136. Sheffel, J.A.; Tsapatsis, M. A model for the performance of microporous mixed matrix membranes with oriented selective flakes. *J. Membr. Sci.* **2007**, *295*, 50–70. [CrossRef]
137. Myles, T.D.; Peracchio, A.A.; Chiu, W.K.S. Extension of anisotropic effective medium theory to account for an arbitrary number of inclusion types. *J. Appl. Phys.* **2015**, *117*, 25101. [CrossRef]
138. Myles, T.D.; Peracchio, A.A.; Chiu, W.K.S. Effect of orientation anisotropy on calculating effective electrical conductivities. *J. Appl. Phys.* **2014**, *115*, 203503. [CrossRef]
139. Ordóñez-Miranda, J.; Alvarado-Gil, J.J.; Medina-Ezquivel, R. Generalized Bruggeman Formula for the Effective Thermal Conductivity of Particulate Composites with an Interface Layer. *Int. J. Thermophys.* **2010**, *31*, 975–986. [CrossRef]
140. Falla, W.R.; Mulski, M.; Cussler, E.L. Estimating diffusion through flake-filled membranes. *J. Membr. Sci.* **1996**, *119*, 129–138. [CrossRef]
141. Chiew, Y.C.; Glandt, E.D. The effect of structure on the conductivity of a dispersion. *J. Colloid Interface Sci.* **1983**, *94*, 90–104. [CrossRef]
142. Jeffrey, D.J. Conduction through a Random Suspension of Spheres. *Proc. R. Soc. Lond. A Math. Phys. Eng. Sci.* **1973**, *335*, 355–367. [CrossRef]
143. Landauer, R.; Garland, J.C.; Tanner, D.B. Electrical Conductivity in inhomogeneous media. *AIP Conf. Proc.* **1978**, *40*, 2–45. [CrossRef]

144. Hashin, Z. Assessment of the Self Consistent Scheme Approximation: Conductivity of Particulate Composites. *J. Compos. Mater.* **1968**, *2*, 284–300. [CrossRef]

145. Gonzo, E.; Parentis, M.; Gottifredi, J. Estimating models for predicting effective permeability of mixed matrix membranes. *J. Membr. Sci.* **2006**, *277*, 46–54. [CrossRef]

146. Rafiq, S.; Maulud, A.; Man, Z.; Mutalib, M.I.A.; Ahmad, F.; Khan, A.U.; Khan, A.L.; Ghauri, M.; Muhammad, N. Modelling in mixed matrix membranes for gas separation. *Can. J. Chem. Eng.* **2015**, *93*, 88–95. [CrossRef]

147. Jeffrey, D.J. Group Expansions for the Bulk Properties of a Statistically Homogeneous, Random Suspension. *Proc. R. Soc. Lond. A Math. Phys. Eng. Sci.* **1974**, *338*, 503–516. [CrossRef]

148. Lu, S.-Y.; Kim, S. Effective thermal conductivity of composites containing spheroidal inclusions. *AIChE J.* **1990**, *36*, 927–938. [CrossRef]

149. Chiew, Y.C.; Glandt, E.D. Effective conductivity of dispersions: The effect of resistance at the particle surfaces. *Chem. Eng. Sci.* **1987**, *42*, 2677–2685. [CrossRef]

150. Acrivos, A.; Chang, E. A model for estimating transport quantities in two-phase materials. *Phys. Fluids* **1986**, *29*. [CrossRef]

151. Monsalve-Bravo, G.M.; Bhatia, S.K. Extending Effective Medium Theory to Finite Size Systems: Theory and Simulation for Permeation in Mixed-Matrix Membranes. *J. Membr. Sci.* **2017**, *531*, 148–159. [CrossRef]

152. Monsalve-Bravo, G.M.; Bhatia, S.K. Comparison of Hollow Fiber and Flat Mixed-Matrix Membranes: Theory and Simulation. *Chem. Eng. Sci.* **2018**, *187*, 174–188. [CrossRef]

153. Monsalve-Bravo, G.M.; Bhatia, S.K. Concentration-dependent transport in finite sized composites: Modified effective medium theory. *J. Membr. Sci.* **2018**, *550*, 110–125. [CrossRef]

154. Bhatia, S.K. Transport in bidisperse adsorbents: Significance of the macroscopic adsorbate flux. *Chem. Eng. Sci.* **1997**, *52*, 1377–1386. [CrossRef]

155. Ash, R.; Barrer, R.M. Mechanisms of surface flow. *Surf. Sci.* **1967**, *8*, 461–466. [CrossRef]

156. Rayleigh, L. LVI. On the influence of obstacles arranged in rectangular order upon the properties of a medium. *Lond. Edinb. Dublin Philos. Mag. J. Sci.* **1892**, *34*, 481–502. [CrossRef]

157. Meredith, R.E.; Tobias, C.W. Resistance to Potential Flow through a Cubical Array of Spheres. *J. Appl. Phys.* **1960**, *31*, 1270–1273. [CrossRef]

158. Petropoulos, J.H.; Papadokostaki, K.G.; Doghieri, F.; Minelli, M. A fundamental study of the extent of meaningful application of Maxwell's and Wiener's equations to the permeability of binary composite materials. Part III: Extension of the binary cubes model to 3-phase media. *Chem. Eng. Sci.* **2015**, *131*, 360–366. [CrossRef]

159. Nielsen, L.E. The Thermal and Electrical Conductivity of Two-Phase Systems. *Ind. Eng. Chem. Fundam.* **1974**, *13*, 17–20. [CrossRef]

160. Maghami, S.; Sadeghi, M.; Mehrabani-Zeinabad, A. Recognition of polymer-particle interfacial morphology in mixed matrix membranes through ideal permeation predictive models. *Polym. Test.* **2017**, *63*, 25–37. [CrossRef]

161. Li, Y.; Chung, T.-S.; Cao, C.; Kulprathipanja, S. The effects of polymer chain rigidification, zeolite pore size and pore blockage on polyethersulfone (PES)-zeolite A mixed matrix membranes. *J. Membr. Sci.* **2005**, *260*, 45–55. [CrossRef]

162. Erucar, I.; Keskin, S. Computational Methods for MOF/Polymer Membranes. *Chem. Rec.* **2016**, *16*, 703–718. [CrossRef] [PubMed]

163. Moore, T.T.; Koros, W.J. Non-ideal effects in organic–inorganic materials for gas separation membranes. *J. Mol. Struct.* **2005**, *739*, 87–98. [CrossRef]

164. Moore, T.T.; Mahajan, R.; Vu, D.Q.; Koros, W.J. Hybrid membrane materials comprising organic polymers with rigid dispersed phases. *AIChE J.* **2004**, *50*, 311–321. [CrossRef]

165. Every, A.G.; Tzou, Y.; Hasselman, D.P.H.; Raj, R. The effect of particle size on the thermal conductivity of ZnS/diamond composites. *Acta Metall. Mater.* **1992**, *40*, 123–129. [CrossRef]

166. Sadeghi, Z.; Omidkhah, M.; Masoumi, M.E.; Abedini, R. Modification of existing permeation models of mixed matrix membranes filled with porous particles for gas separation. *Can. J. Chem. Eng.* **2016**, *94*, 547–555. [CrossRef]

167. Böttcher, C.J.F. The dielectric constant of crystalline powders. *Recueil des Travaux Chimiques des Pays-Bas* **1945**, *64*, 47–51. [CrossRef]

168. Niklasson, G.A.; Granqvist, C.G.; Hunderi, O. Effective medium models for the optical properties of inhomogeneous materials. *Appl. Opt.* **1981**, *20*, 26–30. [CrossRef] [PubMed]

169. Yang, A.-C.; Liu, C.-H.; Kang, D.-Y. Estimations of effective diffusivity of hollow fiber mixed matrix membranes. *J. Membr. Sci.* **2015**, *495*, 269–275. [CrossRef]

170. Minelli, M.; Doghieri, F.; Papadokostaki, K.G.; Petropoulos, J.H. A fundamental study of the extent of meaningful application of Maxwell's and Wiener's equations to the permeability of binary composite materials. Part I: A numerical computation approach. *Chem. Eng. Sci.* **2013**, *104*, 630–637. [CrossRef]

171. Wang, T.; Kang, D.-Y. Highly selective mixed-matrix membranes with layered fillers for molecular separation. *J. Membr. Sci.* **2015**, *497*, 394–401. [CrossRef]

172. Wang, T.; Kang, D.-Y. Predictions of effective diffusivity of mixed matrix membranes with tubular fillers. *J. Membr. Sci.* **2015**, *485*, 123–131. [CrossRef]

173. Andrady, A.L.; Merkel, T.C.; Toy, L.G. Effect of Particle Size on Gas Permeability of Filled Superglassy Polymers. *Macromolecules* **2004**, *37*, 4329–4331. [CrossRef]

174. Petropoulos, J.H.; Papadokostaki, K.G.; Minelli, M.; Doghieri, F. On the role of diffusivity ratio and partition coefficient in diffusional molecular transport in binary composite materials, with special reference to the Maxwell equation. *J. Membr. Sci.* **2014**, *456*, 162–166. [CrossRef]

175. Azimi, H.; Tezel, H.F.; Thibault, J. On the Effective Permeability of Mixed Matrix Membranes. *J. Membr. Sci. Res.* **2018**, *4*, 158–166. [CrossRef]

176. Chen, X.Y.; Hoang, V.-T.; Rodrigue, D.; Kaliaguine, S. Optimization of continuous phase in amino-functionalized metal-organic framework (MIL-53) based co-polyimide mixed matrix membranes for CO_2/CH_4 separation. *RSC Adv.* **2013**, *3*, 24266–24279. [CrossRef]

177. Vu, D.Q. Formation and Characterization of Asymmetric Carbon Molecular Sieve and Mixed-Matrix Membranes for Natural Gas Purification. Ph.D. Thesis, The University of Texas, Austin, TX, USA, 2001.

178. Duval, J.-M.; Folkers, B.; Mulder, M.H.V.; Desgrandchamps, G.; Smolders, C.A. Adsorbent filled membranes for gas separation. Part 1. Improvement of the gas separation properties of polymeric membranes by incorporation of microporous adsorbents. *J. Membr. Sci.* **1993**, *80*, 189–198. [CrossRef]

179. Mahajan, R.; Burns, R.; Schaeffer, M.; Koros, W.J. Challenges in forming successful mixed matrix membranes with rigid polymeric materials. *J. Appl. Polym. Sci.* **2002**, *86*, 881–890. [CrossRef]

180. Dutta, R.C.; Bhatia, S.K. Structure and Gas Transport at the Polymer–Zeolite Interface: Insights from Molecular Dynamics Simulations. *ACS Appl. Mater. Interfaces* **2018**, *10*, 5992–6005. [CrossRef] [PubMed]

181. Low, Z.-X.; Razmjou, A.; Wang, K.; Gray, S.; Duke, M.; Wang, H. Effect of addition of two-dimensional ZIF-L nanoflakes on the properties of polyethersulfone ultrafiltration membrane. *J. Membr. Sci.* **2014**, *460*, 9–17. [CrossRef]

182. Galve, A.; Sieffert, D.; Staudt, C.; Ferrando, M.; Güell, C.; Téllez, C.; Coronas, J. Combination of ordered mesoporous silica MCM-41 and layered titanosilicate JDF-L1 fillers for 6FDA-based copolyimide mixed matrix membranes. *J. Membr. Sci.* **2013**, *431*, 163–170. [CrossRef]

183. Zornoza, B.; Gorgojo, P.; Casado, C.; Téllez, C.; Coronas, J. Mixed matrix membranes for gas separation with special nanoporous fillers. *Desalin. Water Treat.* **2011**, *27*, 42–47. [CrossRef]

184. Kim, S.; Pechar, T.W.; Marand, E. Poly(imide siloxane) and carbon nanotube mixed matrix membranes for gas separation. *Desalination* **2006**, *192*, 330–339. [CrossRef]

185. Ismail, A.F.; Goh, P.S.; Sanip, S.M.; Aziz, M. Transport and separation properties of carbon nanotube-mixed matrix membrane. *Sep. Purif. Technol.* **2009**, *70*, 12–26. [CrossRef]

186. Sedigheh, Z.; Mohammad, A. Improving O_2/N_2 Selective Filtration Using Carbon Nanotube-Modified Mixed-Matrix Membranes. *Chem. Eng. Technol.* **2015**, *38*, 2079–2086. [CrossRef]

187. Galve, A.; Sieffert, D.; Vispe, E.; Téllez, C.; Coronas, J.; Staudt, C. Copolyimide mixed matrix membranes with oriented microporous titanosilicate JDF-L1 sheet particles. *J. Membr. Sci.* **2011**, *370*, 131–140. [CrossRef]

processes

MDPI

Review

Preparation and Potential Applications of Super Paramagnetic Nano-Fe$_3$O$_4$

Hao Zhan [1,2], Yongning Bian [1,2], Qian Yuan [1,2], Bozhi Ren [1,2], Andrew Hursthouse [1,3] and Guocheng Zhu [1,2,*]

[1]   Hunan Provincial Key Laboratory of Shale Gas Resource Utilization and Exploration, Xiangtan 411201, China; zhanhao188@outlook.com (H.Z.); byn1992@outlook.com (Y.B.); superyuanqian@163.com (Q.Y.); xtrbz@sina.com (B.R.); Andrew.Hursthouse@uws.ac.uk (A.H.)
[2]   College of Civil Engineering, Hunan University of Science and Technology, Xiangtan 411201, China
[3]   School of Science & Sport, University of the West of Scotland, Paisley PA1 2BE, UK
*   Correspondence: zhuguoc@hnust.edu.cn

Received: 14 February 2018; Accepted: 3 April 2018; Published: 9 April 2018

Abstract: Ferroferric oxide nanoparticle (denoted as Nano-Fe$_3$O$_4$) has low toxicity and is biocompatible, with a small particle size and a relatively high surface area. It has a wide range of applications in many fields such as biology, chemistry, environmental science and medicine. Because of its superparamagnetic properties, easy modification and function, it has become an important material for addressing a number of specific tasks. For example, it includes targeted drug delivery nuclear magnetic resonance (NMR) imaging in biomedical applications and in environmental remediation of pollutants. Few articles describe the preparation and modification of Nano-Fe$_3$O$_4$ in detail. We present an evaluation of preparation methodologies, as the quality of material produced plays an important role in its successful application. For example, with modification of Nano-Fe$_3$O$_4$, the surface activation energy is reduced and good dispersion is obtained.

Keywords: Nano-Fe$_3$O$_4$; super paramagnetic; water; environment remediation

1. Introduction

Recently, adsorption has become widely used in industrial wastewater treatment technology. In the process, many adsorbents have been synthesized and applied to the treatment of pollutants that contain metallic elements, synthetic dyes and pharmaceutical products [1–4]. However, the removal of suspended adsorbent in wastewater is still a challenge. If it can be handled properly, it allows the effective recycling of the adsorbent with a reduction in operational costs. The addition of magnetic adsorbents has been studied to help efficient removal from wastewater. The use of magnetic Nano-Fe$_3$O$_4$ as an effective means for separating suspended sorbents has been identified [5,6].

As society develops, the water environment and its protection is increasingly more complex and treatment demands are much more changeable. Conventional water treatment materials are not able to satisfy the long term requirements of water treatment processes. Consequently, there is considerable interest in the development of multi-performance materials. Among them, Nanoadsorption materials have been considered to be an effective and environmentally friendly high-performance water treatment material [7]. Compared with atomic or larger scale systems, nanoscale materials have excellent physical and chemical properties from its mesoscopic effect, small objects effect, quantum size and surface effect. Nano-Fe$_3$O$_4$ has attracted wide attention because of its small size, large surface area (BET), super magnetic properties and easy recycling. In addition, it also has the non-toxic and biocompatible characteristics. Nano-Fe$_3$O$_4$ has diverse applications in new biomedical biosensors [8], contrast agent in magnetic resonance imaging [9], magnetic targeting for drug delivery system [10], tissue engineering [11]. New preparation systems and methods of modification have been developed during its application. Prabha et al. [12] studied the preparation of polymer nanocomposite coated magnetic nanoparticles. Compared with the traditional preparation method, this method can increase the reactivity of the nanoferroferric oxide by changing the active groups such as the amino group and the carboxyl group on the surface to increase the surface-active sites.

The application of Nano-Fe_3O_4 has many advantages but it is easy to coalesce and oxidize in the water environment, resulting in limited development. A modification to the Nanomagnetic particles can prevent agglomeration, thus increasing its dispersion in water resulting in broader opportunities for application [13]. To date, superparamagnetic Nano-Fe_3O_4 has been applied to the preparation of magnetic fluid material such as ink. These materials in damping device, rotating seal, magnetic drug target cells, magnetic separation, magnetic card and other fields play an important role [14]. They are also being used in environmental remediation. Given the critical dependence of properties on preparation methodology, it is useful to provide an overview of the Nano-Fe_3O_4 in its preparation, modification and application.

2. Methods for Preparation of Nano-Fe_3O_4

The methods for the preparation of Nano-Fe_3O_4 usually consists of co-precipitation [15], hydrothermal [16], sol-gel [17], micro-emulsion [18], high temperature thermal decomposition [19], solvothermal [20] and a number of less common approaches, which aim to obtain nanometer sized magnetic adsorption material. The performance of Nano-Fe_3O_4 is critically dependent on its particle size and specific surface area. Fuskele et al. [21] focused on the preparation and stability of nanofluids, studied the synthesis methods of various nanoparticles and found that the particle size of ferroferric oxide has a great influence on its performance. As the particle size decreases, the specific surface area increases, the specific saturation magnetization decreases but the coercive force does not change substantially. For example, Dutta et al. [22] proposed a simple method for preparing PEGylated Fe_3O_4 cubic magnetic nanoparticles using iron acetylacetonate thermal decomposition methods and discussed its application in drug release and hyperthermia. The results show that Nano-Fe_3O_4 has good crystallinity and there are active groups such as carboxyl groups on the surface, which provide colloidal stability, low toxicity and anti-protein properties. The modified Nano-Fe_3O_4 has high electrostatic binding affinity with the positively charged anti-cancer drug doxorubicin hydrochloride and which provides useful pH release characteristics.

2.1. Hydrothermal Method

The hydrothermal method—also called hot solvent synthesis—refers to a chemical reaction in a sealed pressure vessel with water as a solvent and under high temperature and pressure. The nanocrystals prepared by the hydrothermal method have relatively advanced development, wide distribution range and do not require high-temperature calcination and treatment. However, because the reaction is carried out at higher temperatures and pressures, the requirements for equipment are more extreme [23,24]. Yang et al. [25] used iron acetylacetonate as the only iron source to prepare magnetite nanoparticles under simple hydrothermal conditions and chose polyacrylic acid as a stabilizer to provide good hydrophilicity. High-water-dispersed Nano-Fe_3O_4 with a particle size of about 50 to 100 nm was obtained. The Nano-Fe_3O_4 was composed of monodispersed magnetite with a size of about 6 nm and had high magnetic properties. The content of magnetite was over 70%. Wu et al. [26] used $FeCl_2$, $FeCl_3$ and glucose as raw materials at 160 °C in water to prepare $Fe_3O_4@C$ composite nanoparticles, as shown in Figure 1. The adsorption behavior of methylene blue on $Fe_3O_4@C$ was studied. The results showed that $Fe_3O_4@C$ prepared by hydrothermal method could adsorb methylene blue effectively and the maximum adsorption capacity was 117 mg/g.

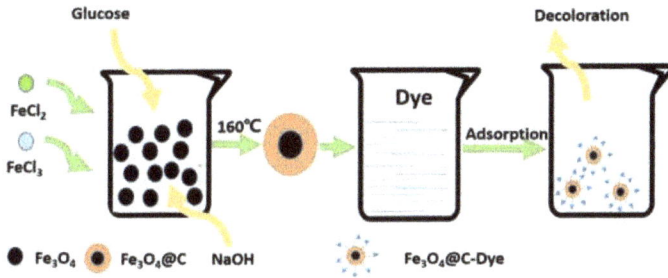

Figure 1. Scheme of the hydrothermal preparation process of Fe_3O_4@C composite particles and its adsorption on dyes.

2.2. Coprecipitation Method

The co-precipitation method involves the mixing of Fe^{2+} and Fe^{3+} in 1:2 proportion, precipitation under alkali conditions followed by filtering, washing and drying to generate Nano-Fe_3O_4. Due to its simple procedure, low cost and rapid reaction co-precipitation methods have been widely used [27,28], (see Figure 2). Qin et al. [29] adopted the chemical coprecipitation method, taking $NH_3 \cdot H_2O$ as the precipitant, which was added to a mixed solution of Fe^{2+} and Fe^{3+}. The particle size of NanoFe$_3$O$_4$ could be controlled within 20 nm under appropriate experimental conditions. Meng et al. [30] used $FeCl_3 \cdot 6H_2O$, $FeCl_2 \cdot 4H_2O$ and deionized water to prepare an iron salt solution and a ferrous salt solution. At the same time, the two solutions were mixed and heated to prepare Nano-Fe_3O_4 with $NH_3 \cdot H_2O$ as a precipitant in the mixed solution. The results showed that the concentration of Fe^{2+}/Fe^{3+} in the mixed solution had the greatest effect on the yield of Nano-Fe_3O_4. In addition, the temperature has an effect on the particle size, indicating that as the temperature increases, the Nano-Fe_3O_4 particle size increases first and then decreases.

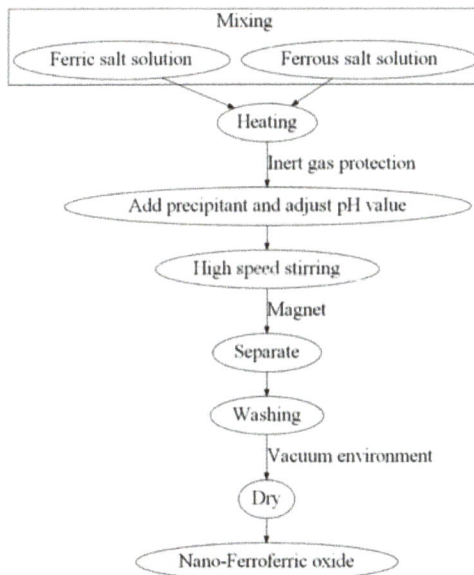

Figure 2. Scheme of co-precipitation synthesis of Nano-Fe_3O_4.

2.3. Sol-Gel Method

The sol-gel method typically utilizes iron sulfate, metal alkoxide hydrolysis and metal oxide or metal hydroxide sol as the raw material and a clean oxide powder [31,32] is then obtained after drying, as shown in Figure 3. The Fe_3O_4 synthesized by the sol-gel method has high purity but the gelation process is slow, the overall period of synthesis is long, with high temperature calcination needed [33]. Zhang et al. [34] prepared Fe_3O_4 aerogels by sol-gel method using $FeCl_3 \cdot 6H_2O$, $FeCl_2 \cdot 4H_2O$ and propylene oxide as precursors and gel promoters, respectively. The results show that Fe_3O_4 aerogel has lower density, larger specific surface area, higher saturation magnetization and its structure and magnetic properties are controlled by factors such as solution concentration, propylene oxide and Fe^{3+} molar ratio and calcining temperature and Fe_3O_4 aerosol is prepared. The gel has certain electromagnetic properties in the 2–18 GHz frequency range. Guo et al. [35] used the sol-gel method for preparation of a Nano-Fe_3O_4 coated with tetraethyl orthosilicate under a constant temperature of 380 °C. It showed a good particle size of 15~20 nm and uniform dispersion.

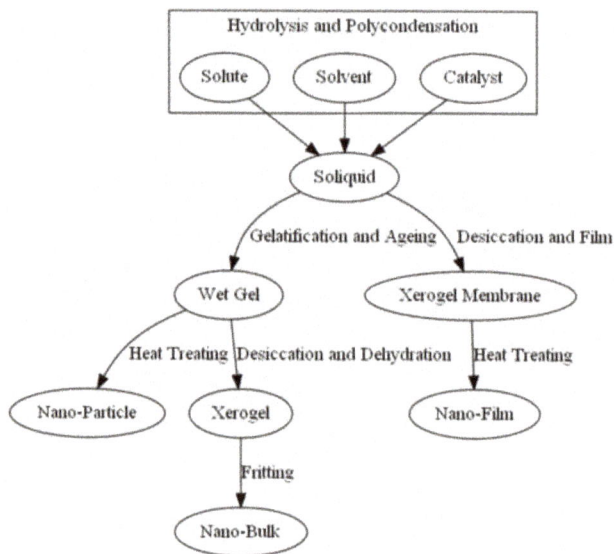

Figure 3. Scheme of sol-gel synthesis of Nano-Fe_3O_4.

2.4. Micro-Emulsion Method

Micro-emulsion method using two kinds of mutually miscible e solvents under the action of surfactant, forms a homogeneous emulsion. The solid phase is subsequently precipitated from the emulsion and the complete process consisting of nucleation, growth, coalescence and agglomeration finally forms spherical particles in a spherical droplet, as shown in Figure 4. The microemulsion method can prevent the agglomeration of particles while synthesizing Fe_3O_4 but the yield of nanoparticles prepared by a single synthesis is low, the separation and purification process of particles is complicated and the water solubility is poor [36,37]. Sun et al. [38] synthesized monodisperse Fe_3O_4 and polyaniline core-shell nanocomposites by microemulsion polymerization. Before the polymerization of aniline, Fe_3O_4 nanoparticles were prepared by thermal decomposition of acetylacetone and oleic acid using benzyl alcohol as solvent. Surface modification was carried out, then sodium dodecylbenzenesulfonate was used as surfactant, ammonium persulfate was used as oxidant and microemulsion was polymerized on the surface of Fe_3O_4 nanoparticles to obtain aniline monomer. Studies have shown that the oleic acid-modified Fe_3O_4 nanoparticles have good dispersion and the particle size is about 10 nm. Lv et al. [39] used the microemulsion hydrothermal method to prepare a new type of tower-like Fe_3O_4 particles. Through experiments, it was found that the reaction time has important influence on the final product. The key factors, such as NaOH concentration and the formation of tower-like Fe_3O_4 microstructures, are very large. The relationship and the micro-scale pagoda-like Fe_3O_4 crystals were grown from the Fe_3O_4 rod structure by an etching process and lithium air cells prepared with a pagoda-type Fe_3O_4 air electrode had a higher specific capacity at 100 mA g^{-1}. The specific capacity is up to 1429 mA h g^{-1}.

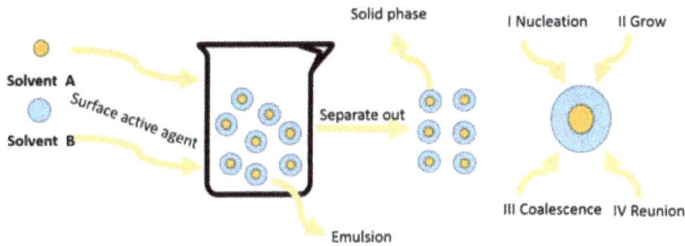

Figure 4. Scheme of Micro-emulsion synthesis of Nano-Fe_3O_4.

2.5. Solvent Heat Method

The solvent heat method is similar to the hydrothermal method but the water is replaced by organic solvents. The solvo-thermal method can precisely regulate the morphology and properties of Fe_3O_4 but it requires stringent preparation conditions [40]. Jiang et al. [41] synthesized monodisperse mesoporous Fe_3O_4 hollow microspheres with a diameter of 220 nm and a shell thickness of 50 nm using the trisodium citrate-assisted solvothermal method. The reaction time was good for this kind of mesoporous structure. The formation method influences the final product. Through experimental studies, it has been found that the synthesized Fe_3O_4 hollow microspheres have superparamagnetism, high saturation magnetization at room temperature, mesopores and high dispersibility. Zeng et al. [42] used ferric chloride hexahydrate as the sole iron source and sodium dodecylbenzenesulfonate as the surfactant. A uniform hollow Fe_3O_4 submicron sphere with a particle size of 350–450 nm was prepared by solvothermal synthesis. And it has high saturation magnetization and very low coercivity. The results show that the addition of sodium dodecylbenzenesulfonate is beneficial to the formation of monodisperse Fe_3O_4 hollow particles with a narrow particle size distribution, a high specific capacitance of 294 F g^{-1} at 0.5 A g^{-1} and good cycle stability. It maintained about 90.8% of the original capacitance after 500 charge and discharge cycles.

Table 1 summarizes the particle sizes of Fe_3O_4 prepared by several commonly used chemical methods; the substances added during the synthesis process and the optimal synthesis conditions; by optimizing the experimental conditions. The adsorption of pollutants and chemical adsorption properties are identified and the difference in the physical and chemical properties of Fe_3O_4 before and after the experimental conditions were optimized are highlighted.

Table 1. Optimization of Experimental Conditions in the Synthesis of Nano-Fe$_3$O$_4$ and Its Effect.

Preparation Methods	Particle Size	Additive	Optimum Ratio	Pollutants	Adsorption Manner	Optimization Effect	References
Coprecipitation method	35 nm	Ethylenediaminetetraacetic acid	15 mol FeCl$_3$·6H$_2$O, 7.5 mol FeCl$_2$·4H$_2$O, 150 mL deionized water	Ag(I), Hg(II), Mn(II), Zn(II), Pb(II) and Cd(II)	Chemical adsorption and physical adsorption	The Fe$_3$O$_4$ modified by ethylenediaminetetraacetic acid had reactive functional groups such as carboxyl groups and amino groups on its surface and could undergo chemical ligand exchange with heavy metal ions in water.	[43]
Coprecipitation method	60 nm	Soluble starch	Reaction time = 2 h, pH = 3, equilibrium concentration = 1.0 mg/L	P	Physical adsorption	The Fe$_3$O$_4$ modified by starch has good dispersion, shown a monolayer disperse. The starch coats on the surface of Fe$_3$O$_4$ nanoparticle through steric hindrance and charge repulsion overcome the van der Waals force and magnetic attraction force.	[44]
Sol-gel method	-	Activated carbon fiber	pH = 2.0–6.0 at room temperature	P	Chemical adsorption	After modification through activated carbon fiber, the Fe$_3$O$_4$ had a significant increase in surface area and total pore volume. The modified Fe$_3$O$_4$ was positively charged and negatively charged by the electrostatic adsorption of phosphate ions. The ion exchange reaction occurred between phosphate and hydroxide. The prepared Fe$_3$O$_4$ particle had a larger BET.	[45]
Coprecipitation method	Around 25 nm	-	pH = 6.0 at 25 °C, 4.0 g/L Fe$_3$O$_4$	Pb and Cr ions	Single phase adsorption Pb ion, multiphase adsorption for Cr ion	Under the van der Waals force and magnetic attraction, it had the link structure of the polygon, which increased the three-dimensional space between the particles. It had a spinel structure and high crystallization degree.	[46]
Solvent heat method	-	Bentonite	Heating for 8 h at 200 °C	Pb^{2+}, Cd^{2+} and Cu^{2+} ions	Chemical adsorption	The composites composed of Fe$_3$O$_4$ and bentonite had reactive functional groups such as hydroxyl and carboxyl groups on the surface, which exhibited better adsorption of heavy metal ions. The specific surface area and total pore volume of composites were larger than those of pure Fe$_3$O$_4$, with higher magnetization saturation and lower remanence.	[47]
Coprecipitation and Sol-gel method	1–100 nm	Cellulose	4.0 g FeCl$_2$·4H$_2$O, 8.0 g FeCl$_3$·6H$_2$O, 150 mL deionized water, Reaction temperature = 60 °C, stirring time = 60 min	Mercury ion	Chemical adsorption	Cellulose slender nanostructures were more likely to adsorb mercury ions. Cellulose-modified Fe$_3$O$_4$ has a reactive functional group such as CH$_2$ on its surface and interacts with mercury ions.	[48]

3. The Modification of Nano-Fe$_3$O$_4$

Magnetic attraction is often present in the Nano-Fe$_3$O$_4$ colloidal solution, which has obvious aggregation effect because of the large BET and small diameter. At the same time, the Nano-Fe$_3$O$_4$ is easily oxidized in the air during the preparation process. It is of great importance to modify the surface of the Nano-Fe$_3$O$_4$ by using physical and chemical methods to solve the problem that agglomeration and oxidation in Nano-Fe$_3$O$_4$. The common surface modification methods include surface chemical reaction [49], surface polymerization reaction [50], ultrasonic chemistry-method [51], surface adsorption deposit [52] and others. According to the different classification of modified raw materials, the package materials on Nano-Fe$_3$O$_4$ are generally classified into 3 kinds [53] including inorganic small molecules [54], organic small molecules [55] and organic polymers [56]. The specific classification was shown in Table 2.

Table 2. Modification of Nano-Fe$_3$O$_4$ by different substances.

Type	Materials	Advantages	References
Inorganic small molecules	(1) SiO$_2$ and other oxides; (2) Au, Co, Ni and other inorganic metals	The modification of SiO$_2$ and other oxides can shield the dipole interaction between the magnetic nanoparticles to prevent the particles from agglomerating and facilitate further functionalization of the particle surface. At the same time, it had a good biocompatibility, hydrophilicity and stability. Encapsulation of inorganic metals can synthesize composite particles of core-shell structure, giving the magnetic nanoparticles rich and excellent physical properties.	[57]
Organic small molecules	(1) Ethanol, organic carboxyl, sulfur and silane coupling agent oil-soluble substances; (2) Sodium oleate, sodium carboxymethylcellulose, β-cyclodextrin, citric acid, amino acids	Particle oil-soluble conversion to water-solubility was achieved by the interaction between the modifier and the stabilizer and the ligand exchange reaction resulting in water-soluble, oil-soluble and amphiphilic nanoparticles. Surfactant modification could control nanoparticle size and shape and changed surface properties of nanoparticle. Modification of silane coupling agents introduces reactive groups on the surface of nanoparticles to provide chemical selectivity for their further functionalization.	[58]
Organic polymers	(1) Glucose, starch, protein, peptides and other natural polymers; (2) Polyethylene glycol, polyvinyl alcohol and other synthetic polymers	Natural biomolecules had a good biodegradability and biocompatibility, greatly improving the biocompatibility of magnetic particles and giving them special biological activity. Synthetic polymer modification could give the material a variety of different properties to meet the actual requirement. Biomacromolecules had excellent bioactivity and were a synthetic polymer-rich chemical-selective organic combination.	[59]

3.1. Modification of Small Inorganic Molecules

Currently, there are two types of materials for Nano-Fe$_3$O$_4$ surface modification. The first includes SiO$_2$ and other oxides and the second is a metal such as Au, Co, Ni. There are also other inorganic materials used to modify Nano-Fe$_3$O$_4$, for example, the introduction of carbon to ferrosoferric oxide can increase its electrochemical performance. Han et al. [60] prepared a series of Fe$_3$O$_4$/C nanocomposites with nanoholes of 5 to 10 nm using a cotton assisted combustion reaction. The particle size of Fe$_3$O$_4$ nanoparticles is less than 10 nm and it is well embedded in the carbon matrix, making Fe$_3$O$_4$/C have good electrochemical performance. At a current of 0.4 A g^{-1}, the presence of carbon favors good dispersion and minimal agglomeration of Fe$_3$O$_4$ nanoparticles, with higher conductivity and buffer volume change.

Coating the Nano-Fe$_3$O$_4$ with a layer of silicon dioxide can effectively improve its corrosion resistance and dispersion. For example, Wang et al. [61] found that Fe$_3$O$_4$ without silica is highly susceptible to corrosion by acidic solutions. However, the silica-coated Fe$_3$O$_4$ forms a shell structure that blocks the contact with the solution. To a large extent, Fe$_3$O$_4$ is protected, which improves its acid resistance. Mostafaei et al. [62] first synthesized Nano-Fe$_3$O$_4$ by chemical precipitation. It was found through analysis that its small particle size was easily modified without precipitation and the dispersion effect was poor before the surface was modified with silica. From the experimental results, it can be seen that the spherical structure did not change. Each of the microspheres covered with SiO$_2$ has a small number of magnetic particles and has a good dispersion. The reason for this phenomenon after modification may be that the SiO_2 core shell blocks the dipole interaction between Nano-Fe_3O_4 to make the particle distribution more uniform.

The surface of mesoporous SiO_2 had Si-OH structure and its surface was easy to be modified after the Nano-Fe_3O_4 coated with SiO_2, further grafting other active functional groups [63–66]. Xue et al. [67] used the hydrothermal method to prepare Nano-Fe_3O_4, compounding a magnetic shell composite microsphere by coating SiO_2 on its surface. It had a unique channel structure and a large BET based on SiO_2 groups in the surface. Finally, through the comparison of its performance in adsorption, they concluded that the adsorption effect for Cr ions by SiO_2 on the Nano-Fe_3O_4 coating was poor. After introducing -NH_2 groups, the attraction of negatively charged CrO_4^{2-} by the positively charged -NH_2 through electrostatic attraction, significantly improved the absorption of Cr ions. Scanone et al. [68] synthesized Nano-Fe_3O_4 by co-precipitating Fe^{2+} and Fe^{3+} in ammonia solution and then modified the surface of SiO_2 with sodium silicate and passed 3-aminopropyltriethoxysilane. After the alkylsilane (APTS) treatment, the amino group was introduced onto the surface of SiO_2 by grafting. Then use 5, 10, 15, 20-tetrakis (4-carboxyphenyl) porphyrin (TCPP) activated by carbondiamine (NHS/EDC) and triiron tetraoxide (MNPNH) and coated with silica. The tri-iron oxide (MNPSINH) is covalently bonded. The photodynamic activities of MNPNH-TCPP and MNPSINH-TCP in the presence of different photo-oxidation substrates and in microbial cell suspensions were compared. The main difference is that the silica coating on the surface of mnpnh-tcpp produces O_2 in water. The photodynamic effect of TCPP linked to MNP is very sensitive to the decomposition of tryptophan. The last study showed that the MNPSINH-TCPP as an antibacterial material can control the proliferation of microorganisms under visible light and can maintain the sterilization state and has potential value in the medical field.

Nano-Fe_3O_4 BET surface area can be controlled in the presence of certain heavy metals. Sun et al. [69] studied composite Fe_3O_4 particles under the influence of Ag. The results of the experiment showed that after coating Ag on Fe_3O_4, the specific surface area had a significant increase, the same as its average pore size and total pore volume. In addition, the modification Fe_3O_4 by Ag provided excellent physical and chemical properties, with good stability even under high temperature and also improved its adsorption capacity in heavy metal removal application.

3.2. Modification of Small Organic Molecules

Inorganic nanomaterials have many advantages in the modification of Nano-Fe_3O_4 but in some systems (water phase system, oil phase system, oil and water system), their impact is limited. The modification of Nano-Fe_3O_4 by small organic molecule can provide good performance in both the water and oil phase systems. Modification by small organic molecules is divided into two, using either coupling agents or surfactants. The most common coupling agent is silane and surface-active agents include β-cyclodextrin [70], sodium carboxymethylcellulose [71], citric acid [71], amino acids [72,73] and other small organic molecules.

Surfactant-treated Nano-Fe_3O_4 often has lipophilic and hydrophilic groups forming micelles in the solution and on this basis, can achieve nanoparticle size and morphology control. The modification by surfactants can be to form a coating layer of long chain hydrocarbon on the Fe_3O_4, with the surface of Fe_3O_4 as the polar end for adsorption [74]. For example, Zhu et al. [75] modified the ferroferric oxide with oleic acid and found that the unmodified Nano-Fe_3O_4 was easy to reunite and the Nano-Fe_3O_4, which was modified by oleic acid, had large particle diameter and better dispersion. The reason for this phenomenon was that oleic acid on the Nano-Fe_3O_4 surface became a layer which effectively prevented the aggregation of particles. The hydrophilic group of oleic acid exposed to air turned the hydrophilic surface of Nano-Fe_3O_4 into a lipophilic surface. In addition, the modified Nano-Fe_3O_4 had more active sites on its surface. Wang et al. [70] synthesized a new cyclodextrin(-CD) polymer adsorbent-cyclodextrin/ethylenediamine/magnetic oxide graphene (CD-E-MgO) to extract Cr(VI) from aqueous solution. The adsorption mechanism was also analyzed: (a) electrostatic attraction; (b) subject-object interaction; (c) Cr (VI), (HCrO$_4^-$) icons and CD-E-MgO combined with hydrogen bonding.

Nano-Fe$_3$O$_4$ after being modified by a silane coupling agent, can undergo further surface modification introducing higher selectivity. Shi et al. [76] used a silane coupling agent as modification material in the process of synthesizing magnetic microspheres. The results showed that the hydroxyl group of the Nano-Fe$_3$O$_4$ and the hydroxyl group were dehydrated, so the coupling agent could be adsorbed on the Nano-Fe$_3$O$_4$ surface. In addition, the surface of the modified Nano-Fe$_3$O$_4$ increases the relative concentration of oxygen atoms on the silane coupling agent, so that it can better integrate with other reactive groups and enhance its surface activity. Gui et al. [77] used a coupling agent to surface-treat Nano-Fe$_3$O$_4$ synthesized by the co-precipitation method and found that the coupling agent bonds well to the surface of Nano-Fe$_3$O$_4$. In addition, studies have shown that lipophilic and hydrophilic amphiphilic Nano-Fe$_3$O$_4$ can be obtained on the surface of Nano-Fe$_3$O$_4$ treated with coupling agent and the coupling agent is chemically bound in the dehydration reaction of the coupling agent and Nano-Fe$_3$O$_4$ to give new surface activity.

The modification of the coupling agent can improve the surface performance of Nano-Fe$_3$O$_4$. Lou et al. [78] compared the properties of the Nano-Fe$_3$O$_4$ had coupling agent on the surface and the unmodified Nano-Fe$_3$O$_4$. The results showed that after modified by coupling agent, the dispersity of Nano-Fe$_3$O$_4$ was greatly improved and the settlement was difficult to occur with the same magnetic responsiveness. The methanol formed during the hydrolysis of coupling agent also increased the hydroxyl groups on the Nano-Fe$_3$O$_4$ surface which is helpful for further modification.

3.3. Modification of Organic Polymers

There are still many deficiencies in aspects of the biocompatibility and stability in the modification of inorganic nanomaterials using small organic molecule in the case of Nano-Fe$_3$O$_4$. Modification of Nano-Fe$_3$O$_4$ by an organic polymer provides a special nucleation, shell structure, which also has the surface functionality and biological activity of the polymer. The modification using an organic polymer is typically divided into two: a natural biological molecule or a synthetic polymer. The natural polymers commonly used include: glucan [79], proteins [80], starch [81] and polypeptides [82]. The synthetic polymers usually used include: polyethylene glycol [83] and polyvinyl alcohol [84,85].

After the modification of natural biomolecules, the biocompatibility of Nano-Fe$_3$O$_4$ can be improved with new biological activity also established. For instance, as a natural, non-toxic protein macromolecule, gelatin has two segments, which can provide Nano-Fe$_3$O$_4$ with micellar encapsulation and water dispersion. The synthesis of magnetic nanoparticles and the stabilizing effect on Fe$_3$O$_4$ nanoparticles were proposed in the polyacrylamide hydrogel gel network by Reddy et al. [86] The study found that the presence of biocompatible gelatin in magnetic hydrogel enhanced the absorption of Nano-Fe$_3$O$_4$ and in vitro blood compatibility by thrombus and hemolysis test to study the biocompatibility of the gel. In addition, drug release was studied under an external magnetic field. The results showed that the hydrogel prepared in this way can be used for magnetic control drug release system and had potential application value in magnetic sensor, actuator and pseudo muscle. The water dispersible Fe$_3$O$_4$ nanoparticles were prepared with the method of gelatin embedding by Cheng et al. [87] and the surface coated with micellar was analyzed. At the same time, because the gelatin contained a large number of active groups, the nanocomposites could be prepared by fluorescent labeling and low-toxic platinum-precursor drugs, which could be applied in anti-cancer therapies.

After the surface modification of Nano-Fe$_3$O$_4$, various copolymers can be introduced to give Nano-Fe$_3$O$_4$ a variety of better and richer properties. For example, Hu et al. [88] looked at polyethylamine (PEI). The Fe$_3$O$_4$ modified by PEI was significantly more dispersed than pure Fe$_3$O$_4$ and the modified Fe$_3$O$_4$ particles were more spherical. The explanation was that the made increased the repulsive effect of charges and increased the steric resistance [89]. He et al. [90] first prepared ferroferric oxide by coprecipitation method at 90 °C, then the ferroferric oxide was coated with sodium polyacrylate by solution dispersion polymerization and surface Ca^{2+} crosslinking. The composite materials prepared had good adsorption properties for Pb^{2+} and Cd^{2+}, which had the highest adsorption rate when the dosage of adsorbent was 1.0 g/L and 1.6 g/L to 200 mg/L Pb^{2+} solution and 100 mg/L Cd^{2+} solution. Yan et al. [91] used catalytic polymerization of aniline to load polyaniline on the surface of Fe_3O_4 to prepare a highly efficient complex catalyst. In a simulated wastewater treatment, COD removal was highly effective. In addition, the hydrogen ions provided by polyaniline can form coordination bonds with iron ions and reduce the loss of iron into the aqueous phase.

3.4. Structure of Nano-Fe_3O_4 Composite Materials

There were four basic sub-structures for ferroferric oxide modified by inorganic or organic substances. These include: core-shell [92], shell-core [93], diffuse [94] and sandwich [95] and are shown in Figure 5a–d. Figure 5a is the core-shell structure, with Nano-Fe_3O_4 as centrally, surrounded by a polymer shell. The Nano-Fe_3O_4 is completely embedded in the polymer. Figure 5b is the shell-core structure, in which the polymer acts as the core and Nano-Fe_3O_4 surrounds as the shell. This kind of composite microsphere is combined with Nano-Fe_3O_4 by complexation or electrostatic adsorption. Figure 5c is the diffuse structure, with the Nano-Fe_3O_4 evenly distributed within the polymer. Figure 5d is the sandwich structure and inner and outer layers of the complex are organic polymers, while the middle layer is Nano-Fe_3O_4. These microspheres are usually coated with organic polymers in the shell-core structure microspheres [96,97]. In addition, magnetic complexes after modification are found to have super paramagnetism, which was faster and easier to separate from the reaction system under the action of magnetic field. Moreover, magnetic compounds can be rapidly dispersed in the reaction system after removing the external magnetic field, which can assist in the recycling of materials and reduce its application costs [98,99].

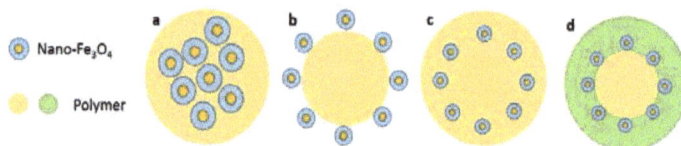

Figure 5. Four different structures of magnetic polymer composite microspheres: (**a**) Core-shell structure; (**b**) Shell-core structure; (**c**) Diffusion structure; (**d**) Sandwich structure.

The magnetic properties of these four different structures varies. The core-shell structure has a number of advantages over the other three structures. It is easier to modify the magnetic complex of the core-shell structure and introduce additional functional groups such as hydroxyl, mercapto and carboxyl group [100,101]. Compared with the diffuse structure and the shell-core structure, the environmental stability and thermal stability of the core-shell structure magnetic complex were improved. At the same time, this structure can prevent Nano-Fe_3O_4 from contacting with air and other substances, which can avoid its core being oxidized and corroded [102]. Compared with the sandwich structure, the core-shell structure magnetic complex was easier to prepare and the surface modification of Nano-Fe_3O_4 can be completed by a one-step method [103], which produced a superparamagnetic product and easier to be separated. For example, Kalska et al. [104] prepared nuclear shell nanoparticles with different magnetic core diameters and different thickness, who used Ag, Au and Cu as spacer metal. The influence of the thickness and composition of the particles in the structure were studied, together with the Cu, Au and Ag on the structure of core crystal magnetite. The results showed that the existence of the precious metal shell and the dipole interaction between the magnetic particles made the magnetic complex separation better. The presence of Cu, Ag and Au in magnetic particles leaded to the addition of superfine magnetic fields, which made them superparamagnetic.

4. Applications of Nano-Fe₃O₄

4.1. Biomedical Sciences

The Nano-Fe_3O_4 had more selective and catalytic activity than general materials because of its characteristics of small particle size, BET and surface activity center [105]. At the same time, the modified Nano-Fe_3O_4 can be used as an anti-tumor drug carrier, which had good specificity and targeting in the field of external magnetic field [106]. The Fe_3O_4 nanoparticles modified by sodium oleate were expected to do drug carrier for osteosarcoma chemotherapy. It could be combined with anticancer drugs targeting tumor body parts through the magnetic field and the basic physiology of the human body with added benefit of low toxicity [107]. For example, Isimian et al. [108] studied the enhanced radio sensitivity of breast cancer cells with the combination of 2-deoxy-D-glucose and doxorubicin plus superparamagnetic Fe_3O_4 nanoparticles. At the same time, it was also found that doxorubicin and 2-deoxy-D-glucose combined with targeted magnetic Fe_3O_4 nanoparticles can promote breast cancer radiotherapy by improving the localization of chemotherapy, increasing the cytotoxicity of tumor cells and reducing the single therapeutic dose.

With good biocompatibility and magnetic effect, Nano-Fe_3O_4 had a wide application in the treatment of tumors and magnetic resonance imaging [109]. Zhang et al. [110] studied the application of Nanoferroferric oxide in rats' hydrocrania CT imaging, which could be used as a photographic developer. They also analyzed the Nano-Fe_3O_4 distribution in rat various organs and found that there was no accumulation in different organs. This phenomenon showed that the Nano-Fe_3O_4 had biocompatibility.

4.2. Removal of Heavy Metals from Aqueous Systems

Modified Nano-Fe_3O_4 surfaces reacts with heavy metal ions and have been studied in the removal for treatment of aqueous systems. For example, Cui et al. [111] synthesized a new magnetic nanocomposite material (MgHAP/Fe_3O_4) by adding Fe_3O_4 into magnesium hydroxyapatite (MgHAP), which can remove the Cu^{2+} from the aqueous solution. The adsorption of Cu^{2+} by MgHAP/Fe_3O_4 was mainly by chemical adsorption, with the sharing or exchange of electrons between adsorbent and Cu^{2+}, in which the valence state is important. Because of this chemical interaction, the combination of MgHAP/Fe_3O_4 and Cu^{2+} had stable chemical and physical properties. In addition, Mg^{2+} and Cu^{2+} in MgHAP had the same charge and similar ionic radius, promoting their exchange in hydroxyapatite. After adsorption, the properties of the material were very stable, which would not cause secondary pollution to the environment, making it advantageous for water treatment.

The modified Nano-Fe_3O_4 can also reduce the reactivity of heavy metals and produce stable hydroxide and iron oxide precipitation on its surface. Tian et al. [112] modified the Nano-Fe_3O_4 by using a kaolin with better functional properties after comparing various modifiers. Based on this, the pH was studied in the process of removing heavy metal chromium. At higher concentrations, the kaolin-modified Nano-Fe_3O_4 forms hydroxides of iron and chromium on the surface, so that the particles are in a stable state and passivated and their surface almost loses reactivity. Chang et al. [113] used coprecipitation to wrap the γ-polyglutamic acid (γ-PGA) on the Nano-Fe_3O_4 surface successfully. The results showed that the γ-PGA/Fe_3O_4 particle size was smaller than that of the Nano-Fe_3O_4, which also had larger surface area. The removal rate of Cr^{3+}, Cu^{2+} and Pb^{2+} by γ-PGA/Fe_3O_4 in the deionized water was more than 99%, which was because of the larger specific surface area of the γ-PGA/Fe_3O_4 mNPs. The experimental results showed that γ-PGA/Fe_3O_4 was better performance than Fe_3O_4 and γ-PGA/Fe_3O_4 in heavy metal removal. The adsorption of heavy metal ions got through the membrane by means of γ-PGA and the unabsorbed ions removal activity was reduced through the membrane.

The heavy metal ions could react with the function group on the modified Nano-Fe_3O_4 surface. For instance, Jin et al. [114] analyzed the process of adsorption of heavy metals by the Nano-Fe_3O_4 surface loaded amino functional group. The study found that the Nano-Fe_3O_4 coated with a single silicon dioxide had a low adsorption capacity of heavy metals and when its surface is modified with amino functional group, faster and more effective removal of heavy metal ions occurred. As a result of the interaction between lewis acid-alkali, amino groups can strengthen the introduction of SiO_2 for heavy metal ions, such as Hg^{2+}, Pb^{2+} and Ag^+. The reason of this phenomenon maybe is the amino changed the silica shell with Nano-Fe_3O_4 nuclear structure, increasing the BET and pore size. Shen et al. [115] investigated the fungi and ferroferric oxide compound material. Using a kind of rice root mildew derived biomaterial to modify Fe_3O_4 and analyzed adsorption mechanism. They concluded that the rhizopus oryzae surface functional groups can readily react with metals, promoting enhanced adsorption of metals from water.

The removal activity of modified Nano-Fe_3O_4 for heavy metals is strongly pH dependent. For example, Zhao et al. [116] successfully synthesized Fe_3O_4-MnO_2 magnetic nanoplates using a simple hydrothermal method to remove divalent heavy metals in water. The modification of amorphous MnO_2 can significantly increase the specific surface area of Fe_3O_4-MnO_2 and reduce the zero-charge point, thus ensuring good adsorption capacity for metal cations. Experiments show that the acid-alkaline environment affects the surface charge of Nano-Fe_3O_4 during the adsorption of divalent heavy metal by Nano-Fe_3O_4. The reason is that at low pH, H^+ ions easily combine with the hydroxyl groups on the surface of Nano-Fe_3O_4, which increases the positive charges and groups on the surface and the negatively charged divalent heavy metal ion complexes are electrostatically attracted to the surface of Nano-Fe_3O_4. At the same time, the ability to adsorb divalent heavy metals is increased. Kilianová et al. [117] reported a simple and cheap synthesis of ultra-fine Nano-Fe_3O_4 with narrow particle size distribution and its application in the field of arsenic removal in a water environment. The study showed that the mesoporous arrangement of nanoparticles in their system enhanced the adsorption capacity, which was due to the strong magnetic interaction between nanoparticles. As (V) would be removed totally when the pH value was in the acid range. At this time, the Zeta potential of Nano-Fe_3O_4 adsorbent was 7.6; pH value was 5–7.6; Fe/As was approximately 20/1 and the balance of arsenic removal was 45 mg/g.

Table 3 lists the modification of Fe_3O_4 by surface chemical physical methods for different substances. After modification, Fe_3O_4 can increase the adsorption efficiency of heavy metal ions and the mechanism of removal of metal ions also differs between them. The three main mechanisms are electrostatic adsorption, surface reaction and chemical ligand exchange.

Table 3. Methods for the modification of Nano-Fe_3O_4 and mechanism for the removal of heavy metals.

Heavy Metal	Methods	Removal Mechanism	References
Pb	With different amounts of glycerol	Surface coordination, chemical adsorption.	[118]
As	With manganese dioxide and graphene oxide	Manganese dioxide can oxidize arsenic into pentavalent arsenic and graphene oxide can increase adsorption ability.	[119]
Cr	With Reduced graphene oxide	Electrostatic adsorption and acid groups adsorption.	[120]
Hg	With cellulose	The complex of mercury ion and cellulose on the surface of Fe_3O_4.	[48]
Pb and Cr	With micrococcus	Weak electrostatic forces between cadmium ions (II) and carboxyl groups or hydroxyl groups; chemical bonding of lead (II) ions and amino groups.	[121]
U and Cu	With calcium alginate containing-chitosan hydrogel beads	Chemical interaction of NH_3-groups in Chitosan with -COOH groups in calcium alginate. Physical pore adsorption and electrostatic adsorption.	[122]

4.3. Electrochemical Sensor and Energy Storage

Complex Nano-Fe_3O_4 had shown excellent charge-discharge cycling stability [123]. For example, composite material combined by Nano-Fe_3O_4 and Graphene after modification had good electrochemical sense performance. The modified graphene can enhance its stability and conductivity and Fe_3O_4 has a reversible capacity synergistic effect [124]. Peng et al. [125] synthesized a novel multifunctional magnetic biomolecule with Fe_3O_4 as the core, and heme protein and polydopamine as shells, using a one-pot chemical polymerization method. The results showed that the synthesized biomacromolecules not only possess the magnetic properties of Fe_3O_4 but also maintain the native structure of the heme protein under the action of an external magnetic field. Polydopamine and Au nanoparticles have good biocompatibility and conductivity. At the same time, due to the presence of Au nanoparticles, the exchange of electrons between the hemoprotein and the electrode is enhanced. Zheng et al. [126] synthesized magnetic Fe_3O_4 nanoparticles using chemical co-precipitation method and mixed Fe_3O_4 nanoparticles with chitosan to form a matrix of immobilized hemoglobin to prepare a hydrogen peroxide biosensor. In the pH range of 4–10, the potential of the Fe(III)/Fe(II) couples changes linearly with increasing pH, indicating that electron transfer is accompanied by transport of single protons in the electrochemical reaction. At the same time, it also has a certain effect on the storage of electrical energy.

After secondary modification, the dispersion of the composite particles was better and graphene had a stronger protective effect on granules, which was of great practical significance. Zhu et al. [127] studied a lab prepared Nano-Fe_3O_4 compound rich in amino and carboxyl groups on the rheophore. After testing the content of the alpha fetoprotein antibody absorbed by amino and electrostatic forces, they concluded that alpha fetoprotein immune sensor had high sensitivity and stability. In addition, the adsorption capacity of nanometer compound to the antibody was increased compared with the traditional method.

4.4. Chemical Catalysis

Cai et al. [128] found that when Ag was loaded on the surface of Fe_3O_4, the effect on the catalytic reduction of nitrophenol was greater than in its absence and that this effect increased with the increase in Ag concentration. Zou et al. [129] used $PdCl_2$, $SnCl_2 \cdot 2H_2O$ as precursors to prepare different Fe_3O_4 catalysts with different Fe_3O_4 content. They found, using cyclic voltammetry and timing current tests, that the catalyst had a good electrocatalytic activity for ethanol oxidation and the charge transfer resistance had a strong relationship with the Pd/Sn content.

Coupling agents were used to connect the heavy metal bridge on the Nano-Fe_3O_4 surface, effectively inhibiting the agglomeration of Fe_3O_4 and enhance its stability. Gu et al. [130] prepared Au/Fe_3O_4 by ultrasound in the presence of 3-aminopropyl triethoxysilane, which had high catalytic activity and did not agglomerate because of the action of the amino containing 3-aminopropyl triethoxysilane. Its rate constant can reach $0.2256 \ min^{-1}$ in the catalytic reduction of 4-nitrophenol. After nine cyclic reactions, the catalytic conversion rate was still very high. Gao et al. [131] utilized Nano-Fe_3O_4 as a filler for polytetrafluoroethylene and the linear expansion coefficient of composite material was greatly reduced. When the mass fraction of Fe_3O_4 was 15%, the linear expansion coefficient decreased by $40.1 \times 10^{-6}/°C$ compared with the single polytetrafluoroethylene.

4.5. Others

The use of Nano-Fe_3O_4 has great practical value in other applications (as shown in Table 4). In addition, to the water environment, a number of studies had shown that Nano-Fe_3O_4 can effectively remove heavy metals from sewage and also remove organic, inorganic compounds, dyes, algae [132] and other environmental pollutants in water. Ito et al. [133] studied the effect of $ZrFe_2(OH)_8$ as adsorbent on phosphorus removal in wastewater. According to the study, $ZrFe_2(OH)_8$ was a good phosphate adsorbent. More than 90% of the phosphate can be removed within 5 min. It can effectively prevent the eutrophication of polluted water and is straight forward to recycle. Phenol was widely used in rubber, pesticides, dyes, plastics and other fields but its toxicity led to significant environmental pollution and subsequent restriction. Nano-Fe_3O_4 shows good adsorption for phenol, for example, Jiang et al. [134] in studies of fungal degradation, compared Nano-Fe_3O_4 immobilized cells to isolated fungal cells without nucleus, free floating cells and calcium alginate immobilized cells without nanoparticles. The results showed that the immobilized dedoxycycline strain had better biodegradability than the free cells and the Nano-Fe_3O_4 immobilized cells had the highest rate of removal of phenol. In addition, when the initial phenol concentration was higher than 900 mg L^{-1}, the Nano-Fe_3O_4 immobilized cells could degrade over 99.9 percent of phenol in 80 h and had good stability in the saline environment. On the contrary, the free cells removal rate of phenol was 34.5% and immobilized cells without Nano-Fe_3O_4 was 81.3%.

Table 4. The application of modified Nano-Fe_3O_4 in treatment of other environmental pollutants.

Pollutants	Methods for Modification of Nano-Fe_3O_4	Removal Principles	References
Congo red	With hydroxyapatite and zeolite	The interaction of the dye and Nano-Fe_3O_4 through Surface coordination, hydrogen bonding, Lewis acid base reaction.	[135]
Natural rubber	With silane coupling agent	The Fe_3O_4 has high binding energy after modification and it was much easier to bond with rubber.	[136]
Rhodamine B	With Fenton reaction in the presence of H_2O_2	The surface of Fe_3O_4 formed complexes and hydroxyl radicals, resulting of degradation of dye.	[137]
Methylene blue	With natural eloise under vacuum impregnation and high temperature pyrolysis	A large number of hydroxyl groups on the surface of rocky oxidized methylene blue.	[138]
Acid orange	With Chitosan	Interacting of ions of dye with the protonated amino ions of chitosan	[139]
Methylene blue	With Graphene	The positive charged oxygen-containing groups in Graphene attracted a negatively charged methylene blue.	[140]
Bisphenol A	With amine-containing β-cyclodextrin	Hybrid effects of electrostatic, hydrophobic and van der Waals.	[141]
Organics 2,4,6-trinitrophenol	With activated carbon	Porous physical adsorption on the surface of activated carbon; Electrostatic adsorption and hydrogen bonding; Surface reaction; In the presence of dissolved oxygen, the phenols on the surface of activated carbon are gathered.	[142]

Due to having a magnetic property, some materials are likely to combine with it and enhance their functions. They often show a combination performance other than individual. For example, in order to short settling time of flocs in coagulation-flocculation, with the help of external magnetic fields, the coagulation by magnetic coagulant consisting of Nano-Fe_3O_4 and conventional coagulants (e.g., polyaluminum chloride [143], chitosan [144,145], polyacrylamide [146] et al.) is easy to achieve the purpose. In fact, it is far more than the effect in settling speed enhancement because the presence of Nano-Fe_3O_4 is likely to enhance original functions of coagulant in charge neutralization, netting adsorption-bridging [147]. Therefore, Nano-Fe_3O_4 is helpful in improving performance of coagulation. Magnetic separation is a green technology and its development is more tempting. Research on the magnetic coagulant is, however, quite deficient. How to ensure its stability and effectiveness is an important problem. In on another part, the Nano-Fe_3O_4 has also received wide attention. A typical example is the metal-organic frameworks (MOFs) being powder materials means that they have a small Nanosize, which means they are often difficult to separate from liquid phase. With the help of Nano-Fe_3O_4, this difficulty can be overcome. The hybridized materials of Nano-Fe_3O_4 and MOFs were able to show more special functions than their individual [148]. Overall, the composite material of Nano-Fe_3O_4 and other functional materials may be a better solution in overcoming their shortcomings or obtaining their expected functions.

5. Conclusions

Because Nano-Fe_3O_4 has the excellent chemical properties, it has been an important research focus in recent years. It is different from general magnetic materials due to unique features which include simple preparation and lower requirements for equipment compared with the traditional adsorbents. Although Nano-Fe_3O_4 has been widely applied in many fields, some problems still need to be addressed. For example, its dispersion and stability. Through modification, iron oxide has multidimensional functional properties such as oxidation, adsorption, catalysis, magnetic separation.

If the Nano-Fe$_3$O$_4$ was applied to different fields, the morphology, structure and surface properties are the main factors controlling their successful application. Further research on their modification is still needed. Finding a more convenient and more economical way to prepare and modify good Nano-Fe$_3$O$_4$ should be addressed to fully exploit their potential in the diverse fields of application.

Acknowledgments: This work is financially supported by National Natural Science Foundation of China (No. 51408215), National Natural Science Foundation of Hunan Province of China (No. 2018JJ2128), China Postdoctoral Science Foundation funded project (No. 2017M622578), Research Foundation of Hunan University of Science and Technology (Nos. E51508 and KJ1808) and Hunan Province graduate research and innovation projects in China (CX2017B638).

Author Contributions: Guocheng Zhu led the academic direction and developed the manuscript. Hao Zhan and Yongning Bian have collected materials and written the paper under academic direction. Qian Yuan translated the paper into English. Andrew Hursthouse further checked English style and language throughout the paper, and also gave some suggestions. Bozhi Ren gave us some suggestions on editing this manuscript.

Conflicts of Interest: The authors declare no conflict of interest.

References

1. Salem, I.A.; Salem, M.A.; El-Ghobashy, M.A. The dual role of ZnO nanoparticles for efficient capture of heavy metals and Acid blue 92 from water. *J. Mol. Liquids* **2017**, *248*, 527–538. [CrossRef]
2. Iannazzo, D.; Pistone, A.; Ziccarelli, I.; Espro, C.; Galvagno, S.; Giofré, S.V.; Romeo, R.; Cicero, N.; Bua, G.D.; Lanza, G.; et al. Removal of heavy metal ions from wastewaters using dendrimer-functionalized multi-walled carbon nanotubes. *Environ. Sci. Pollut. Res.* **2017**, *24*, 14735–14747. [CrossRef] [PubMed]
3. Chaudhary, S.; Sharma, J.; Kaith, B.S.; Yadav, S.; Sharma, A.K.; Aayushi, G. Gum xanthan-psyllium-clpoly (acrylic acid-co-itaconic acid) based adsorbent for effective removal of cationic and anionic dyes: Adsorption isotherms, kinetics and thermodynamic studies. *Ecotoxicol. Environ. Saf.* **2018**, *149*, 150–158. [CrossRef] [PubMed]
4. Islam, M.T.; Hernandez, C.; Ahsan, M.A.; Pardo, A.; Wang, H.; Noveron, J.C. Sulfonated resorcinol-formaldehyde microspheres as high-capacity regenerable adsorbent for the removal of organic dyes from water. *J. Environ. Chem. Eng.* **2017**, *5*, 5270–5279. [CrossRef]
5. Kim, Y.S.; Kim, Y.H. Application of ferro-cobalt magnetic fluid for oil sealing. *J. Magn. Magn. Mater.* **2003**, *267*, 105–110. [CrossRef]
6. Beydoun, D.; Amal, R.; Low, G.K.C.; McEvor, S. Novel Photocatalyst: Titania-Coated Magnetite. Activity and Photodissolution. *J. Phys. Chem. B* **2000**, *104*, 4387–4396. [CrossRef]
7. Lu, F.; Astruc, D. Nanomaterials for removal of toxic elements from water. *Coord. Chem. Rev.* **2018**, *356*, 147–164. [CrossRef]
8. Chen, M.; Hou, C.; Huo, D.; Fa, H.; Zhao, Y.; Shen, C. A sensitive electrochemical DNA biosensor based on three-dimensional nitrogen-doped graphene and Fe$_3$O$_4$ nanoparticles. *Sens. Actuators B Chem.* **2017**, *239*, 421–429. [CrossRef]
9. Cao, J.; Wang, Y.; Yu, J.; Xla, J.; Zhang, C.; Yin, D.; Häfeli, U.O. Preparation and radiolabeling of surface-modified magnetic nanoparticles with rhenium-188 for magnetic targeted radiotherapy. *J. Magn. Magn. Mater.* **2004**, *277*, 165–174. [CrossRef]
10. Huang, Y.S.; Lu, Y.J.; Chen, J.P. Magnetic graphene oxide as a carrier for targeted delivery of chemotherapy drugs in cancer therapy. *J. Magn. Magn. Mater.* **2017**, *427*, 34–40. [CrossRef]
11. Pistone, A.; Iannazzo, D.; Panseri, S.; Montesi, M.; Tampieri, A.; Galvagno, S. Hydroxyapatite-magnetite-MWCNT nanocomposite as a biocompatible multifunctional drug delivery system for bone tissue engineering. *Nanotechnology* **2014**, *25*, 425701. [CrossRef] [PubMed]
12. Prabha, G.; Raj, V. Preparation and characterization of polymer nanocomposites coated magnetic nanoparticles for drug delivery applications. *J. Magn. Magn. Mater.* **2016**, *408*, 26–34. [CrossRef]
13. Ji, F.; Ceng, K.; Zhang, K.; Li, J.; Zhang, J. Synthesis and Properties of PEG Modifiers of Fe$_3$O$_4$ Magnetic Nanoparticles. *Acta Polym. Sin.* **2016**, 1704–1709. [CrossRef]
14. Gupta, J.; Prakash, A.; Jaiswal, M.K.; Agarrwal, A.; Bahadur, D. Superparamagnetic iron oxide-reduced graphene oxide nanohybrid-a vehicle for targeted drug delivery and hyperthermia treatment of cancer. *J. Magn. Magn. Mater.* **2018**, *448*, 332–338. [CrossRef]

15. Butter, K.; Kassapidou, K.; Vroege, G.J.; Philipse, A.P. Preparation and properties of colloidal iron dispersions. *J. Colloid Interface Sci.* **2005**, *287*, 485–495. [CrossRef] [PubMed]

16. Giri, S.; Samanta, S.; Maji, S.; Ganguli, S.; Bhaumik, A. Magnetic properties of α-Fe_2O_3 nanoparticle synthesized by a new hydrothermal method. *J. Magn. Magn. Mater.* **2005**, *285*, 296–302. [CrossRef]

17. Eken, A.E.; Ozenbas, M. Characterization of nanostructured magnetite thin films produced by sol-gel processing. *J. Sol-Gel Sci. Technol.* **2009**, *50*, 321–327. [CrossRef]

18. Liang, X.; Jia, X.; Cao, L.; Sun, J.; Yang, Y. Microemulsion Synthesis and Characterization of Nano-Fe_3O_4 Particles and Fe_3O_4 Nanocrystalline. *J. Dispers. Sci. Technol.* **2010**, *31*, 1043–1049. [CrossRef]

19. Zhu, H.; Han, T.; Zhang, J.; Zhou, Z.; Yang, S. The Synthesis and Application of Water Soluble Fe_3O_4 Nanoparticles. *Guangzhou Chem. Ind.* **2014**, *42*, 88–89.

20. Niu, W.; Shen, Y.; Xu, J.; Ma, L.; Zhao, Y.; Shen, M. Solvothermal Synthesis of Fe_3O_4 Nanospheres and Study on the Catalytic Degradation of Xylenol Orange. *Chin. J. Inorg. Chem.* **2013**, *29*, 2110–2118.

21. Fuskele, V.; Sarviya, R.M. Recent developments in Nanoparticles Synthesis, Preparation and Stability of Nanofluids. *Mater. Today Proc.* **2017**, *4*, 4049–4060. [CrossRef]

22. Dutta, B.; Shetake, N.G.; Gawali, S.L.; Barick, B.K.; Barick, K.C.; Babu, P.D.; Pandey, B.N.; Priyadarsini, K.I.; Hassan, P.A. PEG mediated shape-selective synthesis of cubic Fe_3O_4 nanoparticles for cancer therapeutics. *J. Alloys Compd.* **2018**, *737*, 347–355. [CrossRef]

23. Nadimpalli, N.K.V.; Bandyopadhyaya, R.; Runkana, V. Thermodynamic analysis of hydrothermal synthesis of nanoparticles. *Fluid Phase Equilib.* **2018**, *456*, 33–45. [CrossRef]

24. Ge, S.; Shi, X.; Sun, K.; Li, C.; Baker, J.R., Jr.; Banaszak Holl, M.M.; Orr, B.G. A Facile Hydrothermal Synthesis of Iron Oxide Nanoparticles with Tunable Magnetic Properties. *J. Phys. Chem. C Nanomater. Interfaces* **2009**, *113*, 13593–13599. [CrossRef] [PubMed]

25. Yang, X.; Jiang, W.; Liu, L.; Chen, B.; Wu, S.; Sun, D.; Li, F. One-step hydrothermal synthesis of highly water-soluble secondary structural Fe_3O_4 nanoparticles. *J. Magn. Magn. Mater.* **2012**, *324*, 2249–2257. [CrossRef]

26. Wu, R.; Liu, J.H.; Zhao, L.; Zhang, X.; Xie, J.; Yu, B.; Ma, X.; Yang, S.T.; Wang, H.; Liu, Y. Hydrothermal preparation of magnetic $Fe_3O_4@C$ nanoparticles for dye adsorption. *J. Environ. Chem. Eng.* **2014**, *2*, 907–913. [CrossRef]

27. Xia, T.; Xu, X.; Wang, J.; Xu, C.; Meng, F.; Shi, Z.; Lian, J.; Bassat, J. Facile complex-coprecipitation synthesis of mesoporous Fe_3O_4 nanocages and their high lithium storage capacity as anode material for lithium-ion batteries. *Electrochim. Acta* **2015**, *160*, 114–122. [CrossRef]

28. Wu, J.H.; Ko, S.P.; Liu, H.L.; Kim, S.; Ju, J.S.; Kim, Y.K. Sub 5 nm magnetite nanoparticles: Synthesis, microstructure, and magnetic properties. *Mater. Lett.* **2007**, *61*, 3124–3129. [CrossRef]

29. Qin, R.; Jiang, W.; Liu, H.; Li, F. Preparation and Characterization of Nanometer Magnetite. *Mater. Rev.* **2003**, *17*, 66–68.

30. Meng, H.; Zhang, Z.; Zhao, F.; Qiu, T.; Yang, J. Orthogonal optimization design for preparation of Fe_3O_4 nanoparticles via chemical coprecipitation. *Appl. Surf. Sci.* **2013**, *280*, 679–685. [CrossRef]

31. Lemine, O.M.; Omri, K.; Zhang, B.; Mir, L.; Sajieddine, M.; Alyamani, A.; Bououdina, M. Sol-gel synthesis of 8 nm magnetite (Fe_3O_4) nanoparticles and their magnetic properties. *Superlattices Microstruct.* **2012**, *52*, 793–799. [CrossRef]

32. Chen, G.; Yang, S.; Wang, D.; Zhao, L.; Zhou, T.; Jiang, J. Review of the Preparation of Fe_3O_4 Based on Sol-gel Method. *Guangdong Chem. Ind.* **2017**, *44*, 41–42.

33. Cai, W.; Wan, J. Facile synthesis of superparamagnetic magnetite nanoparticles in liquid polyols. *J. Colloid Interface Sci.* **2007**, *305*, 366–370. [CrossRef] [PubMed]

34. Zhang, Y.; Chai, C.P.; Luo, Y.J.; Wang, L.; Li, G.P. Synthesis, structure and electromagnetic properties of mesoporous Fe_3O_4 aerogels by sol–gel method. *Mater. Sci. Eng. B* **2014**, *188*, 13–19. [CrossRef]

35. Guo, Y.; Xiao, Z. Preparation of magnetic Fe_3O_4/SiO_2 Nano-composite Particles by the Methods of Sol-Gel. *J. Jilin Inst. Archit. Civ. Eng.* **2011**, *28*, 78–80.

36. Lu, T.; Wang, J.; Yin, J.; Wang, A.; Wang, X.; Zhang, T. Surfactant effects on the microstructures of Fe_3O_4 nanoparticles synthesized by microemulsion method. *Colloids Surf. A Physicochem. Eng. Asp.* **2013**, *436*, 675–683. [CrossRef]

37. Hao, J.J.; Chen, H.L.; Ren, C.L.; Yan, N.; Geng, H.J.; Chen, X.G. Synthesis of superparamagnetic Fe_3O_4 nanocrystals in reverse microemulsion at room temperature. *Mater. Res. Innov.* **2010**, *14*, 324–326. [CrossRef]

38. Sun, L.; Zhan, L.; Shi, Y.; Chu, L.; Ge, G.; He, Z. Microemulsion synthesis and electromagnetic wave absorption properties of monodispersed Fe_3O_4/polyaniline core–shell nanocomposites. *Synth. Met.* **2014**, *187*, 102–107. [CrossRef]

39. Lv, H.; Jiang, R.; Li, Y.; Zhang, X.; Wang, J. Microemulsion-mediated hydrothermal growth of pagoda-like Fe_3O_4 microstructures and their application in a lithium–air battery. *Ceram. Int.* **2015**, *41*, 8843–8848. [CrossRef]

40. Abdullaeva, Z.; Kelgenbaeva, Z.; Nagaoka, S.; Matsuda, M.; Masayuki, T.; Koinuma, M.; Nishiyama, T. Solvothermal Synthesis of Surface-Modified Graphene/C and Au-Fe_3O_4 Nanomaterials for Antibacterial Applications. *Mater. Today Proc.* **2017**, *4*, 7044–7052. [CrossRef]

41. Jiang, X.; Wang, F.; Cai, F.; Zhang, X. Trisodium citrate-assisted synthesis of highly water-dispersible and superparamagnetic mesoporous Fe_3O_4 hollow microspheres via solvothermal process. *J. Alloys Compd.* **2015**, *636*, 34–39. [CrossRef]

42. Zeng, X.; Yang, B.; Li, X.; Li, R.; Yu, R. Solvothermal synthesis of hollow Fe_3O_4 sub-micron spheres and their enhanced electrochemical properties for supercapacitors. *Mater. Des.* **2016**, *101*, 35–43. [CrossRef]

43. Ghasemi, E.; Heydari, A.; Sillanpää, M. Superparamagnetic Fe_3O_4@EDTA nanoparticles as an efficient adsorbent for simultaneous removal of Ag(I), Hg(II), Mn(II), Zn(II), Pb(II) and Cd(II) from water and soil environmental samples. *Microchem. J.* **2017**, *131*, 51–56. [CrossRef]

44. Ding, C.; Pan, G.; Zhang, M. Study on preparation of starch-coated Fe_3O_4 and its phosphate removal properties. *Chin. J. Environ. Eng.* **2011**, *5*, 2167–2172.

45. Zhou, Q.; Wang, X.; Liu, J.; Zhang, L. Phosphorus removal from wastewater using nano-particulates of hydrated ferric oxide doped activated carbon fiber prepared by Sol–Gel method. *Chem. Eng. J.* **2012**, *200–202*, 619–626. [CrossRef]

46. Wang, T.; Gao, Y.; Jin, X.; Chen, Z. Simultaneous removal of Pb (II) and Cr(III) from wastewater by magnetite nanoparticles. *Chin. J. Environ. Eng.* **2013**, *7*, 3476–3482.

47. Yan, L.; Li, S.; Yu, H.; Shan, R.; Du, B.; Liu, T. Facile solvothermal synthesis of Fe_3O_4/bentonite for efficient removal of heavy metals from aqueous solution. *Powder Technol.* **2016**, *301*, 632–640. [CrossRef]

48. Zarei, S.; Niad, M.; Raanaei, H. The removal of mercury ion pollution by using Fe_3O_4-nanocellulose: Synthesis, characterizations and DFT studies. *J. Hazard. Mater.* **2018**, *344*, 258–273. [CrossRef] [PubMed]

49. Khanjanzadeh, H.; Behrooz, R.; Bahramifar, N.; Gindl-Altmutter, W.; Bacher, M.; Edler, M.; Griesser, T. Surface chemical functionalization of cellulose nanocrystals by 3-aminopropyltriethoxysilane. *Int. J. Biol. Macromol.* **2018**, *106*, 1288–1296. [CrossRef] [PubMed]

50. Yang, P.; Moloney, M.G.; Zhang, F.; Ji, W. Surface Hydrophobic Modification of Polymers with Fluorodiazomethanes. *Mater. Lett.* **2018**, *210*, 295–297. [CrossRef]

51. Qi, D.; Zhang, H.; Tang, J.; Deng, C.; Zhang, X. Facile Synthesis of Mercaptophenylboronic Acid-Functionalized Core−Shell Structure Fe_3O_4@C@Au Magnetic Microspheres for Selective Enrichment of Glycopeptides and Glycoproteins. *J. Phys. Chem. C* **2010**, *114*, 9221–9226. [CrossRef]

52. Yu, X.; Tian, X.; Wang, S. Adsorption of Ni, Pd, Pt, Cu, Ag and Au on the Fe_3O_4 (111) surface. *Surf. Sci.* **2014**, *628*, 141–147. [CrossRef]

53. Laurent, S.; Forge, D.; Port, M.; Roch, A.; Robic, C.; Elst, LV.; Muller, R.N. Magnetic iron oxide nanoparticles: Synthesis, stabilization, vectorization, physicochemical characterizations, and biological applications. *Chem. Rev.* **2008**, *108*, 2064–2110. [CrossRef] [PubMed]

54. Li, H.; Zhang, Y.; Wang, S.; Wu, Q.; Liu, C. Study on nanomagnets supported TiO_2 photocatalysts prepared by a sol-gel process in reverse microemulsion combining with solvent-thermal technique. *J. Hazard. Mater.* **2009**, *169*, 1045–1053. [CrossRef] [PubMed]

55. Liu, T.Y.; Hu, S.H.; Liu, K.H.; Liu, D.M.; Chen, S.Y. Study on controlled drug permeation of magnetic-sensitive ferrogels: Effect of Fe_3O_4 and PVA. *J. Control. Release* **2008**, *126*, 228–236. [CrossRef] [PubMed]

56. Abu-Much, R.; Meridor, U.; Frydman, A.; Gedanken, A. Formation of a three-dimensional microstructure of Fe_3O_4-poly(vinyl alcohol) composite by evaporating the hydrosol under a magnetic field. *J. Phys. Chem. B* **2006**, *110*, 8194–8203. [CrossRef] [PubMed]

57. Luo, S.; Liu, Y.; Rao, H.; Wang, Y.; Wang, X. Fluorescence and magnetic nanocomposite Fe_3O_4@SiO_2@Au MNPs as peroxidase mimetics for glucose detection. *Anal. Biochem.* **2017**, *538*, 26–33. [CrossRef] [PubMed]

58. Ga, Y.; Zhu, G.; Ma, T. Progress in Fe_3O_4 magnetic nanoparticles and its application in biomedical fields. *Chem. Ind. Eng. Prog.* **2017**, *36*, 973–980.

59. Jiao, L.; He, X.; Wang, L.; Zhang, L.; Ma, Y. Preparation of Fe_3O_4 and Their Modification. *Guangdong Chem. Ind.* **2016**, *43*, 127–128.

60. Han, C.J.; Sheng, N.; Zhu, C.; Akiyama, T. Cotton-assisted combustion synthesis of Fe_3O_4/C composites as excellent anode materials for lithium-ion batteries. *Mater. Today Proc.* **2017**, *5*, 187–195.

61. Wang, P.; Wang, X.; Yu, S.; Zou, Y.; Wang, J.; Chen, Z.; Alharbi, N.S.; Ahmed, A.; Hayat, T.; Chen, Y.; et al. Silica coated Fe_3O_4 magnetic nanospheres for high removal of organic pollutants from wastewater. *Chem. Eng. J.* **2016**, *306*, 280–288. [CrossRef]

62. Mostafaei, M.; Hosseini, S.N.; Khatami, M.; Javidanbardan, A.; Sepahy, A.A.; Asadi, E. Isolation of recombinant Hepatitis B surface antigen with antibody-conjugated superparamagnetic Fe_3O_4/SiO_2 core-shell nanoparticles. *Protein Expr. Purif.* **2018**, *145*, 1–6. [CrossRef] [PubMed]

63. Nie, L.; Yang, H.; Jiang, P. The Research Progress of Magnetic Fe_3O_4/Mesoporous Silica Composite Microspheres. *Chin. J. Biomed. Eng.* **2017**, *36*, 348–353.

64. Gao, F.; Botella, P.; Corma, A.; Blesa, J.; Dong, L. Monodispersed Mesoporous Silica Nanoparticles with Very Large Pores for Enhanced Adsorption and Release of DNA. *J. Phys. Chem. B* **2009**, *113*, 1796–1804. [CrossRef] [PubMed]

65. Wang, Y.; Li, B.; Zhang, L.; Peng, L.; Wang, L.; Zhang, J. Multifunctional magnetic mesoporous silica nanocomposites with improved sensing performance and effective removal ability toward Hg(II). *Langmuir* **2012**, *28*, 1657–1662. [CrossRef] [PubMed]

66. Chen, H.; Xu, X.; Yao, N.; Deng, C.; Yang, P.; Zhang, X. Facile synthesis of C8-functionalized magnetic silica microspheres for enrichment of low-concentration peptides for direct MALDI-TOF MS analysis. *Proteomics* **2008**, *8*, 2778–2784. [CrossRef] [PubMed]

67. Xue, J.; Xu, S.; Zhu, Q.; Qiang, L.; Ma, J. Preparation and Adsorption Properties of Amino Modified Magnetic Mesoporous Microsphere Fe_3O_4@SiO_2@$mSiO_2$. *Chin. J. Inorg. Chem.* **2016**, *32*, 1503–1511.

68. Scanone, A.C.; Gsponer, N.S.; Alvarez, M.G.; Durantini, E.N. Photodynamic properties and photoinactivation of microorganisms mediated by 5,10,15,20-tetrakis(4-carboxyphenyl)porphyrin covalently linked to silica-coated magnetite nanoparticles. *J. Photochem. Photobiol. A Chem.* **2017**, *346*, 452–461. [CrossRef]

69. Sun, Q.; Huang, Y.; Wang, J.; Guan, Z.; Li, M.; Zhou, J.; Wang, Y. Experimental study on mercury removal efficiencies of magnetic Fe_3O_4-Ag composite nanoparticles. *Chem. Ind. Eng. Prog.* **2017**, *36*, 1101–1106.

70. Wang, H.; Liu, Y.G.; Zeng, G.M.; Hu, X.; Hu, X.; Li, T.; Li, H.; Wang, Y.; Jiang, L. Grafting of β-cyclodextrin to magnetic graphene oxide via ethylenediamine and application for Cr(VI) removal. *Carbohydr. Polym.* **2014**, *113*, 166–173. [CrossRef] [PubMed]

71. Mallick, N.; Asfer, M.; Anwar, M.; Kumar, A.; Samim, M.; Talegaonkar, S.; Ahmad, F.J. Rhodamine-loaded, cross-linked, carboxymethyl cellulose sodium-coated super-paramagnetic iron oxide nanoparticles: Development and in vitro localization study for magnetic drug-targeting applications. *Colloids Surf. A Physicochem. Eng. Asp.* **2015**, *481*, 51–62. [CrossRef]

72. Li, L.; Mak, K.Y.; Leung, C.W.; Chan, W.K.; Zhong, W.; Pong, P.W.T. Effect of synthesis conditions on the properties of citric-acid coated iron oxide nanoparticles. *Microelectron. Eng.* **2013**, *110*, 329–334. [CrossRef]

73. Ebrahiminezhad, A.; Ghasemi, Y.; Rasoul-Amini, S.; Barar, J.; Davaran, S. Preparation of novel magnetic fluorescent nanoparticles using amino acids. *Colloids Surf. B Biointerfaces* **2013**, *102*, 534–539. [CrossRef] [PubMed]

74. Ren, H.; Zhuang, H.; Liu, Y. Surface Modification of Fe_3O_4 Nanoparticles. *Chem. Res.* **2003**, *14*, 11–13.

75. Zhu, C.; Song, J.; Qiu, L.; Zhang, Q. Preparation of magnetic Fe_3O_4 nano-particles modified with oleic acid at low-temperature and washed with distilled water. *Chem. Ind. Eng. Prog.* **2011**, *30*, 1552–1555.

76. Shi, W.; Yang, J.; Wang, T.; Jin, Y. Magnetic Fe_3O_4 particles surface organic modification. *Acta Phys.-Chim. Sin.* **2001**, *17*, 507–510.

77. Gui, S.; Shen, X.; Lin, B. Surface organic modification of Fe_3O_4 nanoparticles by silane-coupling agents. *Rare Metals* **2006**, *25*, 426–430. [CrossRef]

78. Lou, M.; Wang, D.; Huang, W.; Zhao, H. The preparation and characterization of the MNPs mondified by silane-coupling agents. *Shanghai J. Biomed. Eng.* **2004**, *25*, 14–19.

79. Portet, D.; Denizot, B.; Rump, E.; Lejeune, J.J.; Jallet, P. Nonpolymeric Coatings of Iron Oxide Colloids for Biological Use as Magnetic Resonance Imaging Contrast Agents. *J. Colloid Interface Sci.* **2001**, *238*, 37–42. [CrossRef] [PubMed]

80. Berry, C.C.; Wells, S.; Charles, S.; Curtis, A.S.G. Dextran and albumin derivatised iron oxide nanoparticles: Influence on fibroblasts in vitro. *Biomaterials* **2003**, *24*, 4551–4557. [CrossRef]

81. Kim, D.H.; Lee, S.H.; Im, K.H.; Kim, K.M.; Shim, I.B.; Lee, M.H.; Lee, Y.K. Surface-modified magnetite nanoparticles for hyperthermia: Preparation, characterization, and cytotoxicity studies. *Curr. Appl. Phys.* **2006**, *6*, e242–e246. [CrossRef]

82. Lewin, M.; Carlesso, N.; Tung, C.H.; Tang, X.W.; Cory, D.; Scadden, D.T.; Weisslder, R. Tat peptide-derivatized magnetic nanoparticles allow in vivo tracking and recovery of progenitor cells. *Nat. Biotechnol.* **2000**, *18*, 410–414. [CrossRef] [PubMed]

83. Martina, M.S.; Nicolas, V.; Wilhelm, C.; Ménager, C.; Barratt, G.; Lesieur, S. The in vitro kinetics of the interactions between PEG-ylated magnetic-fluid-loaded liposomes and macrophages. *Biomaterials* **2007**, *28*, 4143–4153. [CrossRef] [PubMed]

84. Uner, B.; Ramasubramanian, M.K.; Zauscher, S.; Kadla, J.F. Adhesion interactions between poly (vinyl alcohol) and iron-oxide surfaces: The effect of acetylation. *J. Appl. Polym. Sci.* **2006**, *99*, 3528–3534. [CrossRef]

85. Godovsky, D.Y.; Varfolomeev, A.V.; Efremova, G.D.; Moskvina, M.A. Magnetic properties of polyvinyl alcohol-based composites containing iron oxide nanoparticles. *Adv. Funct. Mater.* **1999**, *9*, 87–93. [CrossRef]

86. Reddy, N.N.; Varaprasad, K.; Ravindra, S.; Reddy, G.V.S.; Reddy, K.M.S.; Reddy, K.M.M.; Raju, K.M. Evaluation of blood compatibility and drug release studies of gelatin based magnetic hydrogel nanocomposites. *Colloids Surf. A Physicochem. Eng. Asp.* **2011**, *385*, 20–27. [CrossRef]

87. Cheng, Z.; Dai, Y.; Kang, X.; Li, C.; Huang, S.; Lian, H.; Hou, Z.; Ma, P.; Lin, J. Gelatin-encapsulated iron oxide nanoparticles for platinum (IV) prodrug delivery, enzyme-stimulated release and MRI. *Biomaterials* **2014**, *35*, 6359–6368. [CrossRef] [PubMed]

88. Hu, X.; Tang, W.; He, S.; Yang, L.; Zhang, W.; Xiu, R. Preparation of Fe_3O_4/polyethyleneimine and its application for phosphate adsorption. *Acta Sci. Circumst.* **2017**, *37*, 4129–4138.

89. Zhang, W.; Shen, H.; Xia, J.; Zhang, X.; He, X. Absorption Properties of Aqueous Ferrofluid Modified by Polyelthylenemine. *Nanotechnol. Precis. Eng.* **2007**, *5*, 125–128.

90. He, S.; Zhang, F.; Cheng, S.; Wang, W. Preparation and Pb^{2+}/Cd^{2+} adsorption of encapsulated Fe_3O_4/sodium polyacrylate magnetic crosslinking polymer. *CIESC J.* **2016**, *67*, 4290–4299.

91. Yan, M.; Zhang, Q.; Xie, H.; Kong, J.; Qu, H. Load of PANI on nano-Fe_3O_4 and synergy catalytic degradation of dyes. *China Environ. Sci.* **2017**, *37*, 1394–1400.

92. Cendrowski, K.; Sikora, P.; Zielinska, B.; Horszczaruk, E.; Mijowska, E. Chemical and thermal stability of core-shelled magnetite nanoparticles and solid silica. *Appl. Surf. Sci.* **2017**, *407*, 391–397. [CrossRef]

93. Li, C.L.; Chang, C.J.; Chen, J.K. Fabrication of sandwich structured devices encapsulating core/shell SiO_2/Fe_3O_4 nanoparticle microspheres as media for magneto-responsive transmittance. *Sens. Actuators B Chem.* **2015**, *210*, 46–55. [CrossRef]

94. Feng, W.; Zhou, X.; Nie, W.; Chen, L.; Qiu, K.; Zhang, Y.; He, C. Au/polypyrrole@ Fe_3O_4 nanocomposites for MR/CT dual-modal imaging guided-photothermal therapy: An in vitro study. *ACS Appl. Mater. Interfaces* **2015**, *7*, 4354–4367. [CrossRef] [PubMed]

95. Pan, M.; Sun, Y.; Zheng, J.; Yang, W. Boronic acid-functionalized core-shell-shell magnetic composite microspheres for the selective enrichment of glycoprotein. *ACS Appl. Mater. Interfaces* **2013**, *5*, 8351–8358. [CrossRef] [PubMed]

96. Snoussi, Y.; Bastide, S.; Abderrabba, M.; Chehimi, M. Sonochemical synthesis of Fe_3O_4@NH_2-mesoporous silica@Polypyrrole/Pd: A core/double shell nanocomposite for catalytic applications. *Ultrason. Sonochem.* **2018**, *41*, 551–561. [CrossRef] [PubMed]

97. Sun, C.; Sun, K.; Tang, K. Extended Stöber method to synthesize core-shell magnetic composite catalyst Fe_3O_4@C-Pd for Suzuki coupling reactions. *Mater. Chem. Phys.* **2018**, *207*, 181–185. [CrossRef]

98. Baskakov, A.O.; Solov Eva, A.Y.; Ioni, Y.V.; Starchikov, S.S.; Lyubutin, I.S.; Khodos, I.I.; Avilov, A.S.; Gubin, S.P. Magnetic and interface properties of the core-shell Fe_3O_4/Au nanocomposites. *Appl. Surf. Sci.* **2017**, *422*, 638–644. [CrossRef]

99. Peng, Z.; Fang, X.; Yan, G.; Gao, M.; Zhang, X. Highly efficient enrichment of low-abundance intact proteins by core-shell structured Fe_3O_4-chitosan@graphene composites. *Talanta* **2017**, *174*, 845–852.

100. Kandibanda, S.R.; Gundeboina, N.; Das, S.; Sunkara, M. Synthesis, characterisation, cellular uptake and cytotoxicity of functionalised magnetic ruthenium(II) polypyridine complex core-shell nanocomposite. *J. Photochem. Photobiol. B Biol.* **2018**, *178*, 270–276. [CrossRef] [PubMed]

101. Meng, C.; Zhikun, W.; Qiang, L.; Chunling, L.; Shuangqing, S.; Song, H. Preparation of amino-functionalized Fe$_3$O$_4$@mSiO$_2$ core-shell magnetic nanoparticles and their application for aqueous Fe^{3+} removal. *J. Hazard. Mater.* **2018**, *341*, 198–206. [CrossRef] [PubMed]

102. Khashan, S.; Dagher, S.; Tit, N.; Alazzam, A.; Obaidat, I. Novel method for synthesis of Fe$_3$O$_4$@TiO$_2$ core/shell nanoparticles. *Surf. Coat. Technol.* **2017**, *322*, 92–98. [CrossRef]

103. Murata, J.; Ueno, Y.; Yodogawa, K.; Sugiura, T. Polymer/CeO$_2$–Fe$_3$O$_4$ multicomponent core–shell particles for high-efficiency magnetic-field-assisted polishing processes. *Int. J. Mach. Tools Manuf.* **2016**, *101*, 28–34. [CrossRef]

104. Kalska-Szostko, B.; Wykowska, U.; Satuła, D. Magnetic nanoparticles of core-shell structure. *Colloids Surf. A Physicochem. Eng. Asp.* **2015**, *481*, 527–536. [CrossRef]

105. Zhang, M.; Chen, H.; Yang, X.; Lu, L.; Wang, X. The Development and Prospect of Nanometer Materials. *Missiles Space Veh.* **2000**, *3*, 11–16.

106. Ramasamy, T.; Ruttala, H.B.; Gupta, B.; Poudel, B.K.; Choi, H.G.; Yong, C.S.; Kim, J.O. Smart chemistry-based nanosized drug delivery systems for systemic applications: A comprehensive review. *J. Control. Release* **2017**, *258*, 226–253. [CrossRef] [PubMed]

107. Gong, J.; Tu, Z.; Duan, H.; Zhou, S. Study on the Histocompatibility and Tissue Distribution of Fe$_3$O$_4$ Nanoparticles. *Chin. J. Tissue Eng. Res.* **2016**, *20*, 7872–7877.

108. Isimian, J.P.; Hatamian, M.; Aval, N.A.; Rashidi, M.R.; Mesbahi, A.; Mohammadzadeh, M.; Jafarabadi, M.A. Targeted superparamagnetic nanoparticles coated with 2-deoxy-D-gloucose and doxorubicin more sensitize breast cancer cells to ionizing radiation. *Breast* **2017**, *33*, 97–103.

109. Fakhri, A.; Tahami, S.; Nejad, P.A. Preparation and characterization of Fe$_3$O$_4$-Ag$_2$O quantum dots decorated cellulose nanofibers as a carrier of anticancer drugs for skin cancer. *J. Photochem. Photobiol. B Biol.* **2017**, *175*, 83–88. [CrossRef] [PubMed]

110. Zhang, L.; Lin, G.; Fan, X.; Wang, H.; Wang, J.; Gao, Y.; Xie, B.; Chen, Y.; Song, Z. Superparamagnetic Fe$_3$O$_4$ Nanoparticles in the Application of Water Model of Magnetic Resonance Imaging and CT Imaging in Rat Brain. *Guangdong Trace Elem. Sci.* **2017**, *24*, 10–16.

111. Cui, L.; Hu, L.; Guo, X.; Zhang, Y.; Wang, Y.; Wei, Q.; Du, B. Kinetic, isotherm and thermodynamic investigations of Cu^{2+} adsorption onto magnesium hydroxyapatite/ferroferric oxide nano-composites with easy magnetic separation assistance. *J. Mol. Liquids* **2014**, *198*, 157–163. [CrossRef]

112. Tian, X.; Wang, W.; Tian, N.; Zhou, C.; Yang, C.; Komarneni, S. Cr(VI) reduction and immobilization by novel carbonaceous modified magnetic Fe$_3$O$_4$/halloysite nanohybrid. *J. Hazard. Mater.* **2016**, *309*, 151–156. [CrossRef] [PubMed]

113. Chang, J.; Zhong, Z.; Xu, H.; Yao, Z.; Chen, R. Fabrication of Poly(γ-glutamic acid)-coated Fe$_3$O$_4$ Magnetic Nanoparticles and Their Application in Heavy Metal Removal. *Chin. J. Chem. Eng.* **2013**, *21*, 1244–1250. [CrossRef]

114. Jin, S.; Park, B.C.; Ham, W.S.; Pan, L.; Kim, K.Y. Effect of the magnetic core size of amino-functionalized Fe$_3$O$_4$-mesoporous SiO$_2$ core-shell nanoparticles on the removal of heavy metal ions. *Colloids Surf. A Physicochem. Eng. Asp.* **2017**, *513*, 133–140. [CrossRef]

115. Shen, S.; Wen, G.; Zheng, Y. Preparation of Rhizopus oryzae-Fe$_3$O$_4$ Composites and Their Cu^{2+} Adsorption Experiments. *Anhui Agric. Sci. Bull.* **2016**, *22*, 15–16.

116. Zhao, J.; Liu, J.; Li, N.; Wang, W.; Nan, J.; Zhao, Z.; Cui, F. Highly efficient removal of bivalent heavy metals from aqueous systems by magnetic porous Fe$_3$O$_4$-MnO$_2$: Adsorption behavior and process study. *Chem. Eng. J.* **2016**, *304*, 737–746. [CrossRef]

117. Kilianová, M.; Prucek, R.; Filip, J.; Kolařík, J.; Kvítek, L.; Panáček, A.; Tuček, J.; Zbořil, R. Remarkable efficiency of ultrafine superparamagnetic iron(III) oxide nanoparticles toward arsenate removal from aqueous environment. *Chemosphere* **2013**, *93*, 2690–2697. [CrossRef] [PubMed]

118. Chen, Z.; Li, Y. Preparation of magnetic ferriferous oxide and its application in purification of heavy metal ions in waste water. *Inorg. Chem. Ind.* **2015**, *47*, 20–22.

119. Luo, X.; Wang, C.; Luo, S.; Dong, R.; Tu, X.; Zeng, G. Adsorption of As (III) and As (V) from water using magnetite Fe$_3$O$_4$-reduced graphite oxide-MnO$_2$ nanocomposites. *Chem. Eng. J.* **2012**, *187*, 45–52. [CrossRef]

120. Zhou, L.; Deng, H.; Wan, J.; Shi, J.; Su, T. A solvothermal method to produce RGO-Fe$_3$O$_4$ hybrid composite for fast chromium removal from aqueous solution. *Appl. Surf. Sci.* **2013**, *283*, 1024–1031. [CrossRef]

121. Gupta, P.L.; Jung, H.; Tiwari, D.; Kong, S.H.; Lee, S.M. Insight into the mechanism of Cd(II) and Pb(II) removal by sustainable magnetic biosorbent precursor to Chlorella vulgaris. *J. Taiwan Inst. Chem. Eng.* **2017**, *71*, 206–213.

122. Yi, X.; He, J.; Guo, Y.; Han, Z.; Yang, M.; Jin, J.; Gu, J.; Ou, M.; Xu, X. Encapsulating Fe_3O_4 into calcium alginate coated chitosan hydrochloride hydrogel beads for removal of Cu (II) and U (VI) from aqueous solutions. *Ecotoxicol. Environ. Saf.* **2018**, *147*, 699–707. [CrossRef] [PubMed]

123. Ge, K.; Zhang, X.; Zhang, Y.; Liu, J.; Bi, H. Preparation and Capacitance Properties of Fe_3O_4-ODA/GO/PANI Nanocomposites. *J. Mater. Sci. Eng.* **2017**, *35*, 475–479.

124. Guan, G.; Zou, M.; Feng, Q.; Lin, J.; Huang, Z.; Yan, G. Synthesis of Fe_3O_4/RGO composites and their electrochemical performance. *J. Fuel Chem. Technol.* **2017**, *45*, 362–369.

125. Peng, H.; Liang, R.; Zhang, L.; Qiu, J.D. General preparation of novel core-shell heme protein-Au-polydopamine-Fe_3O_4 magnetic bionanoparticles for direct electrochemistry. *J. Electroanal. Chem.* **2013**, *700*, 70–76. [CrossRef]

126. Zheng, N.; Zhou, X.; Yang, W.; Li, X.; Yuan, Z. Direct electrochemistry and electrocatalysis of hemoglobin immobilized in a magnetic nanoparticles-chitosan film. *Talanta* **2009**, *79*, 780–786. [CrossRef] [PubMed]

127. Zhu, Y.; Tan, S.; Tang, R. AFP immunosensor based on Fe_3O_4-ferrocene nanocomposites study. *Chin. J. Anal. Lab.* **2017**, *36*, 477–480.

128. Cai, K.; Zhang, Y.; Xu, Y.; Ma, F.; Zhang, L.; Zhou, H. One-pot Synthesis of Silver/Fe_3O_4 Particles and Its Catalytic Performances for the Reduction of 4-Nitrophenol. *Ind. Saf. Environ. Prot.* **2017**, *43*, 21–23.

129. Zou, T.; Yi, Q.; Zhang, Y.; Xiang, B.; Zhou, X. Preparation of PdSn/Fe_3O_4-C Catalysts and Their Electro-catalytic Activities for The Oxidation of Ethyl Alcohol. *Chin. J. Synth. Chem.* **2017**, *25*, 480–486.

130. Gu, Y.; Liu, G.; Zou, C.; Zhang, Z.; Liu, J.; Sun, M.; Cheng, G.; Yu, L. Ultrasonic Synthesis and Application in Catalytic 4-Nitrophenols Reduction of Au/Fe_3O_4. *Chin. J. Inorg. Chem.* **2017**, *33*, 787–795.

131. Gao, X. Effect of nano Fe_3O_4 on properties of PTFE-based high temperature resistant sealing material. *China Synth. Resin Plast.* **2016**, *33*, 38–40.

132. Gao, Z.; Peng, X.; Zhang, H.; Luan, Z.; Fan, B. Montmorillonite-Cu(II)/Fe(III) oxides magnetic material for removal of cyanobacterial Microcystis aeruginosa and its regeneration. *Desalination* **2009**, *247*, 337–345. [CrossRef]

133. Ito, D.; Nishimura, K.; Miura, O. Removal and recycle of phosphate from treated water of sewage plants with zirconium ferrite adsorbent by high gradient magnetic separation. *J. Phys. Conf. Ser.* **2009**, *156*, 012033. [CrossRef]

134. Jiang, Y.; Deng, T.; Shang, Y.; Yang, K.; Wang, H. Biodegradation of phenol by entrapped cell of *Debaryomyces* sp. with nano-Fe_3O_4 under hypersaline conditions. *Int. Biodeterior. Biodegrad.* **2017**, *123*, 37–45. [CrossRef]

135. Fang, Q.; Lin, J.; Zhan, Y.; Yang, M.; Zheng, W. Synthesis of Hydroxyapatite/Magnetite/Zeolite Composite for Congo Red Removal from Aqueous Solution. *Environ. Sci.* **2014**, *35*, 2992–3001.

136. Yuan, J.; Mao, M.; Cao, Y.; Wang, J. Preparation and properties of Fe_3O_4 nanoparticles/natural rubber composites. *New Chem. Mater.* **2015**, *43*, 33–36.

137. Liu, H.; Li, X.; Wei, B.; Xu, F. Synthesis of super-paramagnetic Fe_3O_4 nanoparticles and its Fenton-like properties. *Chin. J. Environ. Eng.* **2017**, *11*, 3525–3531.

138. Guo, B.; Ouyang, J.; Yang, H. Adsorption Performance to Methylene Blue by Nano-Fe_3O_4 Assembled in Lumen of Halloysite Nanotubes. *J. Chin. Ceram. Soc.* **2016**, *44*, 1655–1661.

139. Zhou, L.; Jin, J.; Liu, Z.; Liang, X.; Shang, C. Adsorption of acid dyes from aqueous solutions by the ethylenediamine-modified magnetic chitosan nanoparticles. *J. Hazard. Mater.* **2011**, *185*, 1045–1052. [CrossRef] [PubMed]

140. Ai, L.; Zhang, C.; Chen, Z. Removal of methylene blue from aqueous solution by a solvothermal-synthesized graphene/magnetite composite. *J. Hazard. Mater.* **2011**, *192*, 1515–1524. [CrossRef] [PubMed]

141. Ghosh, S.; Badruddoza, A.Z.M.; Hidajat, K.; Uddin, M.S. Adsorptive removal of emerging contaminants from water using superparamagnetic Fe_3O_4 nanoparticles bearing aminated β-cyclodextrin. *J. Environ. Chem. Eng.* **2013**, *1*, 122–130. [CrossRef]

142. Mohan, D.; Sarswat, A.; Singh, V.K.; Alexandre-Franco, M.; Pittman, C.U., Jr. Development of magnetic activated carbon from almond shells for trinitrophenol removal from water. *Chem. Eng. J.* **2011**, *172*, 1111–1125. [CrossRef]

143. Zhang, M.; Xiao, F.; Xu, X.Z.; Wang, D.S. Novel ferromagnetic nanoparticle composited PACls and their coagulation characteristics. *Water Res.* **2012**, *46*, 127–135. [CrossRef] [PubMed]

144. Liu, B.; Chen, X.; Zheng, H.; Wang, Y.; Sun, Y.; Zhao, C.; Zhang, S. Rapid and efficient removal of heavy metal and cationic dye by carboxylate-rich magnetic chitosan flocculants: Role of ionic groups. *Carbohydr. Polym.* **2018**, *181*, 327–336. [CrossRef] [PubMed]
145. Xu, B.; Zheng, H.; Wang, Y.; An, Y.; Luo, K.; Zhao, C.; Xiang, W. Poly (2-acrylamido-2-methylpropane sulfonic acid) grafted magnetic chitosan microspheres: Preparation, characterization and dye adsorption. *Int. J. Biol. Macromol.* **2018**, *112*, 648–655. [CrossRef] [PubMed]
146. Zhou, Z.; Shan, A.; Zhao, Y. Synthesis of a novel magnetic polyacrylamide coagulant and its application in wastewater purification. *Water Sci. Technol.* **2017**, *75*, 581–586. [CrossRef] [PubMed]
147. Liu, D.; Wang, P.; Wei, G.; Dong, W.; Hui, F. Removal of algal blooms from freshwater by the coagulation–magnetic separation method. *Environ. Sci. Pollut. Res.* **2013**, *20*, 60–65. [CrossRef] [PubMed]
148. Lu, A.H.; Salabas, E.L.; Schüth, F. Magnetic nanoparticles: Synthesis, protection, functionalization, and application. *Angew. Chem. Int. Ed.* **2007**, *46*, 1222–1244. [CrossRef] [PubMed]

processes

MDPI

Review

A Review on the Separation of Lithium Ion from Leach Liquors of Primary and Secondary Resources by Solvent Extraction with Commercial Extractants

Thi Hong Nguyen [1,2] and Man Seung Lee [1,*]

[1]   Department of Advanced Materials Science & Engineering, Institute of Rare Metal, Mokpo National University, Jeollanamdo 534-729, Korea; nthong43@ctu.edu.vn

[2]   College of Natural Sciences, Can Tho University, Can Tho City 900000, Viet Nam

*   Correspondence: mslee@mokpo.ac.kr; Tel.: +82-61-450-2492

Received: 10 April 2018; Accepted: 9 May 2018; Published: 12 May 2018

Abstract: The growing demand for lithium necessitates the development of an efficient process to recover it from three kinds of solutions, namely brines as well as acid and alkaline leach liquors of primary and secondary resources. Therefore, the separation of lithium(I) from these solutions by solvent extraction was reviewed in this paper. Lithium ions in brines are concentrated by removing other metal salts by crystallization with solar evaporation. In the case of ores and secondary resources, roasting followed by acid/alkaline leaching is generally employed to dissolve the lithium. Since the compositions of brines, alkaline and acid solutions are different, different commercial extractants are employed to separate and recover lithium. The selective extraction of Li(I) over other metals from brines or alkaline solutions is accomplished using acidic extractants, their mixture with neutral extractants, and neutral extractants mixed with chelating extractants in the presence of ferric chloride ($FeCl_3$). Among these systems, tri-n-butyl phosphate (TBP)- methyl isobutyl ketone (MIBK)-$FeCl_3$ and tri-n-octyl phosphine oxide (TOPO)- benzoyltrifluoroacetone (HBTA) are considered to be promising for the selective extraction and recovery of Li(I) from brines and alkaline solutions. By contrast, in the acid leaching solutions of secondary resources, divalent and trivalent metal cations are selectively extracted by acidic extractants, leaving Li(I) in the raffinate. Therefore, bis-2,4,4-trimethyl pentyl phosphinic acid (Cyanex 272) and its mixtures are suggested for the extraction of metal ions other than Li(I).

Keywords: lithium resources; lithium; solvent extraction; commercial extractants; separation

---

1. Introduction

1.1. Applications and Resources

Lithium is an indispensable element in the manufacture of the electrode materials for batteries. It is also widely used in the fields of ceramic glass, enamels, adhesive, lubricant greases, metal alloys, air-conditioning and dyeing [1]. Therefore, the demand for lithium metal and its compounds has significantly increased. Albema [2] reported that lithium consumption reached 150,000 t in 2012 and is expected to increase by 50% by 2020. The increase in the demand of lithium and its compounds is related to their applications to nuclear energy and lithium-ion batteries [3]. The global market of lithium products for tradition and energy usage is shown in Table 1. Lithium-ion batteries represent about 37% of the rechargeable battery world market [1].

Lithium is found in several primary resources such as different ores, clays, brines and seawater [4]. A distribution of lithium among these resources shows that continental brines are the biggest resources (59%) of lithium, followed by pegmatite and spodumene (25%), hectorite (7%) and geothermal brines, oilfield brines and jaderite (3%) [4]. Liu et al. [5] reported that brines are rich in lithium and account for over 80% of lithium reserves in the world. There are about 20 minerals containing lithium but only four minerals, namely lepidolite ($KLi_{1.5}Al_{1.5}[Si_3O_{10}][F,OH]_2$), spodumene ($LiO_2 \cdot Al_2O_3 \cdot 4SiO_2$), petalite ($LiO_2 \cdot Al_2O_3 \cdot 8SiO_2$) and amblygonite ($LiAl[PO_4][OH,F]$), occur in sufficient quantities for commercial interest as well as industrial importance [4,6]. The geological features of pegmatite, brine and other types of lithium deposits and their potential for large scale and long-term production were reviewed by Kesler et al. [7]. The authors reported that total amount of lithium in these deposits is enough to meet the estimated lithium demand for the next century. The increase in demand for lithium-ion batteries results in a massive amount of scrap during the manufacture of lithium-ion batteries [8]. Therefore, the recycling of the scrap and spent lithium-ion batteries is important to recover lithium and other valuable metals [9]. In particular, spent lithium-ion batteries contain 2–7 wt % lithium and other valuable metals [10].

Table 1. Applications in traditional uses and energy of lithium and lithium compounds [2].

	Applications	Market Size	Lithium and Lithium Compounds
Traditional uses	Glass/ceramics	46 kt	• Spodumene • Li_2CO_3
	Greases/lubricants	18 kt	• LiOH
	Chemical synthesis	11 kt	• Li organometallics fed by Li metal LiCl
Energy	Portable electronics and other handheld devices	48 kt	• BG Li_2CO_3 • BG LiOH • BG Li metal • BG electrolyte salts • BG LiCl • BG alloys • BG specialty compounds
	Plug-in hybrid and hybrid electric vehicles		
	Battery electric vehicles (BEVs)		
	Grid and other power storage applications		

Total Global Lithium Carbonate Equivalent (LCE) Market: 160 kt (2014); kt: kiloton; BG: battery grade.

1.2. Lithium Recovery

Lithium demand is predominantly driven by the expansion of the battery industry. Therefore, the recovery of lithium from primary and secondary resources has attracted much attention to either meet lithium requirements or minimize waste disposal problems [4]. Pyrometallurgy and hydrometallurgy processes are commonly employed in the recovery of lithium from primary and secondary resources [4,11,12]. Although pyrometallurgical processes are techno-economically feasible, they require intensive investment and cause environment pollution [4]. Hydrometallurgical processes including acid/alkaline leaching followed by solvent extraction, ion exchange, and precipitation are considered to be promising methods, for the recovery of lithium in a pure lithium carbonate (Li_2CO_3) and lithium hydroxide (LiOH) form due to technological advantages such as smaller scale, minimal energy investment, minimal toxic gas emission and waste management efficiency [1]. In leaching solutions, Li(I) exists as a cationic species with other metal ions such as Na(I), K(I), Ca(II), Mg(II), etc. These metals have nearly identical ionic radii and thus the separation of Li(I) from the leaching solutions becomes more difficult [13–15]. For the purpose of the separation and recovery of lithium from solutions in the presence of impurities, solvent extraction, ion exchange, and precipitation have been widely employed [16–21]. In the precipitation method, impurities such as Mg(II), Ca(II) and Ni(II) should be removed before the production of lithium [16]. Precipitation has several drawbacks, such as the lower purity of the products due to the co-precipitation of other metals and slow kinetics. Ion exchange offers high separation efficiency of Li(I) from brines but the application of this method is limited in the large scale owing to the low loading capacity of resins [17,18]. Considering the high separation and recovery efficiency, low cost and easy operation, solvent extraction is commonly regarded as a favorable method to recover Li(I) from lithium resources [19–21]. Some commercial extractants have been employed for the recovery of Li(I) from the leaching solutions of primary and secondary resources [19–21]. Crown ethers and their derivatives employ ether oxygens as donor atoms ("hard base"), which coordinate well with alkali metal cations ("hard acids") on the basis of the hard-soft acid-base (HSAB) principle [1]. Although the highly selective extraction efficiency of lithium can be obtained from the solution containing K(I), Na(I), Rb(I) and Cs(I) using crown ethers, the application of these extractants is limited due to their high cost [1]. Therefore, finding commercial extractants with a low cost and high extraction efficiency is needed for the recovery of Li(I) from lithium resources. Some review papers have been published on the recovery of lithium from primary and secondary resources by hydrometallurgy and pyrometallurgy. However, little data has been reported in these review papers on the solvent extraction of Li(I) with promising extractants [4,12]. Swain [1] reviewed the separation and purification of lithium from brines and alkaline solutions by solvent extraction and supported liquid membrane, while physical and chemical processes for the recycling of spent lithium-ion batteries were summarized by Ordonez et al. [11]. Generally, the comparison of the extraction performance of commercial extractants for the separation of lithium from different leach liquors of primary and secondary resources is scarce in the reported literature. For this purpose, the present work reviewed the separation of lithium from leach liquors of primary and secondary resources by solvent extraction with commercial extractants and their mixtures.

2. Pretreatment and Leaching of Primary and Secondary Resources

In brines, the weight percentages of Li(I) are generally 0.01–0.2 wt %, while large amounts of chloride salts of sodium, potassium, calcium, and magnesium are also present [16]. Therefore, in recovery of lithium from brines, a process of solar evaporation is widely employed to remove Na(I), K(I), Ca(II) and Mg(II) through the evaporation and crystallization of these salts [16,22]. The salts of Na(I), K(I), Ca(II) and Mg(II) are crystallized in the sun while lithium remains in the solution. After liquid-solid separation between crystallized salts and the aqueous solution containing lithium, this process is repeated several times to remove water and salts until the concentration of lithium in the solution reaches a required concentration [16,19,22]. Then the concentrated brines are fed into further purification steps such as solvent extraction, ion exchange, and precipitation to produce pure lithium products [6].

Alkaline and sulfuric acid processes are widely used in the recovery of lithium from ores and clays [6,23–30]. In alkaline processes, the roasting of the ores such as lepidolite, zinnwaldite, spodumene, and montmorillonite with $Na_2SO_4/CaSO_4/CaCO_3/(CaSO_4 + Ca(OH)_2)$ at 850–1100 °C followed by water leaching is commonly employed to recover lithium [23–26,29,30]. Most of Li(I) was leached by this method and pure lithium carbonate products were obtained by evaporation and precipitation with sodium carbonate [23–26,29,30]. Several researchers suggested a possible process to recover lithium from petalite concentrate as a solution of lithium sulfate by following three main steps: (i) calcination; (ii) the roasting of the calcines with sulfuric acid; and (iii) water leaching [6,27,28]. The solutions of lithium sulfate were subsequently converted to lithium carbonate as a final lithium product by the addition of sodium carbonate to the solutions after pH adjustment, purification and evaporation [12]. However, disadvantages of sulfuric acid processes are the requirement of a strong acid concentration and complicated purification processes. The recovery efficiency of Li(I) from ores and clays by calcination and roasting followed by water leaching is summarized in Table 2.

Spent lithium-ion batteries are made up of valuable metals (Co(II), Ni(II) and Li(I)), organic chemical products and plastics. Thus, preliminary mechanical separation processes and thermal treatments are carried out to treat the outer cases and shells and to concentrate the metallic fraction before applying hydrometallurgical processes (acid/alkaline leaching, solvent extraction, precipitation and electrochemical processes) [11]. Lithium cobalt oxide ($LiCoO_2$) is commonly used as an active cathode material, which is very difficult to dissolve by common leaching reagents. Various inorganic acids (HCl, HNO_3 and H_2SO_4) and organic acids (citric acid, oxalic acid, ascorbic acid and 2-hydroxybutanedioic acid (DL-malic acid)) have been employed to dissolve the active cathode materials [11]. In order to enhance the leaching efficiency of metals in spent lithium-ion batteries and to reduce acid consumption, the addition of reducing agents such as H_2O_2 and $NaHSO_3$ is required in leaching processes [31–36]. Among the inorganic acids, HCl leaching offers a higher leaching efficiency of Co(II), Li(I) and Ni(II) than that of H_2SO_4 and HNO_3 systems [33]. Organic acids are found to be more effective in dissolving Li(I) and Co(II) from the spent lithium-ion batteries and to release lower emissions of toxic gases than inorganic acids. However, the main disadvantage of organic acids is their high cost [34]. A summary of the operational conditions for the leaching of Li(I), Co(II) and other metals from spent lithium-ion batteries using inorganic and organic acids is shown in Table 3.

Table 2. Summary of the pretreatment and leaching of ores and clays by alkaline processes.

Ores/Clays	Pretreatment and Leaching Condition		Li Leaching, %	Ref.
	Calcination and Roasting	Water Leaching		
Petalite	Calcination: 1050–1100 °C Roasting: 93% H_2SO_4; 250 °C; 1 h	-	85	[6]
	Calcination: 1100 °C; 2 h Roasting: H_2SO_4; 300 °C; 1 h	S/L = 1/7.5; 320 rpm; 1 h; 50 °C	97	[27]
Spodumene	Calcination: 1050–1090 °C, 0.5 h Roasting: H_2SO_4	S/L = 4; 225 °C; 1 h	96	[28]
Zinnwaldite	Roasting: $CaSO_4$ + $Ca(OH)_2$; 950 °C; 1 h	S/L:1/10; 10 min; 90 °C	96	[23]
	Roasting: $CaCO_3$; 825 °C; 1 h	S/L: 1/5; 1 h; 90–95 °C	85	[29]
	Roasting: $CaSO_4$ + $Ca(OH)_2$; 975 °C	S/L:1/5; 1 h; 90 °C	93	[24]
	Roasting: $CaCO_3$; 825 °C; 1 h	S/L: 1/10, 400 rpm; 4 h, 95 °C	84	[30]
Clay	Roasting: $CaSO_4$; 1050 °C, 1 h Roasting: Na_2SO_4; 850 °C, 1 h	S/L: 1/10; 10 min; 85 °C	8497	[25]
Lepidolite	Roasting: Na_2SO_4 + K_2SO_4 + CaO; 850 °C, 0.5 h	S/L: 1/2.5; 0.5 h; room temperature	92	[26]

* S/L: solid-liquid ratio.

Table 3. Summary of operational conditions for Li(I), Co(II) and other metals leaching from spent lithium-ion batteries using inorganic and organic acids.

Type of Acid	Leaching Condition	Leaching Efficiency, %		Ref.
		Li(I)	Others	
Inorganic acids	Ultrasonic power: 90 W 2M HCl/H_2SO_4; S/L: 1/40; 5 h; 60 °C	97 (H_2SO_4) 98 (HCl)	Co: 48 (H_2SO_4) Co: 76 (HCl)	[34]
	2M HCl/H_2SO_4/HNO_3; S/L:5%w/v; 18h; 25 °C	80 (HCl) <80 (H_2SO_4) >80 (HNO_3)	Co, Ni, Al: >60 (HCl) Co, Ni, Al: ≤40 (H_2SO_4) Co, Ni, Al: ≤40 (HNO_3)	[33]
	4M HCl; S/L: 1/50; 1 h; 80 °C	>99%	Co, Mn, Ni: >99	[37]
	3M HCl + 3.5%v/v H_2O_2; S/L: 1/20; 1 h; 80 °C	89%	Co: 89%	[38]
	1M H_2SO_4 + 30%v/v H_2O_2; S/L: 1/1.4; 2 h; 80 °C	-	Co: 88	[39]
	2M H_2SO_4 + 6%v/v H_2O_2; S/L:1:10; 1 h; 60 °C; 300 rpm	-	Co: >99%	[40]
	2M H_2SO_4 + 5%v/v H_2O_2; S/L: 1/10; 0.5 h; 75 °C	94	Co: 93	[8]
	6%v/v H_2SO_4 + 5%v/v H_2O_2; S/L: 3/10; 1 h; 65 °C	95	Co: 80 Al: 55	[41]
	2M H_2SO_4 + 15%v/v H_2O_2; S/L: 1:20; 10 min; 75 °C; 300 rpm 4M H_2SO_4 + 10%v/v H_2O_2; S/L: 1/10; 2 h; 85 °C	100 96	Co: 95 Co: 95	[42] [31]
	2M H_2SO_4 + 4%v/v H_2O_2; S/L: 1:10; 2 h; 70 °C	99	Co: 100 Ni: 99 Mn: 98	[36]
	1M H_2SO_4 + 0.0075M NaHSO3; S/L: 1/50; 4 h; 95 °C	97	Co: 92 Ni:96 Mn:88	[35]
	1M HNO_3 + 1.7%v/v H_2O_2; S/L: 1:50; 1 h; 75 °C	95	Co: 95	[43]

Table 3. *Cont.*

Type of Acid	Leaching Condition	Leaching Efficiency, %		Ref.
		Li(I)	Others	
Organic acids	2M citric acid + 1.25%v/v H_2O_2 S/L: 3/100; 2 h; 60 °C	92	Co: 81	[32]
	Ultrasonic power: 90 W 2M citric acid + 0.55M H_2O_2; S/L: 1/40; 5 h; 60 °C	98	Co: 96	[34]
	1.25M citric acid + 1%v/v H_2O_2; S/L: 1/40; 0.5 h; 90 °C; 300 rpm	100	Co: >90	[44]
	1.25M ascorbic acid; S/L: 1/40; 20 min; 70 °C	99	Co: 95	[45]
	1.5M DL-malic acid+ 2%v/v H_2O_2; S/L: 1/40; 40 min; 90 °C	100	Co: 90	[46]
	1M oxalic acid; S/L: 1/66.7; 2.5 h; 95 °C; 400 rpm	98	Co: 97	[47]

3. Separation of Li(I) from Leach Liquors of Primary and Secondary Resources by Solvent Extraction

3.1. Selective Extraction of Li(I) from Brines/Alkaline Solutions

Lithium exists in leach liquors as a cationic species, Li^+, which is difficult to selectively extract due to its strong tendency to be hydrated [20]. Various acidic and neutral commercial extractants (see Table A1) have been used to extract Li(I) from brines, sea waters and alkaline solutions. The extraction and separation of Li(I) from brines and alkaline solutions by commercial extractants are summarized in Table 4. Hano et al. [21] reported the extraction of Li(I) from geothermal water containing Na(I), K(I), Mg(II) and Ca(II) using single D2EHPA (di-(2-ethylhexyl)phosphoric acid) and MEHPA (mono-2-ethylhexyl-phosphonic acid). The highly selective extraction of Li(I) over Na(I) and K(I) was obtained, while Mg(II) and Ca(II) were well extracted by these extractants compared to Li(I), Na(I) and K(I). The authors also found that the addition of TBP to D2EHPA/MEHPA led to the selective extraction of Li(I) from the geothermal water. Although commercial acidic extractants offer a high separation factor between lithium and other monovalent metal cations from brines and seawaters, the application of these extractants is limited due to their low extraction efficiency [21].

Table 4. Summary of the extraction and separation of Li(I) from brines and alkaline solutions by commercial extractants.

Extractants	Condition	Remarks	Ref.
HBTA-TOPO	Li(I): 0.14 g/L pH = 11.2	97% of Li(I) was extracted with form complexes of Li.2BTA.TOPO; scrubbing with 0.5M HCl; stripping of Li(I) with 2.5M HCl; regeneration of the organic phase was achieved by washing with NaOH	[48]
Thenoyltrifluoroacetone (TTA)-TOPO in kerosene	Li(I): 1 mg/L pH = 10.6	Mg^{2+} had a strong effect on Li(I) extraction; 70% Li(I) was extracted from Mg(II)-free aqueous solution	[48]
TTA-1, 10-phenanthroline (Phen) in chlorobenzene	Li(I): 0.01–0.1 mol/L pH = 6.5–11.6	Li(I) was extracted in the wide phen concentration while the extraction of K(I) and Na(I) was only possible in a high phen concentration	[49]
TTA-TOPO in m-xylene/MIBK/n-henxane/ benzene/chloroform	Li(I): 5.8.10^{-4}M NH_4Cl: 0.1M, pH = 9	Extraction efficiency of Li(I) followed the sequence: m-xylene > benzene > MIBK > n-hexane > chlorofrom; extracted sepecies were Li.TTA.2TOPO	[50]
β-carbonyl amide (NB2EHOTA)-TBP-FeCl$_3$	HCl: 0.05M; Li(I): 2 g/L Fe(III)/Li(I):1.3 MgCl$_2$: 4.8M	Separation factor of Li(I)/Mg(II) was higher than 450; extracted sepecies were (LiFeCl$_4$.2TBP.NB2EHOTA).4TBP.NB2EHOTA	[51]
Dioctyl phthalate (DOP)/ acetyl tributyl citrate(ATBC)/ tri-n-butyl citrate(TBC)-TBP-FeCl$_3$	HCl: 0.05 Li(I): 1.86 g/L Fe(III)/Li(I):1.3 MgCl$_2$: 4.8	Li(I) extraction efficiency was in the order of DOP > ATBC > TBP;after three stages, 99.5% Li was extracted with extracted species of LiFeCl$_4$.2TBP.0.1DOP; separation factors of Li(I)/Mg(II), Li(I)/Na(I) and Li(I)/K(I) were 31,458, 1259 and 16,508, respectively	[15]

Table 4. *Cont.*

Extractants	Condition	Remarks	Ref.
TBP/MIBK-FeCl₃-keosene	LiCl: 0.025–0.05 mol/L MgCl₂: 3.5–4 mol/L FeCl₃: 0.025–0.09 mol/L	Extracted species were LiFeCl₄.TBP and LiFeCl₄.2MIBK	[52]
TBP -FeCl₃-keosene	Li(I): 0.2 mol/L Fe(III)/Li(I): 1.0–1.9 MgCl₂/CaCl₂/NH₄Cl	Fe(III) extraction was a precondition of Li(I) extraction; the extraction efficiency of Li(I) followed the sequence: MgCl₂ > CaCl₂ > NH₄Cl; MgCl₂ at Fe(III)/Li(I) = 1.9 was the optimum condition for Li(I) extraction	[3]
TBP-MIBK-FeCl₃	Li(I): 0.05 mol/L Mg(II): 4.74 mol/L SO₄²⁻: 0.12 mol/L Cl– 9.43 mol/L	98% Li(I) was extracted at a high Mg(II)/Li(I) molar ratio; Mg(II) scrubbing with LiCl/NaCl; Li(I) stripping with HCl/NaCl; regeneration of the organic phase was obtained by washing with NaOH/NaCl	[53]
α-acetyl-m-dodecylacetophenone (LIX 54) - a mixture of four trialkylphosphine oxides (Cyanex 923)	pH = 11 Li(I): 1 g/L Na(I): 20–80 g/L	High separation of Li(I) at a high Na(I) concentration (SF = 110–1500); 95% of Li(I) was extracted after three stages; extracted species was LiR Cyanex 923	[54]

Several authors have reported the extraction of Li(I) using a mixture of chelating and neutral extractants such as TBP and TOPO in kerosene [29,54,55]. Neither Li(I) nor Na(I) was extracted by LIX 54, TOPO and Cynanex 923, while some mixed systems consisting of LIX 54 and neutral extractants (TOPO and Cyanex 923) showed synergism for the selective extraction of Li(I) from Na(I) and K(I) in sulfate or chloride solutions [29,54,55]. The extraction efficiency of Li(I) by the mixture of LIX 54 and Cyanex 923 was higher than that by the mixture of LIX 54 and TOPO because the solubility of Cyanex 923 was higher in organic diluents than that of TOPO. In the extraction with the mixture of LIX54 and Cyanex 923, LIX 54 played the role of extractant and Cyanex 923 acted as a synergist. The extracted species of Li(I) by the mixture of LIX 54 and Cyanex 923 were found to be LiR(Cynanex 923), where R denotes the deprotonated LIX54. HCl solutions with moderate acidity can strip Li(I) from the loaded organic mixtures. The recovered LiCl in the HCl stripping processes is one of the products for the market and an intermediate for the production of either lithium hydroxide or carbonate [54].

The extraction efficiency of Li(I) from brines/alkaline solutions was enhanced by the employment of neutral extractants such as TBP and MIBK dissolved in kerosene in the presence of ferric chloride (FeCl₃) [3,52,56,57]. In these extraction processes, FeCl₃ plays the role as a co-extracting agent, which leads to a great increase in the extraction of lithium [49]. The stepwise extraction reactions can be represented by Equations (1) and (2) [3,52,56,57]. In concentrated chloride solutions, ferric chloride exists as $FeCl_4^-$, which is extracted by neutral extractants to form extracted species (HFeCl₄nL) through an ion association mechanism (see Equation (1)). Then ion exchange reaction occurs between the hydrogen in HFeCl₄ nL and the Li(I) in the aqueous phase, as represented by Equation (2). Zhou et al. [52] reported that the extraction capacity of TBP for Li(I) is much higher than that of MIBK. The difference between the extraction of Li(I) by TBP and that by MIBK might be related to the interaction performance between $FeCl_4^-$ and an effective functional group P = O in TBP or C = O in MIBK [52].

$$FeCl_{4(aq)}^- + H_{(aq)}^+ + nL_{(org)} = HFeCl_4 \, nL_{(org)} \tag{1}$$

$$Li_{(aq)}^+ + HFeCl_4 \, nL_{(org)} = LiFeCl_4 \, nL_{(org)} + H_{(aq)}^+ \tag{2}$$

where L denotes the neutral extractants (TBP/MIBK) and the subscripts (aq) and (org) denote the aqueous and organic phases, respectively.

Equation (1) indicates that the formation of $FeCl_4^-$ is a prerequisite for the extraction of Li(I) to occur; thus a certain concentration of chloride ions is required for $FeCl_4^-$ to form. As chloride ion sources, MgCl₂, CaCl₂ and NH₄Cl were tested for the extraction of Li(I) by TBP [3]. The extraction efficiency of Li(I) was in the order of MgCl₂ > CaCl₂ > NH₄Cl due to the competitive effect of Mg²⁺, Ca²⁺, and NH₄⁺ with Li⁺ and the salting out effect of the three salts [3]. In fact, when salting out agents (MgCl₂, CaCl₂ and NH₄Cl) are added to the solution, some of water molecules are attracted by the salt ions, resulting in a decrease in the amount of free water molecules. The increase in the fraction of hydrogen ions accelerates the extractability of iron (see Equation (1)) and thus the extraction efficiency of Li(I) is improved with MgCl$_2$, CaCl$_2$ and NH$_4$Cl added as chloride sources (see Equation (2)). Zhou et al. [3] reported that MgCl$_2$ has a stronger salting-out effect than CaCl$_2$ and NH$_4$Cl, so MgCl$_2$ is suggested as a promising chloride resource for the extraction of lithium by TBP/kerosene/FeCl$_3$. The main disadvantage of using the TBP-FeCl$_3$ system in kerosene as a diluent is a significant loss of the extractant to the aqueous phase during extraction at high TBP concentrations [15]. Moreover, the formation of a third phase occurs at low TBP concentrations due to the low solubility of the extracted species in an inert diluent, such as kerosene [3]. It has been demonstrated that 2-octanol, a polar diluent, has strong intermolecular forces with the extracted complexes, while MIBK has low density and viscosity. Thus, MIBK and 2-octanol were used in TBP-FeCl$_3$ systems as diluents to prevent the formation of the third phase [22,56]. According to the obtained results from the reported literature, the extraction efficiency of Li(I) was in the order of TBP-FeCl$_3$-MIBK > TBP-FeCl$_3$-kerosene > TBP-FeCl$_3$-2-octanol [22,56]. This means that the use of MIBK as a diluent in the TBP/FeCl$_3$ system not only prevents the formation of a third phase but also enhances the extraction efficiency of Li(I) from chloride solutions by synergistic extraction with TBP and MIBK [22].

In the extraction process of Li(I) from salts containing high Mg(II)/Li(I) ratios, the co-extraction of Mg(II) by TBP necessitates the employment of a scrubbing step to remove Mg(II) from the loaded organic phase [22]. The stripping process of TBP-FeCl$_3$-MIBK consists of three steps: (i) the scrubbing of Mg(II) from the loaded organic phase using LiCl + NaCl solution; (ii) the stripping of Li(I) using HCl + NaCl solution; and (iii) the regeneration of the organic phase using NaOH + NaCl [22]. Ji et al. [15,51] developed synergistic extraction systems to enhance the extraction efficiency of Li(I) from saturated MgCl$_2$ solutions using TBP-dioctyl phthalate/β-carbonyl amide-FeCl$_3$. The advantages of dioctyl phthalate/β-carbonyl amide in TBP-FeCl$_3$ systems are good stability, low to negligible solubility and low corrosiveness towards instruments. Dioctyl phthalate reacts with LiFeCl$_4$, thus reducing the polarity of Li(I) complex and increasing the solubility of these molecules [15]. An almost complete extraction of Li(I) was obtained without phase separation problems by controlling the concentration of TBP and dioctyl phthalate in the mixture of TBP-dioctyl phthalate-FeCl$_3$ [51]. In these extraction processes, the saturated concentration of MgCl$_2$ in the aqueous phase is required to form FeCl$_4^-$ [51]. However, the aqueous phase with saturated MgCl$_2$ would be viscous, which could lead to a kinetic problem in mixer-settler operations.

TTA forms ion-pairs with TOPO to offer a synergistic effect for the extraction of Li(I) from seawater in the absence of Mg(II), which reacts with TTA to form magnesium chelate that is insoluble in solvent [58]. However, the extraction and separation efficiency of Li(I) by TTA-TBP/TOPO is still low because of the high co-extraction of other metals like Na(I), K(I) and Ca(II). Strong selectivity for Li(I) over Na(I) and K(I) was obtained by employing TTA in the presence of phen [49]. The extracted species of Li(I) can be represented as Li(TTA)(phen), and the separation factor between Li(I) and Na(I) by TTA-phen systems was much higher than that by TTA-TOPO systems [49]. Although the employment of the TTA-phen system resulted in the selective extraction of Li(I) from the aqueous solutions containing Na(I) and K(I), the application of this system is limited due to the high toxicity, high price, high water solubility, and poor solubility in conventional diluents [1]. Zhang et al. [48] reported an innovative application of synergistic extraction system HBTA and TOPO to extract Li(I) from alkaline brine. Most of Li(I) was extracted over Na(I) by this system without the formation of a third phase or emulsification. The extraction reactions of Li(I) by HBTA-TOPO can be represented by Equations (3)–(5). With high capacity and stability, a simple extraction process, and good regeneration efficiency, HBTA-TOPO is recognized as a promising system for Li(I) extraction from alkaline brine [48]. A process for the recovery of Li(I) as LiCl from alkaline brine with HBTA-TOPO is represented in Figure 1.

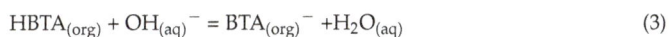

$$HBTA_{(org)} + OH_{(aq)}^- = BTA_{(org)}^- + H_2O_{(aq)} \tag{3}$$

$$BTA_{(org)}^- + Na_{(aq)}^+ + mTOPO_{(org)} = NaBTA \cdot mTOPO_{(org)} \tag{4}$$

$$\text{NaBTA mTOPO}_{(\text{org})} + \text{Li}_{(\text{aq})}^+ = \text{LiBTA TOPO}_{(\text{org})} + (\text{m-1})\text{TOPO}_{(\text{org})} \qquad (5)$$

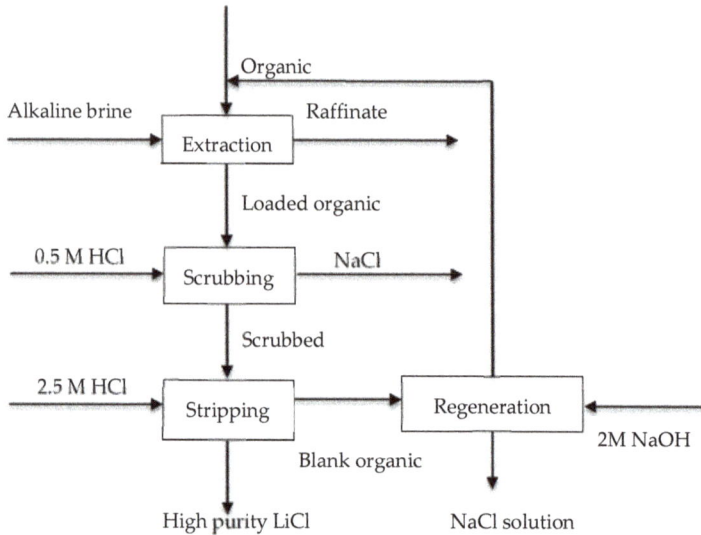

Figure 1. Flow-sheet of lithium extraction from alkaline brine with the HBTA-TOPO-kerosene system [48].

3.2. Selective Extraction and Recovery of Li(I) from Leach Liquors of Secondary Resources

The leach liquors of spent lithium-ion batteries contain large amounts of Co(II), Li(I) and small amounts of Ni(II), Cu(II), Al(III) and Fe(III) as cationic metal ions, which can be extracted by commercial acidic extractants (see Table A1). Swain [1] reported that divalent/trivalent metal cations have stronger affinities for acidic extractants than lithium ions; thus, it is difficult to selectively extract Li(I) over the divalent/trivalent metals by acidic extractants. Various kinds of single acidic extractants and their mixtures have been used to separate and recover Co(II) and Li(I) from leach liquors of spent lithium-ion batteries [8–10,59–64]. Zhang et al. [65] used D2EHPA and PC88A (2-ethylhexyl 2-ethylhexyphosphonic acid) in kerosene to separate Co(II) and Li(I) from HCl leaching solutions. PC88A was found to be more effective in selectively extracting Co(II) over Li(I) than D2EHPA in terms of lower co-extraction efficiency of Li(I), but low phase disengagement occurred at high PC88A concentrations [65]. Other researchers found that Cyanex 272 can selectively extract Co(II) over Li(I) from either acidic or alkaline solutions [8,59–61]. Although the extraction efficiency of Co(II) was enhanced with Cyanex 272 concentration, the increase in the viscosity of the organic phase caused some problems in phase disengagement [8]. Moreover, high co-extraction of Li(I) at high Cyanex 272 concentrations led to great difficulty in the scrubbing process. Since the increase in the viscosity ultimately decreased the rate of mass transfer, the Cyanex 272 extraction system should be operated at the maximum loading capacity to avoid phase disengagement problems [62,63]. On the other hand, the saponification of Cyanex 272 was found to be helpful in maintaining the extraction rate [62]. Most of Co(II) was selectively extracted over Li(I) and Ni(II) from reductive leaching solutions (H_2SO_4 + H_2O_2) of spent lithium-ion batteries using saponified Cyanex 272 without phase disengagement problems [40]. Nayl et al. [10] reported that Cyanex 272 existed as a dimer, while its saponified form existed as a monomer. Therefore, the saponification reaction of Cyanex 272 and the extraction reactions between monovalent/divalent metal ions and saponified Cyanex 272 can be represented as follows:

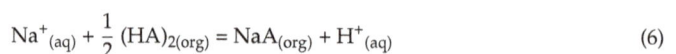

$$\text{Na}^+_{(\text{aq})} + \frac{1}{2}\,(\text{HA})_{2(\text{org})} = \text{NaA}_{(\text{org})} + \text{H}^+_{(\text{aq})} \qquad (6)$$

$$M^{2+}_{(aq)} + A^-_{(org)} + 2(HA)_{2(org)} = (MA_2 \, 3HA)_{(org)} + H^+_{(aq)} \qquad (7)$$

$$M^+_{(aq)} + A^-_{(org)} + 2(HA)_{2(org)}) = (MA \, 2HA)_{(org)} \qquad (8)$$

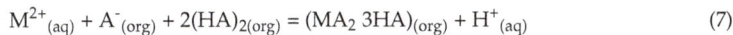

Generally, the recovery of pure Li(I) and Co(II) from leach liquors containing impurities such as Al(II), Fe(III), Cu(II), Ni(II) and Mn(II) consists of the following steps: (i) the elimination of some impurities such as Fe(III), Cu(II) and Al(III) by solvent extraction or precipitation; (ii) the selective extraction of Mn(II) and Co(II) by solvent extraction; (iii) the separation of Ni(II) by ion exchange; and (iv) the precipitation of Li(I) from the raffinate as lithium carbonate [9,10,64,66]. A conceptual process flowsheet for the recovery of Li(I) and other metals from leach solutions of spent lithium-ion batteries is presented in Figure 2. The mixture of 5-nonylsalicylaldoxime (Acorga M5640) and 2-ethylhexyl phosphonic acid mono-2-ethylhexyl ester (Ionquest 801) has a synergistic effect of selectively extracting Cu(II), Al(III) and Fe(III), leaving Co(II), Ni(II) and Li(I) in the raffinate. After extracting Al(III), Cu(II) and Fe(III), the selective extraction of Co(II) over Ni(II) and Li(I) was obtained by employing Cyanex 272/Na-Cyanex 272. A small amount of the co-extracted Li(I) (<20%) into the organic phase was scrubbed using Na$_2$CO$_3$ solution and then the complete stripping of Co(II) was achieved using acidic solutions. Finally, an ion-exchange resin such as Dowex M4195 was employed to load Ni(II), leaving Li(I) in the effluent [66]. Zhao et al. [63] reported that the mixture of Cyanex 272 and PCC8A has synergistic effect on the selective extraction of Co(II) and Mn(II) over Li(I) from simulated sulfuric acid. The addition of EDTA (ethylenediaminetetraacetic acid) to the mixtures of Cyanex 272 and PC88A suppressed the extraction efficiency of Co(II), while the extractability of Mn(II) was slightly increased. Therefore, the mixture of Cyanex 272-PC88A-EDTA was considered to be a promising system for the separation of Co(II) and Mn(II) in terms of extraction efficiency and stripping properties [63]. A process for the recovery of Li(I), Co(II) and Mn(II) from spent lithium-ion batteries is shown in Figure 3. According to this process, Li(I) can be recovered as a precipitate of Li$_2$CO$_3$ by adding sodium carbonate after the separation of all of the metals from the leach liquors.

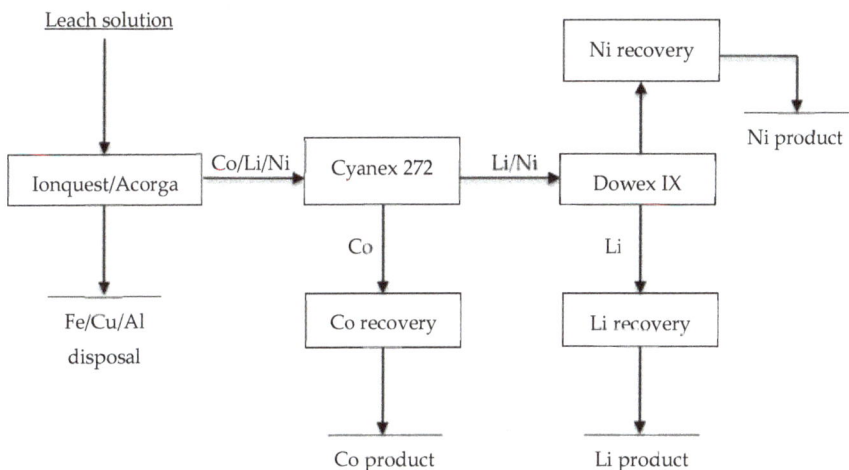

Figure 2. A conceptual process flow-sheet for the recovery of Co(II), Ni(II) and Li(I) from spent battery leach solutions [66].

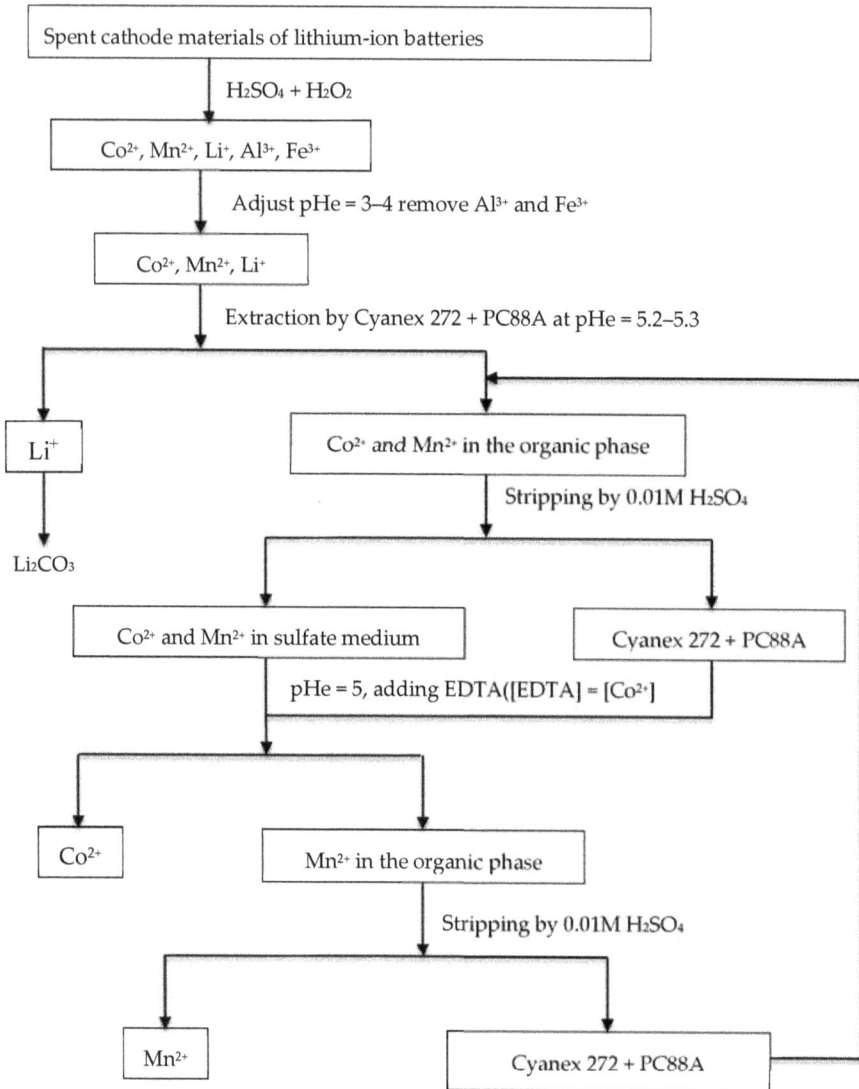

Figure 3. Flow-sheet for the separation of Co^{2+}, Mn^{2+} and Li^+ from each other in order to recycle the spent cathode materials of lithium-ion batteries (LIBs)[63].

4. Conclusions

This paper presents a review of the separation of lithium from leach liquors of primary and secondary resources by solvent extraction with commercial extractants and their mixtures. Solar evaporation is employed to concentrate lithium ions in brines by removing the salts of Na(I), Ca(II), Mg(II) and K(I), while acid and alkaline leaching processes are employed to dissolve lithium and other metals from primary and secondary resources. Several commercial extractants are then employed to separate lithium ions from the leach liquors of primary and secondary resources. Li(I) was selectively extracted over monovalent metal cations from brines or alkaline solutions by single acidic extractants and the mixture of acidic and neutral extractants. However, the extraction efficiency of lithium by these extractants was low. The mixture of neutral extractants, TBP/TOPO, and chelating extractants in the presence of FeCl₃ had a synergistic effect on the extraction and separation efficiency of Li(I) from chloride solutions containing Na(I), Ca(II), K(I) and Mg(II). The small amount of co-extracted divalent metal cations in the TBP/TOPO system was scrubbed and then LiCl was obtained by the stripping process. In the acidic leaching solutions of secondary resources, divalent/trivalent metal cations are selectively extracted over Li(I) by single acidic extractants, which renders the recovery process of Li(I) complicated. Therefore, these divalent/trivalent metal cations should be separated before Li(I) purification. Among the acidic extractants and their mixtures, the Cyanex 272 system showed a high extraction performance for these divalent and trivalent metals cations. From the raffinate, Li(I) can be recovered as Li₂CO₃ by adding sodium carbonate.

Author Contributions: M.S.L. participated in valuable discussion and gave insightful comments on a draft of the manuscript. T.H.N. wrote the paper.

Acknowledgments: This work was supported by the Global Excellent Technology Innovation of the Korea Institute of Energy Technology Evaluation and Planning (KETEP), granted financial resource from the Ministry of Trade, Industry and Energy, Republic of Korea (No. 20165010100880).

Conflicts of Interest: The authors declare no conflict of interest.

Appendix A

Table A1. Structure of commercial extractants used for the extraction and recovery of lithium in the reported literature.

Extractants	Structure of the Compound
D2EHPA Di-2-ethylhexyl phosphoric acid	
PC88A 2-Ethylhexyl phosphonic acid mono-2-ethylhexyl ester	
MEHPA Mono-2-ethylhexyl phosphoric acid	
Cyanex 272 Bis-2,4,4-trimethyl pentyl phosphinic acid	
LIX 54 α-acetyl-m-dodecylacetophenone	

Table A1. *Cont.*

Extractants	Structure of the Compound
MIBK Methyl isobutyl ketone	
TBP Tri-n-butyl phosphate	
TOPO Tri-n-octyl phosphine oxide	
Cyanex 923 Mixture of main trialkyl phosphine oxides	$R'R_2P{=}O$ (31%) $R_2'RP{=}O$ (42%) $R_3'P{=}O$ (14%) $R_3P{=}O$ (8%) $R,R' = [CH_3(CH_2)_7]-$normal octyl

References

1. Swain, B. Separation and purification of lithium by solvent extraction and supported liquid membrane, analysis of their mechanism: A review. *J. Chem. Technol. Biotechnol.* **2016**, *91*, 2549–2562. [CrossRef]
2. Albema. Global Lithium Market Outlook. In Proceedings of the Goldman Sachs Houston Chemical Intensity Days Conference, Houston, TX, USA, 15 March 2016. Available online: https://www.scribd.com/document/341213533/HCID-Conference-ALB-Lithium-Presentation-v1-1. (accessed on 15 March 2016).
3. Zhou, Z.; Qin, W.; Liu, Y.; Fei, W. Extraction equilibria of lithium with tributyl phosphate in kerosene and FeCl₃. *J. Chem. Eng. Data* **2011**, *57*, 82–86. [CrossRef]
4. Swain, B. Recovery and recycling of lithium: A review. *Sep. Purif. Technol.* **2017**, *172*, 88–403. [CrossRef]
5. Liu, X.; Zhong, M.; Chen, X.; Zhao, Z. Separating lithium and magnesium in brine by aluminum-based materials. *Hydrometallurgy* **2018**, *176*, 73–77. [CrossRef]
6. Wietelmann, U.; Steinbild, M. Lithium and Lithium Compounds. *Ullmann's Encycl. Ind. Chem.* **2014**, 1–38. [CrossRef]
7. Kesler, S.E.; Gruber, P.W.; Medina, P.A.; Keoleian, G.A.; Everson, M.P.; Wallington, T.J. Global lithium resources: Relative importance of pegmatite, brine and other deposits. *Ore Geol. Rev.* **2012**, *48*, 55–69. [CrossRef]
8. Swain, B.; Jeong, J.; Lee, J.C.; Lee, G.H.; Sohn, J.S. Hydrometallurgical process for recovery of cobalt from waste cathodic active material generated during manufacturing of lithium ion batteries. *J. Power Sources* **2007**, *167*, 536–544. [CrossRef]
9. Nan, J.; Han, D.; Zuo, X. Recovery of metal values from spent lithium-ion batteries with chemical deposition and solvent extraction. *J. Power Sources* **2005**, *152*, 278–284. [CrossRef]
10. Nayl, A.A.; Hamed, M.M.; Rizk, S.E. Selective extraction and separation of metal values from leach liquor of mixed spent Li-ion batteries. *J. Taiwan Inst. Chem. Eng.* **2015**, *55*, 119–125. [CrossRef]

11. Ordoñez, J.; Gago, E.J.; Girardm, A. Processes and technologies for the recycling and recovery of spent lithium-ion batteries. *Renew. Sustain. Energy Rev.* **2016**, *60*, 195–205. [CrossRef]

12. Meshram, P.; Pandey, B.D.; Mankhand, T.R. Extraction of lithium from primary and secondary sources by pre-treatment, leaching and separation: A comprehensive review. *Hydrometallurgy* **2014**, *150*, 192–208. [CrossRef]

13. Maraghechi, H.; Rajabipour, F.; Pantano, C.G.; Burgos, W.D. Effect of calcium on dissolution and precipitation reactions of amorphous silica at high alkalinity. *Cem. Concr. Res.* **2016**, *87*, 1–13. [CrossRef]

14. Harvianto, G.R.; Kim, S.H.; Ju, C.S. Solvent extraction and stripping of lithium ion from aqueous solution and its application to seawater. *Rare Met.* **2016**, *35*, 948–953. [CrossRef]

15. Ji, L.; Hu, Y.; Li, L.; Shi, D.; Li, J.; Nie, F. Lithium Extraction with a Synergistic System of Dioctyl Phthalate and Tributyl Phosphate in Kerosene and FeCl₃. *Hydrometallurgy* **2016**, *162*, 71–78. [CrossRef]

16. An, J.W.; Kang, D.J.; Tran, K.T.; Kim, M.J.; Lim, T.; Tran, T. Recovery of lithium from Uyuni salar brine. *Hydrometallurgy* **2012**, *117*, 64–70. [CrossRef]

17. Bukowsky, H.; Uhlemann, E.; Steinborn, D. The recovery of pure lithium chloride from "brines" containing higher contents of calcium chloride and magnesium chloride. *Hydrometallurgy* **1991**, *27*, 317–325. [CrossRef]

18. Chitrakar, R.; Makita, Y.; Ooi, K.; Sonoda, A. Lithium recovery from salt lake brine by H₂TiO₃. *Dalt. Trans.* **2014**, *43*, 8933–8939. [CrossRef] [PubMed]

19. Works, D.S. Extraction of lithium from the Dead Sea. *Hydrometallurgy* **1981**, *6*, 269–275.

20. El-Eswed, B.; Sunjuk, M.; Al-Degs, Y.S.; Shtaiwi, A. Solvent Extraction of Li⁺ using Organophosphorus Ligands in the Presence of Ammonia. *Sep. Sci. Technol.* **2014**, *49*, 1342–1348. [CrossRef]

21. Hano, T.; Matsumoto, M.; Ohtake, T.; Egashira, N.; Hori, F. Recovery of lithium from geothermal water by solvent extraction technique. *Solvent Extr. Ion Exch.* **1992**, *10*, 195–206. [CrossRef]

22. Xiang, W.; Liang, S.; Zhou, Z.; Qin, W.; Fei, W. Lithium recovery from salt lake brine by counter-current extraction using tributyl phosphate/FeCl₃ in methyl isobutyl ketone. *Hydrometallurgy* **2017**, *171*, 27–32. [CrossRef]

23. Jandová, J.; Vu, H.N.; Belková, T.; Dvorák, P.; Kondás, J. Obtaining Li₂CO₃ from Zinnwaldite Wastes. *Ceram-Silikáty* **2009**, *53*, 108–112.

24. Kondás, J.; Jandová, J. Lithium extraction from zinnwaldite wastes after gravity dressing of Sn-W ore. *Acta Metall. Slovaca* **2006**, *12*, 197–202.

25. Siame, E.; Pascoe, R.D. Extraction of lithium from micaceous waste from china clay production. *Miner. Eng.* **2011**, *24*, 1595–1602. [CrossRef]

26. Yan, Q.; Li, X.; Wang, Z.; Wu, X.; Wang, J.; Guo, H. Extraction of lithium from lepidolite by sulfation roasting and water leaching. *Int. J. Miner. Process.* **2012**, *110*, 1–5. [CrossRef]

27. Sitando, O.; Crouse, P.L. Processing of a Zimbabwean petalite to obtain lithium carbonate. *Int. J. Miner. Process.* **2012**, *102*, 45–50. [CrossRef]

28. Clarke, G.M. Lithium-ion batteries: Raw material considerations. *Amer. Inst. Chem. Eng.* **2013**, 44–52.

29. Jandová, J.; Dvorák, P.; Vu, H.N. Processing of zinnwaldite waste to obtain Li₂CO₃. *Hydrometallurgy* **2010**, *103*, 12–18. [CrossRef]

30. Vu, H.; Bernardi, J.; Jandová, J.; Vaculíková, L.; Goliáš, V. Lithium and rubidium extraction from zinnwaldite by alkali digestion process: Sintering mechanism and leaching kinetics. *Int. J. Miner. Process.* **2013**, *123*, 9–17. [CrossRef]

31. Chen, L.; Tang, X.; Zhang, Y.; Li, L.; Zeng, Z.; Zhang, Y. Process for the recovery of cobalt oxalate from spent lithium-ion batteries. *Hydrometallurgy* **2011**, *108*, 80–86. [CrossRef]

32. Golmohammadzadeh, R.; Rashchi, F.; Vahidi, E. Recovery of lithium and cobalt from spent lithium-ion batteries using organic acids: Process optimization and kinetic aspects. *Waste Manag.* **2017**, *64*, 244–254. [CrossRef] [PubMed]

33. Joulié, M.; Laucournet, R.; Billy, E. Hydrometallurgical process for the recovery of high value metals from spent lithium nickel cobalt aluminum oxide based lithium-ion batteries. *J. Power Sources* **2014**, *247*, 551–555. [CrossRef]

34. Li, L.; Zhai, L.; Zhang, X.; Lu, J.; Chen, R.; Wu, F. Recovery of valuable metals from spent lithium-ion batteries by ultrasonic-assisted leaching process. *J. Power Sources* **2014**, *262*, 380–385. [CrossRef]

35. Meshram, P.; Pandey, B.D.; Mankhand, T.R. Hydrometallurgical processing of spent lithium ion batteries (LIBs) in the presence of a reducing agent with emphasis on kinetics of leaching. *Chem. Eng. J.* **2015**, *281*, 418–427. [CrossRef]

36. Nayl, A.A.; Elkhashab, R.A.; Badawy, S.M.; El-Khateeb, M.A. Acid leaching of mixed spent Li-ion batteries. *Arab. J. Chem.* **2017**, *10*, S3632–S3639. [CrossRef]

37. Wang, R.C.; Lin, Y.C.; Wu, S.H. A novel recovery process of metal values from the cathode active materials of the lithium-ion secondary batteries. *Hydrometallurgy* **2009**, *99*, 194–201. [CrossRef]

38. Shuva, M.A.H.; Kurny, A. Hydrometallurgical Recovery of Value Metals from Spent Lithium Ion Batteries. *Am. J. Mater. Eng. Technol.* **2013**, *1*, 8–12. [CrossRef]

39. Yong-jia, L.; Ting, L. Hydrometallurgical Process for Recovery and Synthesis of $LiCoO_2$ from Spent Lithium-ion Batteries. In Proceedings of the 2011 International Conference on Electric Technology and Civil Engineering (ICETCE), Lushan, China, 22–24 April 2011; pp. 6009–6011.

40. Kang, J.; Senanayake, G.; Sohn, J.; Shin, S.M. Recovery of cobalt sulfate from spent lithium ion batteries by reductive leaching and solvent extraction with Cyanex 272. *Hydrometallurgy* **2010**, *100*, 168–171. [CrossRef]

41. Dorella, G.; Mansur, M.B. A study of the separation of cobalt from spent Li-ion battery residues. *J. Power Sources* **2007**, *170*, 210–215. [CrossRef]

42. Shin, S.M.; Kim, N.H.; Sohn, J.S.; Yang, D.H.; Kim, Y.H. Development of a metal recovery process from Li-ion battery wastes. *Hydrometallurgy* **2005**, *79*, 172–181. [CrossRef]

43. Lee, C.K.; Rhee, K.I. Reductive leaching of cathodic active materials from lithium ion battery wastes. *Hydrometallurgy* **2003**, *68*, 5–10. [CrossRef]

44. Li, L.; Ge, J.; Wu, F.; Chen, R.; Chen, S.; Wu, B. Recovery of cobalt and lithium from spent lithium ion batteries using organic citric acid as leachant. *J. Hazard. Mater.* **2010**, *176*, 288–293. [CrossRef] [PubMed]

45. Li, L.; Lu, J.; Ren, Y.; Zhang, X.X.; Chen, R.J.; Wu, F. Ascorbic-acid-assisted recovery of cobalt and lithium from spent Li-ion batteries. *J. Power Sources* **2012**, *218*, 21–27. [CrossRef]

46. Li, L.; Ge, J.; Chen, R.; Wu, F.; Chen, S.; Zhang, X. Environmental friendly leaching reagent for cobalt and lithium recovery from spent lithium-ion batteries. *J. Hazard. Mater.* **2010**, *30*, 2615–2621. [CrossRef] [PubMed]

47. Zeng, X.; Li, J.; Shen, B. Novel approach to recover cobalt and lithium from spent lithium-ion battery using oxalic acid. *J. Hazard. Mater.* **2015**, *295*, 112–118. [CrossRef] [PubMed]

48. Zhang, L.; Li, L.; Shi, D.; Li, J.; Peng, X.; Nie, F. Selective extraction of lithium from alkaline brine using HBTA-TOPO synergistic extraction system. *Sep. Purif. Technol.* **2017**, *188*, 167–173. [CrossRef]

49. Ishimori, K.; Imura, H.; Ohashi, K. Effect of 1,10-phenanthroline on the extraction and separation of lithium(I), sodium(I) and potassium(I) with thenoyltrifluoroacetone. *Anal. Chim. Acta* **2002**, *454*, 241–247. [CrossRef]

50. Kim, Y.S.; In, G.; Choi, J.M. Chemical Equilibrium and Synergism for Solvent Extraction of Trace Lithium with Thenoyltrifluoroacetone in the Presence of Trioctylphosphine Oxide. *Bull. Korean Chem. Soc.* **2003**, *24*, 1495–1500. [CrossRef]

51. Ji, L.; Li, L.; Shi, D.; Li, J.; Liu, Z.; Xu, D. Extraction equilibria of lithium with *N,N*-bis(2-ethylhexyl)-3-oxobutanamide and tributyl phosphate in kerosene and $FeCl_3$. *Hydrometallurgy* **2016**, *164*, 304–312. [CrossRef]

52. Zhou, Z.; Qin, W.; Fei, W.; Li, Y. A study on stoichiometry of complexes of tributyl phosphate and methyl isobutyl ketone with lithium in the presence of $FeCl_3$. *Chin. J. Chem. Eng.* **2012**, *20*, 36–39. [CrossRef]

53. Zhou, Z.; Qin, W.; Liu, Y.; Fei, W. Extraction equilibria of lithium with tributyl phosphate in Three Diluents. *J. Chem. Eng. Data* **2011**, *56*, 3518–3522. [CrossRef]

54. Pranolo, Y.; Zhu, Z.; Cheng, C.Y. Separation of lithium from sodium in chloride solutions using SSX systems with LIX 54 and Cyanex 923. *Hydrometallurgy* **2015**, *154*, 33–39. [CrossRef]

55. Kinugasa, T.; Nishibara, H.; Murao, Y.; Kawamura, Y.; Watanabe, K.; Takeuchi, H. Equilibrium and Kinetics of Lithium Extraction by a Mixture of LIX54 and TOPO. *J. Chem. Eng. Jpn.* **1994**, *27*, 815–818. [CrossRef]

56. Zhou, Z.; Qin, W.; Chu, Y.; Fei, W. Elucidation of the structures of tributyl phosphate/Li complexes in the presence of $FeCl_3$ via UV-visible, Raman and IR spectroscopy and the method of continuous variation. *Chem. Eng. Sci.* **2013**, *101*, 577–585. [CrossRef]

57. Zhou, Z.; Qin, W.; Fei, W.; Liu, Y. A study on stoichiometry of complexes of tributyl phosphate and metyl isobutyl ketone with lithium in the presence of $FeCl_3$. *Chin. J. Chem. Eng.* **2012**, *20*, 36–39. [CrossRef]

58. Harvianto, G.R.; Jeong, S.G.; Ju, C.S. The effect of dominant ions on solvent extraction of lithium ion from aqueous solution. *Korean J. Chem. Eng.* **2014**, *31*, 828–833. [CrossRef]

59. Lupi, C.; Pasquali, M.; Dell'Era, A. Nickel and cobalt recycling from lithium-ion batteries by electrochemical processes. *Waste Manag.* **2005**, *25*, 215–220. [CrossRef] [PubMed]

60. Mantuano, D.P.; Dorella, G.; Elias, R.C.A.; Mansur, M.B. Analysis of a hydrometallurgical route to recover base metals from spent rechargeable batteries by liquid-liquid extraction with Cyanex 272. *J. Power Sources* **2006**, *159*, 1510–1518. [CrossRef]

61. Swain, B.; Mishra, C.; Jeong, J.; Lee, J.C.; Hong, H.S.; Pandey, B.D. Separation of Co(II) and Li(I) with Cyanex 272 using hollow fiber supported liquid membrane: A comparison with flat sheet supported liquid membrane and dispersive solvent extraction process. *Chem. Eng. J.* **2015**, *271*, 61–70. [CrossRef]

62. Devi, N.B.; Nathsarma, K.C.; Chakravortty, V. Sodium salts of D2EHPA, PC-88A and Cyanex-272 and their mixtures as extractants for cobalt(II). *Hydrometallurgy* **1994**, *34*, 331–342. [CrossRef]

63. Zhao, J.M.; Shen, X.Y.; Deng, F.L.; Wang, F.C.; Wu, Y.; Liu, H.Z. Synergistic extraction and separation of valuable metals from waste cathodic material of lithium ion batteries using Cyanex272 and PC-88A. *Sep. Purif. Technol.* **2011**, *78*, 345–351. [CrossRef]

64. Chen, X.; Chen, Y.; Zhou, T.; Liu, D.; Hu, H.; Fan, S. Hydrometallurgical recovery of metal values from sulfuric acid leaching liquor of spent lithium-ion batteries. *Waste Manag.* **2015**, *38*, 349–356. [CrossRef] [PubMed]

65. Zhang, P.; Yokoyama, T.; Itabashi, O.; Wakui, Y.; Suzuki, T.M.; Inoue, K. Hydrometallurical process for recovery of metal values from spent nickel-metal hydride secondary batteries. *Hydrometallurgy* **1998**, *50*, 61–75. [CrossRef]

66. Pranolo, Y.; Zhang, W.; Cheng, C.Y. Hydrometallurgy Recovery of metals from spent lithium-ion battery leach solutions with a mixed solvent extractant system. *Hydrometallurgy* **2010**, *102*, 37–42. [CrossRef]

processes

MDPI

Article

The Influence of Cation Treatments on the Pervaporation Dehydration of NaA Zeolite Membranes Prepared on Hollow Fibers

Xuechao Gao *,†, Bing Gao †, Xingchen Wang, Rui Shi, Rashid Ur Rehman [ORCID] and Xuehong Gu *

State Key Laboratory of Materials-Oriented Chemical Engineering, College of Chemical Engineering, Nanjing Tech University, 5 Xinmofan Road, Nanjing 210009, China; gb123456789@njtech.edu.cn (B.G.); xingchenwang@njtech.edu.cn (X.W.); 15062281631@njtech.edu.cn (R.S.); dr.rehman@njtech.edu.cn (R.U.R.)
* Correspondence: xuechao.gao@njtech.edu.cn (X.G.); xhgu@njtech.edu.cn (X.G.); Tel.: +86-25-8317-2268 (X.Gu)
† The authors contribute equally to this work.

Received: 21 April 2018; Accepted: 23 May 2018; Published: 1 June 2018

Abstract: NaA zeolite membrane is an ideal hydrophilic candidate for organic dehydrations; however, its instability in salt solutions limits its application in industries as the membrane intactness was greatly affected due to the replacement of cation ions. In order to explore the relationship between the structural variation and the cation types, the obtained NaA zeolite membranes were treated by various monovalent and divalent cations like Ag^+, K^+, Li^+, NH_4^+, Zn^{2+}, Mg^{2+}, Ba^{2+} and Ca^{2+}. The obtained membranes were subsequently characterized by contact angle, scanning electron microscopy (SEM), pervaporation (PV), and vapor permeation (VP). The results showed that all of the hydrophilicities of the exchanged membrane were reduced, and the membrane performance varied with cation charges and sizes. For the monovalent cations, the membrane performance was largely determined by the cation sizes, where the membrane remained intact. On the contrary, for the divalent cation treatments, the membrane separation was generally reduced due to the presence of cation vacancies, resulting in some unbalanced stresses between the dispersive interaction and electrostatic forces, thereby damaging the membrane intactness. In the end, a set of gas permeation experiments were conducted for the two selected cation-treated membranes (K^+ and Ag^+) using H_2, CO_2, N_2 and CH_4, and a much higher decreasing percentage (90% for K^+) occurred in comparison with the permeation drop (10%) in the PV dehydration, suggesting that the vaporization resistance of phase changing for the PV process was more influential than the water vapor transport in the pore channel.

Keywords: NaA zeolite membrane; pervaporation; cation treatments; membrane separation; hollow fibers

1. Introduction

Fuel ethanol is considered as a promising alternative to the conventionally used fossils due to its significant advantages in terms of environment protection, energy intensity, and resource abundancy [1–3]. At present, microbial fermentations are the mainstream route that is used in the industry to produce fuel ethanol; however, by-products and water generated during the fermenting activities [4] should be removed simultaneously in order to maintain the production efficiency. Among the diverse separation strategies, the membrane-based pervaporation (PV) process, as a newly emerging technique, ha attracted considerable attention due to its significant advantages of less energy consumption and operational convenience over the traditional separation methods, such as extraction and distillation.

In general, the membranes that are used for organic dehydrations can be classified into two categories, which include (i) polymeric and (ii) inorganic membranes. Although the polymer is easy to fabricate due to its high flexibility, it suffers from weak mechanical strengths and swelling problems, thereby limiting its application [5–7]. Not only has the inorganic membrane demonstrated good mechanical strengths, but also strong chemical resistance and high thermostability were also derived, so it is more promising to be used in harsh solutions containing additives, like dimethylformamide (DMF), tetrahydrofuan (THF) and salts. Among the various types of inorganic membranes, zeolite membrane, making use of its unique pore structure and surface properties, are widely deployed for organic dehydrations [8]. In the past two decades, many types of zeolites have been fabricated as the supported zeolite membranes [7,9], including LTA [10,11], FAU [12], T [13], MOR [14] and MFI [15–17]. Among these candidates, The tubular NaA zeolite membrane has been widely applied in pharmaceutical and food industries for solvent recoveries [18]. NaA zeolite contains sodium cations, and the chemical formula is $Na_{12}[(AlO_2)_{12}(SiO_2)_{12}]\cdot 27H_2O$ [19], where the sodalite cages are connected through double four-membered rings (4MR) in order to form a large cavity. For the NaA zeolite membrane with a Si/Al ratio of 1.0, high permselectivities can be achieved due to its strong hydrophilicity. Since the first commercial application of tubular NaA membranes in practical factories by Mitsui Engineering and Shipping in the 1990s, over 200 sets of the membrane plants have been established in the world for various purposes [20–22]. However, the tubular NaA membrane often contains large diameters and thick walls, so the packing density and flux in a certain membrane module are limited, thereby increasing the membrane facility investment.

To further reduce the membrane cost, the packing density of the membrane module should be significantly increased. Therefore, there is an on-going effort to prepare the NaA membrane on hollow fibers with a much smaller diameter and thinner walls. However, at present, the NaA membranes prepared on the hollow fiber are still unsatisfactory due to their low selectivity. This could be caused by the high curvature of the substrate surface, which induces stresses between the membrane and substrate surface, thereby leading to the presence of the inter-crystalline defects [23,24]. Therefore, it is necessary to use chemical approaches to improve the membrane performance in order to increase the adhesive forces between the membrane layer and substrate. To fix the membrane defects derived during the preparation, cations exchanges could be employed to cover interstitial gaps by changing the crystal cell parameters, which resulted in the contraction and expansion of the zeolite particles. Further, inorganic salts are generally presented in organic solvents used in the fields of organic chemicals, fine chemicals, and pharmaceutical chemicals. The salts inevitably interacted with the NaA zeolite and thus affected the separation performance of the NaA zeolite membrane by changing the zeolite structure and surface hydrophilicity. According to the work by Breck and co-workers [25], the pore size of LTA zeolite could be tailored using various cation treatments. For the zeolite that was treated by potassium, the effective pore size was reduced to 0.28 nm; on the contrary, the calcium salt produced a larger pore size of 0.51 nm. Francisco et al. [26] investigated several ion-exchanged NaA/carbon membranes for hydrogen purification, and it had found that the Rb-LTA and Cs-LTA/carbon membranes provided better separation performance. Shirazian et al. [27]. Prepared the K-exchanged LTA zeolite membranes for the dehydration of natural gas; however, the separation factors were still close to the prediction of Knudsen diffusion. Yang et al. studied the influence of several inorganic salts on the PV performance for the ethanol/H_2O/salt mixtures, and a significant flux drop was found, which was assumed to be caused by the pore blocking and surface precipitation of salts [28–31]. However, the above research only involved a limit number of cations, and the membrane structural changes caused by the cation exchanges were not explicitly correlated with the ionic size and charge, but were lumped with the driving force variations and transport resistance increments during the PV characterization.

Based on above discussion, it is necessary to scientifically and systematically explore the relationship of separation variations with cation sizes and charges using both PV dehydration and gas permeation techniques, so as to provide guidance in order to broaden the industrial application of hollow fibered supported NaA zeolite membranes. In this paper, we prepared hollow fiber supported NaA zeolite membranes in a batch mode with acceptable separation performances, which was later treated by different cations solutions (monovalent and divalent cations) to derive ion-exchanged NaA zeolite membranes. Then, the ion-exchanged membranes were characterized by PV for 90 wt %

ethanol/water solution. The variations of permeation fluxes and selectivities were studied and were used to correlate the coordinated effect between the cation size and charge. In the end, the PV decrement for several cation treatments was compared with that for the gas permeation and vapor permeation (VP) tests to evaluate the importance of the surface hydrophilicity and the vaporization resistance of the NaA zeolite membranes for water molecules.

2. Experimental

2.1. Materials

The monovalent and divalent salts (monovalent: NaCl, LiCl, NH_4Cl, KCl and $AgNO_3$; divalent: $ZnCl_2$ $MgCl_2$ $6H_2O$, $CaCl_2$ and $BaCl_2$) were purchased from commercial companies, and they were used as received without any purification. The silica sol, water glass and $NaAlO_2$ were used as Si and Al sources with industrial purity. The original NaA zeolite seeds (cubic crystals, with a size of 2–4 µm) and deionized water were produced in the laboratory. The alumina four channel hollow fibers (4CHF) were used as the support (homemade, out diameter: 3.4–3.8 mm, length: 70 mm, porosity: 45–55%).

2.2. Synthesis of NaA Zeolite Membranes and Ion-Exchange Processes

The preparation process of NaA crystal suspension: 0.2 g silica sol was added to a certain amount of ball-milled NaA zeolite crystal solution so as to improve the viscosity and to prevent the agglomeration of seeds; after that, the solution was mixed by an ultrasonic cell grinder (power 200 W) for 30 min, in order to obtain a uniformly dispersed seed suspension. Seeding the hollow fiber: the hollow fibers, soaked with 0.1 mol·L^{-1} sodium hydroxide solution for 12 h, were placed in a 60 °C oven and were dried overnight; after that, the hollow fibers were pre-coated by the dip-coating method for 5–15 s.

The preparation process of the NaA zeolite membrane was provided in Figure 1. At the beginning, the sodium hydroxide that was dissolved in deionized water to provide basic conditions, was later mixed with sodium aluminate to derive a clear solution, and then a certain amount of water glass was added into the mixture under vigorously stirring for 1.5 h, where the molar ratio of the synthesis solution is Al_2O_3:SiO_2:NaOH:H_2O = 1:2:2:120. Subsequently, the treated supports that were vertically placed in a Teflon-stainless steel autoclave containing the synthesis sol, were transferred into the oven for 4 h at 100 °C. The as-synthesis membrane was washed with deionized water several times, before drying at 70 °C overnight.

The preparation of ion-exchanged NaA-type zeolite membrane: the 90% ethanol/water solution was mixed with a certain amount of NaCl, LiCl, NH_4Cl, KCl, $AgNO_3$, $ZnCl_2$, $MgCl_2$, $CaCl_2$, or $BaCl_2$, respectively, to derive a mixture with a salt concentration of 0.05 mol/L–0.10 mol/L by sonicating for 30 min; the ion-exchange solution was pumped into the liner that was equipped with the prepared NaA zeolite membrane to exchange for 24 h at 50 °C. After the exchanging, the membrane was immersed in deionized water for 2 h, and was then was dried for characterizations.

The preparation of ion-exchanged NaA zeolite powder: the salt solutions were prepared according to the above method, and then 40 g of the salt solution and 0.6 g NaA crystal were stirringly mixed in a conical flask immersed in a water bath of 50 °C for 4 h. After the exchanging, the suspension was centrifuged for several times and dried.

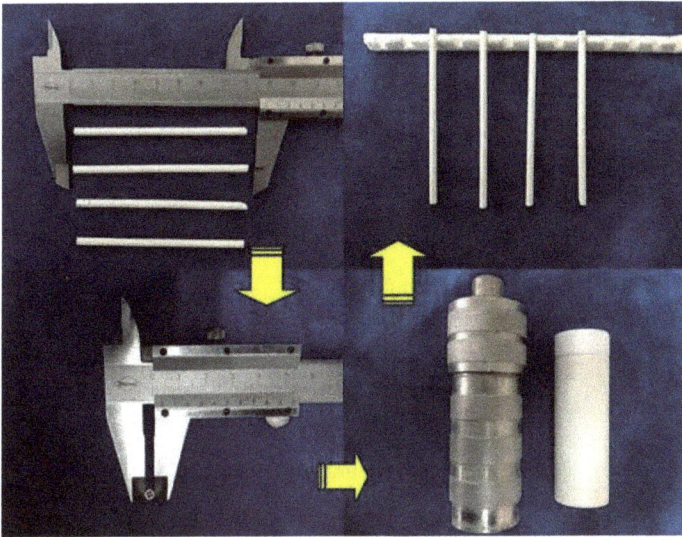

Figure 1. Cation treatment diagram of NaA zeolite membranes.

2.3. Characterization of NaA Zeolite Particles and Membranes

The morphologies of the as-synthesis NaA zeolite powders and membranes were characterized by field emission scanning electron microscopy (FESEM) and installed with a cold field emission gun that was operating at 5 kV and 10 μA (S-4800, Hitachi, Tokyo, Japan). The crystal structure of the sample was analyzed by X-ray diffraction (XRD) at a tube voltage of 40 kV and a tube current of 15 mA (MiniFlex 600, Rigaku, Tokyo, Japan), where the Cu target and the Ni filter were used to generate Kα rays and to remove Kβ rays, respectively. The tests were performed at room temperature with a diffraction range of 5°–50°, and a scan rate of 12°/min. After exchanging, the elemental content of the sample was measured by inductively coupled plasma optical emission spectrometry (ICP-OES, optima 7000DV, PerkinElmer, Waltham, MA, US), where the samples were dissolved in an acidic mixture before testing.

2.4. PV Test and Single Gas Permeation

The separation performances of NaA zeolite membranes were evaluated by dehydrating 90 wt % ethanol-water solution at 75 °C by a PV apparatus, as reported in our previous work [32]. The compositions of the feed and permeate were analyzed by gas chromatography (GC-6890, Ruihong, Tengzhou, China). The PV performance of each membrane was determined by the separation factor (α) and flux (J), which were, respectively, defined as follows:

$$\alpha = \frac{y_w/y_e}{x_w/x_e} \tag{1}$$

$$J = \frac{m}{A \cdot \Delta t} \tag{2}$$

where y_w and y_e are the mass fractions of water and ethanol in the permeate, respectively; x_w and x_e correspond to the fraction values in the feed, respectively; m is the mass (kg) permeated over a time period of Δt (h); and, A is the effective membrane area (m^2).

By excluding the influence of the membrane hydrophilicities, gas permeation at room temperature was used to characterize the importance of the vaporization effect in the NaA zeolite membranes that were treated by different cations. For the permeation experiment, the membrane was sealed by silicone O-rings and the volumetric flow rates of single gases (H_2, CO_2, CH_4 and N_2) were measured using a bubble flow meter. The tested gas on the feed side provided a driving force for gas permeation from the outside of the membrane to the lumen inside, where the gas permeation flux (Pm_i) can be expressed as:

$$Pm_i = \frac{n}{A \cdot \Delta t \cdot \Delta P} \tag{3}$$

where Pm_i represents the molar quantity of the gas flowing through the membrane per unit time and per unit area under per unit pressure difference($mol \cdot m^{-2} \cdot h^{-1} \cdot Pa^{-1}$); n corresponds to the moles of the gas; and, ΔP is the pressure difference (Pa).

3. Results and Discussion

3.1. PV Performances of the Exchanged NaA Zeolite Membrane

Figure 2 depicted the ion exchange model for the NaA zeolite with high aluminum content to show how different cation affects the open pore passages. As suggested, cation charges affect the exchange number of ions during the exchange in order to maintain the valence balance of the zeolite. According to the law of conservation of charge, each divalent cation consumes two replacements of monovalent sodium ion, and vice versa. In the cation treatment, in order to ensure the full exchange of ions with the membrane, an exchanging time of 24 h and a cation concentration of 0.1 mol/L were firstly selected. To initially examine the effect of ion exchanges of different cations on the membrane performance, both the feed and permeate were analyzed by a gas chromatograph, where the PV solution was a 90 wt % ethanol/water mixture.

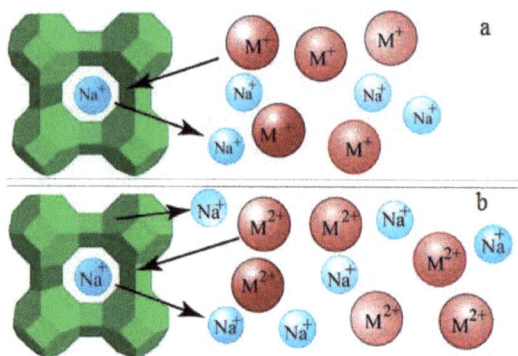

Figure 2. Different cation exchange model of the NaA-Type zeolite. (**a**) Monovalent cation and (**b**) Divalent cation.

Table 1 summarized the PV results for different salts, where the corresponding cation sizes were provided in Table 2. As suggested, the kinetic radii of the monovalent cations, like K^+ and Ag^+ (1.49 Å and 1.26 Å), are greater than that of Na^+ (1.17 Å), and the PV separation factor of the membrane was increased and the permeate flux was decreased. This is probably because K^+ or Ag^+ entered into zeolite cage and replaced the original Na^+, which resulted in the decrease in accessible pore size. Smaller pores yielded lower permeation fluxes, but higher separation factor. For the smallest monovalent cation (Li^+), when the original Na^+ was replaced by Li^+, the pore was significantly opened, which led to a much higher permeate flux, but without any selectivity during the PV test. For the treatment by the acidic cation solution of NH_4Cl, both the permeation flux and the separation factor were considerably reduced, suggesting that the membrane structure was significantly changed. The decrease in the permeation flux could be explained by the ionic size effect (1.48 Å versus 1.17 Å), where the water passage in the membrane was blocked by the larger cation. The decrease in the separation factor was caused by the chemical attack to the zeolite skeleton due to the presence of H proton that was released from the cation for the NH_4Cl solution being acidic. On the other hand, the surface precipitation of salts on the membrane surface could also cause the decrease of permeation flux. To evaluate such an effect, the NaCl solution was used as a reference. As given in the Table 1, before and after the treatment with NaCl solution, the permeation flux and the separation factor of the membrane only change slightly, so the surface precipitation effect could be neglected in the experiment.

For the divalent cations of Mg^{2+} and Zn^{2+} with smaller ion sizes (0.72 Å and 0.74 Å), both the permeation flux and the separation factor of the NaA zeolite membranes dramatically decreased after cation exchanging. This was because of the displacement of Na^+ cations around the six-membered ring-like structures. When the Na^+ ions were deprived from zeolite pores by the smaller divalent cations of Mg^{2+} and Zn^{2+}, the open pore size became larger than the molecular size of ethanol, so the separation factor was reduced. In addition, since the molar ratio of ion exchanging between Na^+ and Mg^{2+}/Zn^{2+} was 2, the remaining cation sites in the zeolite were vacancies, so some unbalanced stresses were induced, where the membrane porosity was reduced due to some collapse. When the effect of porosity decrement outweighed the effect that was caused by the pore enlargement, the overall permeation flux still decreased. For the other two divalent cations (Ca^{2+} and Ba^{2+}) with larger ion sizes (1.0 Å and 1.35 Å), the coordinated effect between open pore size and unbalanced stresses due to cation vacancies were similar except the pore size that was caused by Ba^{2+} treatment became smaller. The decrease in open pore size should cause the increase in the separation factor; however, the unbalanced stresses generated some defects between crystals in the membrane, so the membrane intactness was reduced, which caused the decrease in the separation factor. Another evidence for the presence of defects in the Ba^{2+} treatment was the increase in flux.

Table 1. Pervaporation of 90 wt % ethanol/water mixture through NaA-type zeolite membrane and its cation-exchanged membrane (exchange condition: 12 h at 60 °C and cation concentration of 0.1 M).

Ion Valence State	Material	Processing Conditions	Fluxes/$kg \cdot m^{-2} \cdot h^{-1}$	Separation Factor
1	KCl	Un-exchange	4.05 ± 0.019	449 ± 2.4
		Exchange	3.58 ± 0.023	1062 ± 12.8
	NH_4Cl	Un-exchange	7.42 ± 0.023	306 ± 1.1
		Exchange	1.44 ± 0.005	38 ± 0.1
	LiCl	Un-exchange	4.58 ± 0.019	208 ± 0.5
		Exchange	leaking	1 ± 0.01
	NaCl	Un-exchange	4.75 ± 0.023	449 ± 2.4
		Exchange	4.83 ± 0.018	519 ± 3.1
	$AgNO_3$	Un-exchange	6.51 ± 0.027	548 ± 3.6
		Exchange	5.98 ± 0.025	2136 ± 50.7
2	$MgCl_2$	Un-exchange	5.05 ± 0.025	2167 ± 52.5
		Exchange	1.99 ± 0.014	47 ± 0.1
	$BaCl_2$	Un-exchange	4.80 ± 0.024	208 ± 0.6
		Exchange	5.19 ± 0.025	162 ± 0.4
	$CaCl_2$	Un-exchange	6.45 ± 0.025	306 ± 1.2
		Exchange	6.07 ± 0.024	52 ± 0.1
	$ZnCl_2$	Un-exchange	5.06 ± 0.019	2156 ± 53.4
		Exchange	2.90 ± 0.015	5 ± 0.04

Table 2. The cation radius of the used salts.

Cations	Hydrated Radius, Å [33]	Bare Radius, Å [34]
Li^+	3.82	0.94
Na^+	3.58	1.17
K^+	3.31	1.49
Ag^+	3.41	1.26
NH_4^+	3.31	1.48
Mg^{2+}	4.28	0.72
Ca^{2+}	4.12	1.00
Ba^{2+}	4.04	1.35
Zn^{2+}	4.30	0.74

3.2. Surface Morphology of the Exchanged NaA Zeolite Membrane

To further testify the above structural alternations, which gave rise to the changes in permeation fluxes and separation factors, the surface morphologies of NaA zeolite membranes after various cation treatments were examined by the scanning electron microscopy (SEM) technique. The obtained images for the monovalent cation treatment were depicted in Figure 3. As suggested, no large cracks of crystal particles and interstitial defects were found on the surface of the zeolite membrane after being treated with 0.1 M $AgNO_3$, NaCl, KCl and NH_4Cl solutions, so the membranes intactness was maintained in the cation treatments. When compared to the feature of the original NaA zeolite crystals in Figure 3a, no significant changes occurred to the samples that were treated with KCl and NaCl solutions in Figure 3b,f. Further, for $AgNO_3$ treatment in Figure 3c, a large amount of amorphous substances were found on the membrane surface, which may be caused by the partial destruction of particle due to the crystal cell expansion by the replacement of Ag^+, i.e., the interstitial defects between the particles were largely fixed, which increased the separation factor but decreased the permeation flux. However, for the NH_4Cl treatment in Figure 3d, the zeolite crystals on the membrane surface started to dissolute as the sharp edges of the crystal particles were absent. This was largely due to the dealumination in the zeolite framework under acidic conditions (NH_4Cl solution is acidic), thereby resulting in collapses of the skeleton and the partial dissolution of the membrane. For the smallest monovalent cation treatment of LiCl in Figure 3e, no amorphous substances were found on the membrane surface, and the crystal shapes remained unchanged except with larger interstitial spaces due the crystal cell contraction, which may be responsible for the membrane leaking in the PV tests in Table 1.

The SEM images for divalent cation treatments were provided in Figure 4, where the crystal shapes were unaffected, with larger interstitial spaces. However, for the treatment by $BaCl_2$ solution in Figure 4b, some amorphous substances were found on the membrane surface. Similar to the explanation for $AgNO_3$, the amorphous substances were generated by the cell expansion due to the replacement of larger ionic sizes of Ba^{2+}. Since the interstitial gaps were covered, the separation factor was slightly increased. As the PV results given in Table 1, the coordinated effect between the pore blocking effect and the minimization of intercrystals gaps by the amorphous substance led to the slight variation of PV performances of NaA zeolite membrane treated by Ba^{2+}. However, for other small divalent cations, due to the unbalanced exchange number between Na^+ and M^{2+}, some cation sites in the zeolite structure were left vacant, so some porosities of crystals were lost due to the structural collapse, which resulted in a complex behavior of fluxes and separation factors.

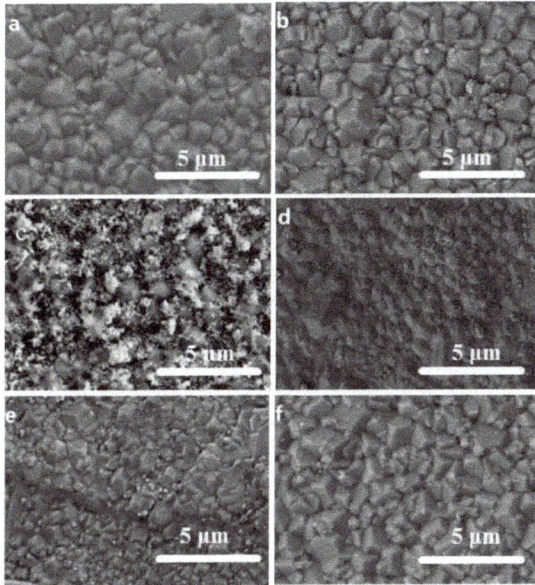

Figure 3. Scanning electron microscopy (SEM) images of NaA zeolite membrane (**a**) and its cation treated surface morphologies by (**b**) KCl, (**c**) AgNO$_3$, (**d**) NH$_4$Cl, (**e**) LiCl and (**f**) NaCl solutions.

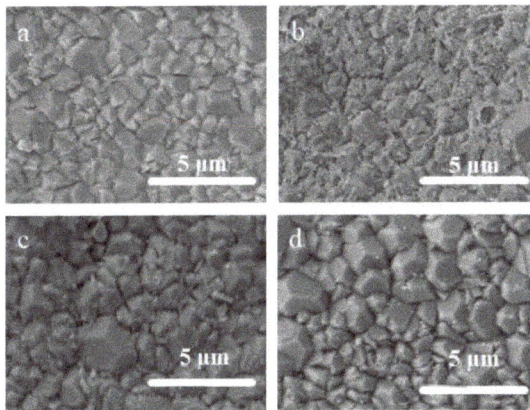

Figure 4. SEM images of NaA zeolite membrane surface morphologies treated by (**a**) CaCl$_2$, (**b**) BaCl$_2$, (**c**) ZnCl$_2$ and (**d**) MgCl$_2$ solutions.

3.3. The Structural Analysis of the Exchanged NaA Zeolite Powders

Due to the replacements of new cations, not only were the cell parameters of NaA zeolite crystals changed to some extent, but also new substances were obtained, which in turn led to the variation of permeation flux and separation factors, and such a variation could be confirmed by the XRD characterization. Figure 5 illustrated the XRD results for the cation treated zeolite particles. As suggested in Figure 5a, the XRD peak intensities of the sample that was treated by AgNO$_3$ were dramatically reduced, suggesting the presence of amorphous substances. In Figure 5b, a similar trend also occurred in the sample that was treated by Ba^{2+}, so the interpretation for the membrane surface morphologies was validated. In general, the PV performances of divalent cation treated NaA zeolite membrane were unsatisfactory. For Mg^{2+}, Ca^{2+} and Zn^{2+} ion treatments, both the permeation flux and the separation factor of the membrane decreased, which had little change in Ba^{2+} treatment. For the monovalent cation treatments, the permeation fluxes of the membrane after the exchange of Ag^+ and K^+ slightly decreased, but the separation factor has been greatly improved. Therefore, the variation trends of the membrane performance after M^{2+} exchange depended on the coordinated effect between porosity losses and ionic sizes, where a smaller ionic size generally reduced the membrane porosity due to structural collapse, and such an effect could be inhibited in the case for larger size cations due to a stronger dispersive interaction. For monovalent cations, in addition to the acidic/basic condition, the ionic size played a dominating role in the predication of membrane performance variations as no cation site was left vacant after exchanging treatments.

By observing the variations of crystal morphologies and the cation contents for the powder sample, a deeper understanding of ion exchanging could be achieved for the membranes. Table 3 illustrated that the ICP characterization results for the treated and original NaA zeolite crystals. It could be found that the content of Na^+ only increased slightly after NaCl solution treatment. However, according to the conservation of charge, the content of Na^+ should be unchanged in theory. The possible reasons were due to the presence of a small amount of residual ions in the crystals. For other cations, the content of Na^+ steadily decreased. For the monovalent salts, about 90%, 16% and 59% of Na^+ was replaced by Ag^+, Li^+ and K^+, respectively. In comparison, for the divalent salts, about 78%, 49%, 66% and 81% of Na^+ were replaced by Zn^{2+}, Mg^{2+}, Ca^{2+} and Ba^{2+}, respectively. In addition, the relative molar content of the positive charge was very close to 1 for all of the samples, matching well with the NaA zeolite structure.

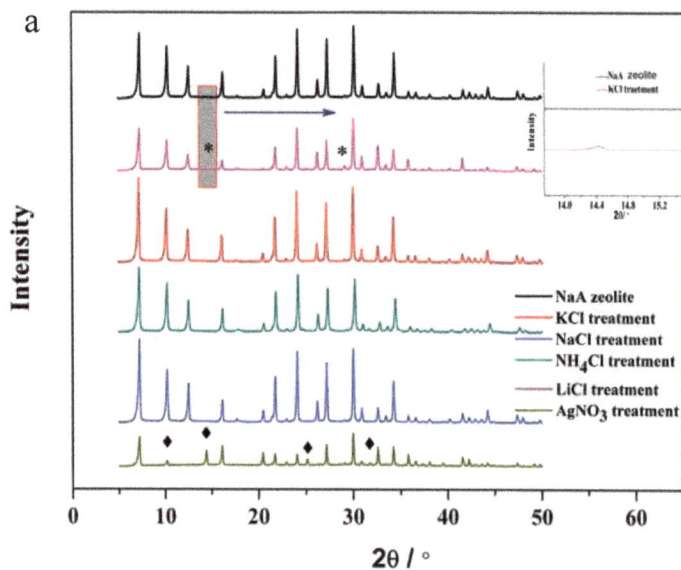

Figure 5. *Cont.*

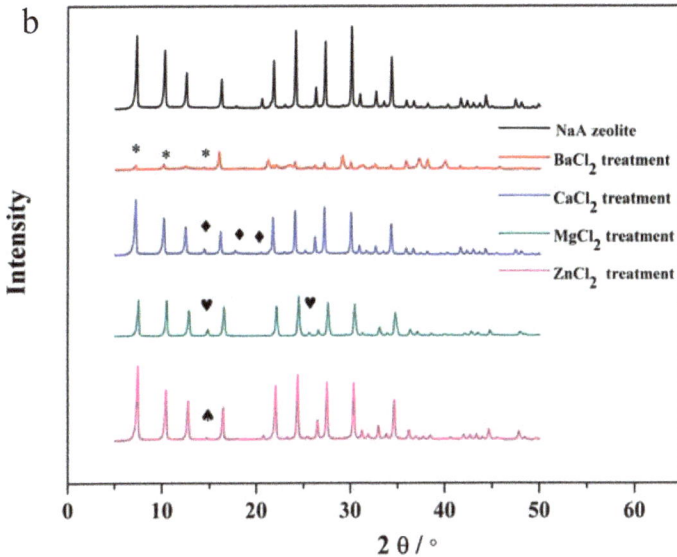

Figure 5. X-ray diffraction (XRD) patterns of NaA zeolite seeds and after treatment seeds with (**a**) monovalent cations and (**b**) divalent cations.

Table 3. Inductively coupled plasma (ICP) tests of NaA zeolites treated by different cations.

Type	Relative Content of Na	Relative Content of Positive Charge [a]
NaA zeolite	1.00	0.96
NaCl treatment	1.12	1.03
LiCl treatment	0.84	0.93
KCl treatment	0.41	0.97
$AgNO_3$ treatment	0.10	0.99
$ZnCl_2$ treatment	0.22	0.97
$MgCl_2$ treatment	0.51	1.06
$CaCl_2$ treatment	0.34	1.02
$BaCl_2$ treatment	0.19	0.99

[a] $[(Na^+ + M^+)/Al \text{ or } (Na^+ + 2M^{2+})/Al]$. Each element represents its molar content.

Since the cations were successfully incorporated in the NaA zeolites, the SEM technique was used to examine the morphology changes of the crystals, with the results being summarized in Figure 6. As suggested in Figure 6b,e,f, the salts of KCl, NaCl and LiCl had no destructive effects on the crystals; on the contrary, the crystals after the $AgNO_3$ treatment were rounded in some way (Figure 6c), as the distinct edges became blurred due to swelling. For the NH_4Cl treatment in Figure 6d, the crystals became less distinctive and smaller, indicating that acidic fluid dissolved the crystalline structures. Based on above discoveries, it was evident that the influence of Ag^+ on the NaA zeolite structure was relatively significant, which was consistent with the discoveries that were based on the SEM image for $AgNO_3$ solution in Figure 4b. Further, after the divalent salt treatments by $CaCl_2$, $BaCl_2$, $ZnCl_2$ and $MgCl_2$ solution in Figure 6g–j, not only did some damages occur to the particles during the replacements of cations, but also a small amount of amorphous substances were found on the crystal surface, which may be related to the flux decrement in the treated membranes, i.e., pore blocking effect due to the porosity loss.

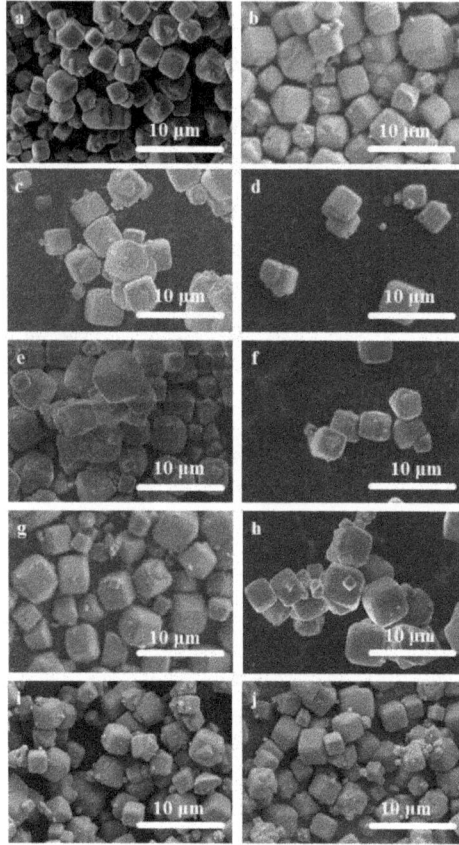

Figure 6. SEM images of NaA zeolite particles (**a**) and ion-exchanged zeolite with (**b**) KCl, (**c**) AgNO$_3$, (**d**) NH$_4$Cl, (**e**) LiCl, (**f**) NaCl, (**g**) CaCl$_2$, (**h**) BaCl$_2$, (**i**) ZnCl$_2$ and (**j**) MgCl$_2$ solution treatment.

3.4. The Concentration Effect on the Exchanged Membrane

Since the PV performance of the NaA zeolite membrane that was treated by divalent cations was not improved, only the concentration effect of monovalent cations on the PV performance was explored for the exchanged membrane. Table 4 summarized that the PV performances of different NaA zeolite membranes that were treated by various monovalent cations with several concentrations. For a high concentration of 0.10 M, it was evident that the separation factor of the NaA zeolite membrane was greatly improved after conducting KCl, AgNO$_3$ exchanging, where the decrease in the flux was caused due to the pore blocking effect of K$^+$ and Ag$^+$ cations. For the lower concentrations of K$^+$ and Ag$^+$ with a value of 0.05 M, the membrane separation factors were still very high (>10,000), and the corresponding fluxes became 4.45 and 5.87 kg·m^{-2}·h^{-1}, respectively, which were slightly lower than that for the original NaA zeolite membrane (4.96 kg·m^{-2}·h^{-1} and 6.51 kg·m^{-2}·h^{-1}, respectively). This was due to the faster exchanging rate with Na$^+$ ions that was caused by the higher concentration. As the highest separation factor occurred to the mild concentration, it was reasonable to conclude that higher exchanging rate may also induce some unbalanced structural variations in the membrane. Similarly, the exchange effect of NH$_4$Cl at a concentration of 0.05 M is better than that of 0.1 M by PV characterization. This was caused by the stronger acidity of the 0.1 M NH$_4$Cl solution, which dissolved the membrane layer to some extent. For LiCl, no differences were obtained for the two conditions as the leakages occurred due to the smaller kinetic diameter of Li^+, which permitted both water and ethanol molecules passing through the membranes. Based on the PV results for the treated NaA-type zeolite membranes, it was evident that the concentration of 0.05 M was more suitable for Ag^+ and K^+ to fix the membrane defects.

Table 4. Pervaporation (PV) performances of the NaA zeolite membranes treated by various concentrations of monovalent cation solutions, by dehydrating 90 wt % ethanol/water mixture.

No.	Cations	Concentration/M	Flux/kg·m^{-2}·h^{-1}	Separation Factor
M1	-	-	4.96 ± 0.03	449 ± 2.4
M1′	K^+	0.05	4.45 ± 0.02	$>10,000 \pm 198.2$
M1″	K^+	0.1	3.58 ± 0.02	1062 ± 12.8
M2	-	-	6.51 ± 0.03	548 ± 3.6
M2′	Ag^+	0.05	5.87 ± 0.03	$>10,000 \pm 198.2$
M2″	Ag^+	0.1	5.98 ± 0.03	2136 ± 50.9
M3	-	-	7.42 ± 0.03	306 ± 1.2
M3′	NH_4^+	0.05	4.95 ± 0.02	107 ± 0.2
M3″	NH_4^+	0.1	1.44 ± 0.01	38 ± 0.1

The SEM images of the NaA membranes that were treated by different concentrations were shown in Figure 7. It was observed that the ion exchanging rate difference due to the cation concentration clearly occurred to Ag^+ and NH_4^+ in Figure 7a,b and Figure 7c,d due to the larger cation sizes, which yielded a limited exchanging rate. For the size identical cation of K^+ in Figure 7e,f, and the difference between the two conditions was negligible due to the close similarity of crystals in two conditions.

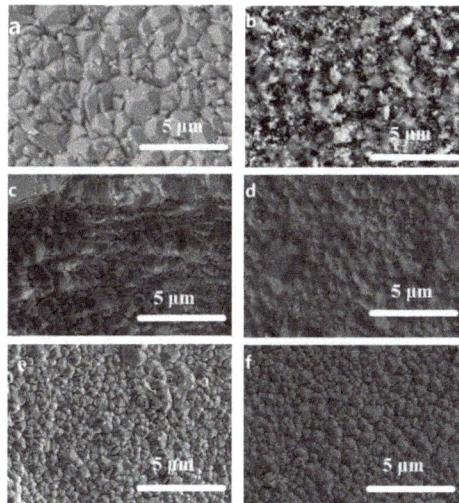

Figure 7. SEM images of NaA zeolite membrane treated by different concentration cations of $AgNO_3$ (**a**) 0.05 M, (**b**) 0.1 M, NH_4Cl (**c**) 0.05 M, (**d**) 0.1 M, KCl (**e**) 0.05 M and (**f**) 0.1 M.

3.5. PV Stability Test of NaA Zeolite Membrane

Since improved performances were found for the monovalent salts of KCl and $AgNO_3$, respectively, the PV stability tests were conducted for the treated NaA zeolite membranes that were derived under a salt concentration of 0.05 M for 24 h at 60 °C. The results were shown in Figure 8. As given in Figure 8a for KCl, the flux increased after the first immersing. This was due to the initial dissolution of some amorphous substance on the membrane surface, where the membrane intactness and the crystal structure were affected. Since a lower transport resistance occurred, the flux was increased without affecting the separation factor. After the secondary immersing, the membrane flux increased from 4.96 kg·m^{-2}·h^{-1} to 7.6 kg·m^{-2}·h^{-1}. This could be explained that the amorphous substances dissolved in the solution exposed the interstitial defects, thereby increasing the flux but reducing the selectivity. After the third immersing, the PV performance tended to be stable, where the flux remained at 6.5 kg·m^{-2}·h^{-1}, with the water content as high as 99.7%. In total, within 220 h, the KCl-exchanged NaA membrane maintained adequate fluxes with enough water purity in the permeate (above 99.6 wt %), and no significant damages occurred to the membrane. For the durability test of Ag$^+$ exchanged zeolite membrane in Figure 8b, the flux tended to decrease significantly with the extension of operation time. This was caused by the preferential adsorption of water molecules on the membrane surface, leading to the blockage of the pore channel, thereby, resulting in a decrease in flux. However, after immersing in the raw material fluid for a long time, the flux increased due to a partial dissolution and destruction of the amorphous material on the membrane surface. Since the membrane intactness was still neat, the permeate purity was hardly affected. In total, the NaA zeolite membrane that was treated with Ag$^+$, could keep a high performance in the PV test in 230 h. Since the amorphous substances were crucial to the performance evolution for the ion-exchanged membranes, the Energy-dispersive X-ray spectrometry (EDX) analysis was used to detect its chemical compositions, and the corresponding results were provided in Figure 9. As suggested, besides the elements (K or Ag) from the salts, the elements of Al, Si and O were also found, so the amorphous substances on the membrane surface were regarded as aluminosilicate.

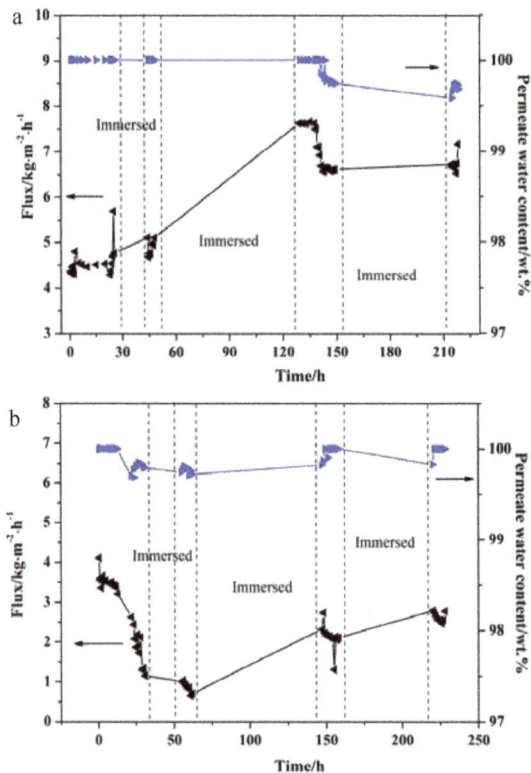

Figure 8. PV dehydration of 90 wt % ethanol/water mixture with different cation-treated NaA zeolite membranes (**a**) KCl solution treatment, (**b**) AgNO$_3$ solution treatment.

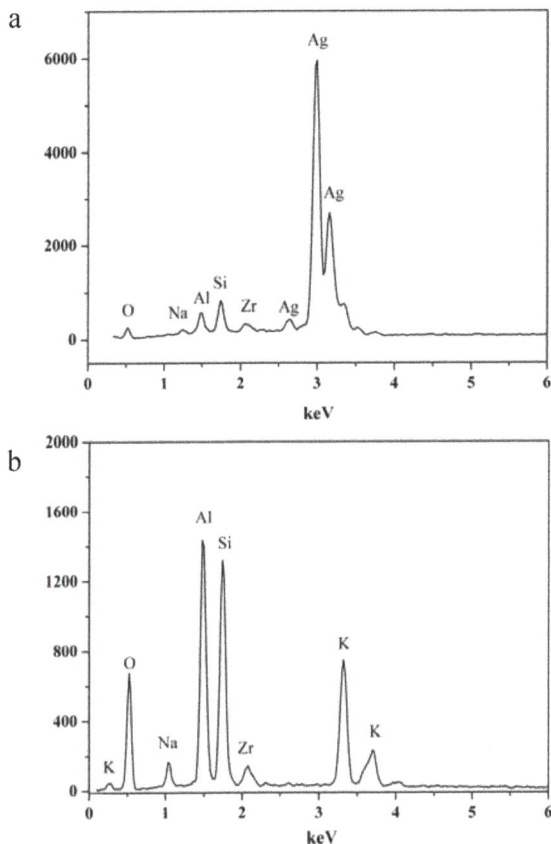

Figure 9. Energy-dispersive X-ray spectrometry (EDX) analysis for different cation-treated NaA zeolite membranes (**a**) AgNO₃ solution treatment, (**b**) KCl solution treatment.

3.6. Gas Permeation of NaA Zeolite Membranes

NaA zeolite membranes have exhibited high performance in liquid separation, but it has to admit that the membrane performance for gas separation is still far below the expectation, as the non-zeolitic pores and defects permit gas molecules passing through the membrane layer via Knudsen diffusion rather than by molecular sieving [22]. However, the gas permeation characterization provides effective insights for the structural changes in the membrane layer by excluding the hydrophilicity variation of the membrane surface due to the cation exchanges. In order to further elucidate the relative importance of the pore channel variation on NaA zeolite membrane due to the cation replacement, the contact angle was first used to inspect the variation of membrane hydrophilicity, and such a result was provided in Table 5. As suggested, all the hydrophilicities of the exchanged NaA zeolite membranes were considerably reduced in comparison with that of the original membrane, suggesting interplay between the pore variation and hydrophilicities in the complex behaviour of the exchanged membrane that was treated by different cations.

The gas permeation test was carried out on NaA ion-exchanged membrane with good PV dehydration performances, and the blocking effect of Ag⁺ and K⁺ was determined by the changes of permeance. Figure 10 illustrated the variation of permeance through the membrane as a function of the kinetic diameter of gases, using a driving pressure of 2.0 bar at 25 °C, where flow rates of single gases

(H_2, CO_2, N_2 and CH_4) were measured by bubble flow meter. As suggested, for the original NaA zeolite membrane, the experimental permeance order followed a Knudsen pattern of $H_2 > CH_4 > N_2 > CO_2$. For the smallest gas of H_2, the permeance reached up to 7.66×10^{-7} mol·m^{-2}·s^{-1}·Pa^{-1}, much higher than that for the rest gases of CO_2, N_2 and CH_4. For the most strongly adsorbed component of CO_2, the molecules were anchored in the zeolitic cavities due to dispersive interactions, and the transport was controlled by the molecular movements, thereby leading to a lower CO_2 permeance than that of N_2 and CH_4, which have higher thermal velocities. After the membrane was treated by $AgNO_3$ solution, despite the same permeance pattern of gases, all of the permeances were decreased by a factor of 35%. This was because the larger size of Ag^+ (1.26 Å) reduced the accessible pore passage. In comparison, for the membrane that was treated with KCl solution with even larger cation size (K^+ = 1.49 Å), the gas permeance decreased more sharply, and a factor of 90% was found for all of the gases. Since the effect of the hydrophilicity was excluded in gas permeation, the pore narrowing effect was further confirmed. On the other hand, it has been noted that the percentage of permeance drop for gases was more significant than that for the PV dehydration in Table 1. For instance, the drop of permeation flux for the KCl exchanges was only 10% in Table 1, which is far below the value of 90% in gas permeation. Clearly, the smaller decreasing was irrelevant of the variation of the membrane surface hydrophilicities, and it could be mainly attributed to the extra vaporing resistance in the PV operation, where the water molecules escape from the liquid feed to the water vapor phase. Since the permeance difference for PV dehydration is much smaller than that for the gas permeation, so it was reasonable to conclude that the vaporization process of phase change is much more influential than the vapor transport in the zeolite pores. A similar result was also found in the vapor permeation (VP) characterization using different feed mixtures of ethanol/water for Ag^+ and K^+, where the variation of the permeance and separation factor was provided in Figure 11. As suggested, a greater decreasing percentage of 60% for the permeance was obtained for the vapor mixture with 10 wt % water content, thus confirming that the importance of vaporization resistance in PV dehydration outweighed that of vapor phase transport in the pore channel under current conditions.

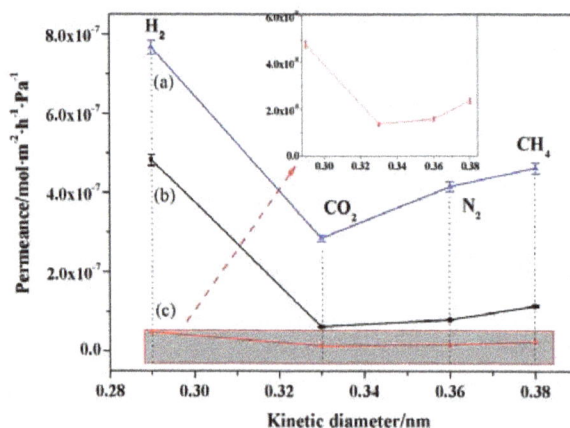

Figure 10. Gas permeance through (a) the original, (b) the $AgNO_3$ treated and (c) the KCl solution treated NaA zeolite membrane as a function of the gas kinetic diameter at 25 °C and 2.0 bar.

Table 5. The contact angles of the NaA zeolite membrane treated by different cations.

Cation Type	Exchanged NaA Zeolite Membranes	Contact Angle/θ
Original	NaA zeolite membrane	6.6°
Monovalent cation	Li$^+$ exchanged NaA zeolite membrane	46.5°
	K$^+$ exchanged NaA zeolite membrane	43.0°
	Ag$^+$ exchanged NaA zeolite membrane	99.6°
Divalent cation	Zn^{2+} exchanged NaA zeolite membrane	72.6°
	Mg^{2+} exchanged NaA zeolite membrane	18.3°
	Ca^{2+} exchanged NaA zeolite membrane	46.2°
	Ba^{2+} exchanged NaA zeolite membrane	33.5°

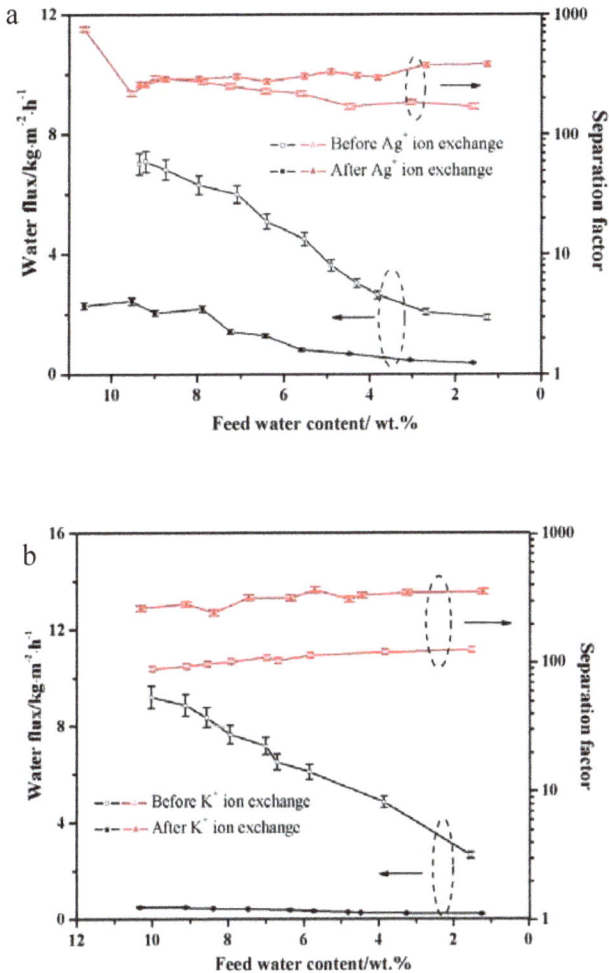

Figure 11. Vapor permeation (VP) dehydration of different ethanol/water mixtures using salt treated NaA zeolite membrane. (**a**) AgNO$_3$ solution and (**b**) KCl solution, where the vapor pressure was 1.4 bar under a temperature of 100 °C.

4. Conclusions

Different cation-exchanged NaA zeolite membranes were successfully derived on hollow fibers, using monovalent and divalent salts. For the monovalent cations, the membrane showed very stable membrane structure and dehydration performances, except for the very acidic salts of NH_4^+, as the dealumination effect occurred in the zeolite frameworks. For the rest monovalent cations of (Li^+, K^+ and Ag^+), it had been found that the permeation flux variations in dehydration largely related to the cations size. For the smaller sizes, like Li^+, a much higher permeation flux was obtained, but the separation factor was completely lost due to the over enlargement of accessible passages that was caused by the replacement of Na^+ by Li^+. For the larger cations, like K^+ and Ag^+, a mild decrement of permeation flux occurred to the treated membrane, but the separation factors became higher for both cations, which was because the accessible passage became narrower. For the divalent cation-treated membrane, the membrane permeance variation was complex, where the permeation flux was generally reduced due to the interplay between the cation size and charges. For larger cation (Ba^{2+}), the PV dehydration performances were slightly reduced for both permeation flux and separation factor, indicating that the membrane intactness was largely maintained and the unbalanced stresses was minimized. For other cations with smaller sizes, both the permeation flux and separation factor were reduced. This was because the number of Na^+ was unevenly exchanged with M^{2+} to balance the charge, so some cation vacancies were left in the treated membrane and thus induced unbalanced stresses in the membrane, thereby damaging the membrane intactness. To further verify the structure variation of the treated membrane by excluding the hydrophilicity factor occurring in the liquid separation, gas permeation using H_2, CO_2, N_2 and CH_4 was conducted for the treated NaA zeolite membrane by Ag^+ and K^+, and all of the permeance of gases was found to be decreased, which was consistent with the result that was obtained in the PV dehydration. Thus, the narrowing of the accessible pore channel due to the larger cation replacement in the membrane was firmly validated. However, it has been found that the decrement percentage (90% for K^+) in the gas permeation was far above that for the PV dehydration (10% for K^+), and this could be explained by the higher importance of vaporization resistance of water molecules from liquid to vapor phases, which outweighs the vapor transport resistance in the pore channels in the membrane.

Author Contributions: X. Gao and B. Gao conceived and designed the experiments; B. Gao and R. Shi and X. Wang performed the experiments; B. Gao and R. Ur Reshman analyzed the data; X. Gao and X. Gu supervised the project and provided research ideas. X. Gao and B. Gao contributed to the drafting of this paper.

Acknowledgments: This work is supported by National Natural Science Foundation of China (51502311), and the Doctorial Programs for Innovation and Entrepreneurship of Jiangsu Province.

Conflicts of Interest: The authors declare no conflict of interest.

References

1. Zeng, L.; Li, Z. A new process for fuel ethanol dehydration based on modeling the phase equilibria of the anhydrous $MgCl_2$ + ethanol + water system. *AIChE J.* **2015**, *61*, 664–676. [CrossRef]
2. Sanchez, O.J.; Cardona, C.A. Trends in biotechnological production of fuel ethanol from different feedstocks. *Bioresour. Technol.* **2008**, *99*, 5270–5295. [CrossRef] [PubMed]
3. Rasmussen, M.L.; Koziel, J.A.; Jane, J.; Pometto, A.L., III. Reducing bacterial contamination in fuel ethanol fermentations by ozone treatment of uncooked corn mash. *J. Agric. Food Chem.* **2015**, *63*, 5239–5248. [CrossRef] [PubMed]
4. Seo, D.; Takenaka, A.; Fujita, H.; Mochidzuki, K.; Sakoda, A. Practical considerations for a simple ethanol concentration from a fermentation broth via a single adsorptive process using molecular-sieving carbon. *Renew. Energ.* **2018**, *118*, 257–264. [CrossRef]
5. Zah, J.; Krieg, H.M.; Breytenbach, J.C. Single gas permeation through compositionally different zeolite NaA membranes: Observations on the intercrystalline porosity in an unconventional semicrystalline layer. *J. Membr. Sci.* **2007**, *287*, 300–310. [CrossRef]

6. Aoki, K.; Kusakabe, K.; Morooka, S. Preparation of oriented A-type zeolite membranes. *AIChE. J.* **2000**, *46*, 221–224. [CrossRef]

7. Caro, J.; Noack, M. Zeolite membranes-Recent developments and progress. *Microporous Mesoporous Mater.* **2008**, *115*, 215–233. [CrossRef]

8. Shao, J.; Zhan, Z.; Li, J.; Wang, Z.; Li, K.; Yan, Y. Zeolite NaA membranes supported on alumina hollow fibers: Effect of support resistances on pervaporation performance. *J. Membr. Sci.* **2014**, *451*, 10–17. [CrossRef]

9. Wenten, I.G.; Dharmawijaya, P.T.; Aryanti, P.T.P.; Mukti, R.R.; Khoiruddin, K. LTA zeolite membranes: Current progress and challenges in pervaporation. *RSC Adv.* **2017**, *7*, 29520–29539. [CrossRef]

10. Kita, H.; Horii, K.; Ohtoshi, Y.; Tanaka, K.; Okamoto, K.I. Synthesis of a zeolite NaA membrane for pervaporation of water-organic liquid mixtures. *J. Mater. Sci. Lett.* **1995**, *14*, 206–208. [CrossRef]

11. Shakarova, D.; Ojuva, A.; Bergstrom, L.; Akhtar, F. Methylcellulose-drected synthesis of nanocrystalline zeolite NaA with high CO_2 uptake. *Materials* **2014**, *7*, 5507–5519. [CrossRef] [PubMed]

12. Nikolakis, V.; Xomeritakis, G.; Abibi, A.; Dickson, M.; Tsapatsis, M.; Vlachos, D.G. Growth of a faujasite-type zeolite membrane and its application in the separation of saturated/unsaturated hydrocarbon mixtures. *J. Membr. Sci.* **2001**, *184*, 209–219. [CrossRef]

13. Cui, Y.; Kita, H.; Okamoto, K.I. Zeolite T membrane: Preparation, characterization, pervaporation of water/organic liquid mixtures and acid stability. *J. Membr. Sci.* **2004**, *236*, 17–27. [CrossRef]

14. Zhou, R.; Li, J.; Zhu, M.; Hu, Z.; Duan, L.; Chen, X. Synthesis and pervaporation performance of mordenite membranes on mullite tubes. *Chin. J. Inorg. Chem.* **2010**, *26*, 469–475.

15. Yang, S.; Arvanitis, A.; Cao, Z.; Sun, X.; Dong, J. Synthesis of silicalite membrane with an aluminum-containing surface for controlled modification of zeolitic pore entries for enhanced gas separation. *Processes* **2018**, *6*, 13. [CrossRef]

16. Dincer, E.; Culfaz, A.; Kalipcilar, H. Effect of seeding on the properties of MFI type zeolite membranes. *Desalination* **2006**, *200*, 66–67. [CrossRef]

17. Korelskiy, D.; Ye, P.; Zhou, H.; Mouzon, J.; Hedlund, J. An experimental study of micropore defects in MFI membranes. *Microporous Mesoporous Mater.* **2014**, *197*, 358. [CrossRef]

18. Liu, Y.; Wang, X.; Zhang, Y.; He, Y.; Gu, X. Scale-up of NaA zeolite membranes on alpha-Al_2O_3 hollow fibers by a secondary growth method with vacuum seeding. *Chin. J. Chem. Eng.* **2015**, *23*, 1114–1122. [CrossRef]

19. Guan, G.; Kusakabe, K.; Morooka, S. Gas permeation properties of ion-exchanged LTA-type zeolite membranes. *Sep. Purif. Technol.* **2001**, *36*, 2233–2245. [CrossRef]

20. Li, Y.; Chen, H.; Liu, J.; Yang, W. Microwave synthesis of LTA zeolite membranes without seeding. *J. Membr. Sci.* **2006**, *277*, 230–239. [CrossRef]

21. Li, Y.; Zhou, H.; Zhu, G.; Liu, J.; Yang, W. Hydrothermal stability of LTA zeolite membranes in pervaporation. *J. Membr. Sci.* **2007**, *297*, 10–15. [CrossRef]

22. Hasegawa, Y.; Nagase, T.; Kiyozumi, Y.; Hanaoka, T.; Mizukami, F. Influence of acid on the permeation properties of NaA-type zeolite membranes. *J. Membr. Sci.* **2010**, *349*, 189–194. [CrossRef]

23. Caro, J.; Albrecht, D.; Noack, M. Why is it so extremely difficult to prepare shape-selective Al-rich zeolite membranes like LTA and FAU for gas separation? *Sep. Purif. Technol.* **2009**, *66*, 143–147. [CrossRef]

24. Huang, A.; Caro, J. Cationic polymer used to capture zeolite precursor particles for the facile synthesis of oriented zeolite LTA molecular sieve membrane. *Chem. Mater.* **2010**, *22*, 4353–4355. [CrossRef]

25. Breck, D.W.; Eversole, W.G.; Milton, R.M.; Reed, T.B.; Thomas, T.L. Crystalline zeolites. I. The properties of a new synthetic zeolite, type A. *J. Am. Chem. Soc.* **1956**, *78*, 5963–5972. [CrossRef]

26. Shirazian, S.; Ashrafizadeh, S.N. LTA and ion-exchanged LTA zeolite membranes for dehydration of natural gas. *J. Ind. Eng. Chem.* **2015**, *22*, 132–137. [CrossRef]

27. Varela-Gandia, F.J.; Berenguer-Murcia, A.; Lozano-Castello, D.; Cazorla-Amoros, D. Hydrogen purification for PEM fuel cells using membranes prepared by ion-exchange of Na-LTA/carbon membranes. *J. Membr. Sci.* **2010**, *351*, 123–130. [CrossRef]

28. Wei, X.; Liang, S.; Xu, Y.; Sun, Y.; An, J.; Chao, Z. Methylcellulose-assisted synthesis of a compact and thin NaA zeolite membrane. *RSC Adv.* **2016**, *6*, 71863–71866. [CrossRef]

29. Xu, K.; Jiang, Z.; Feng, B.; Huang, A. A graphene oxide layer as an acid-resisting barrier deposited on a zeolite LTA membrane for dehydration of acetic acid. *RSC Adv.* **2016**, *6*, 23354–23359. [CrossRef]

30. Zhang, Y.; Avila, A.M.; Tokay, B.; Funke, H.H.; Falconer, J.L.; Noble, R.D. Blocking defects in SAPO-34 membranes with cyclodextrin. *J. Membr. Sci.* **2010**, *358*, 7–12. [CrossRef]

31. Yang, J.; Li, H.; Xu, J.; Wang, J.; Meng, X.; Bai, K.; Lu, J.; Zhang, Y.; Yin, D. Influences of inorganic salts on the pervaporation properties of zeolite NaA membranes on macroporous supports. *Microporous Mesoporous Mater.* **2014**, *192*, 60–68. [CrossRef]

32. Liu, D.; Zhang, Y.; Jiang, J.; Wang, X.; Zhang, C.; Gu, X. High-performance NaA zeolite membranes supported on four-channel ceramic hollow fibers for ethanol dehydration. *RSC Adv.* **2015**, *5*, 95866–95871. [CrossRef]

33. Nightingaljer, E.R. Phenomenological the ion solvation. Effective radii of hydrated ions. *J. Phys. Chem.* **1959**, *63*, 1381–1387. [CrossRef]

34. Shannon, R.D. Revised effective ionic radii and systematic studies of interatomie distances in halides and chaleogenides. *Acta Crystallogr. Sect. A* **1976**, *32*, 751–767. [CrossRef]

processes

MDPI

Article

Estimation of Pore Size Distribution of Amorphous Silica-Based Membrane by the Activation Energies of Gas Permeation

Guozhao Ji [1,2,*], Xuechao Gao [3,4], Simon Smart [1], Suresh K. Bhatia [4], Geoff Wang [4], Kamel Hooman [5] and João C. Diniz da Costa [1]

1   FIM²Lab-Functional Interfacial Materials and Membranes Laboratory, School of Chemical Engineering, the University of Queensland, Brisbane, QLD 4072, Australia; s.smart@uq.edu.au (S.S.); j.dacosta@eng.uq.edu.au (J.C.D.d.C.)
2   School of Environmental Science and Technology, Dalian University of Technology, Dalian 116024, China
3   State Key Laboratory of Materials-Oriented Chemical Engineering, College of Chemical Engineering, Nanjing Tech University, 5 Xinmofan Road, Nanjing 210009, China; xuechao.gao@njtech.edu.cn
4   School of Chemical Engineering, the University of Queensland, Brisbane, QLD 4072, Australia; s.bhatia@uq.edu.au (S.K.B.); gxwang@uq.edu.au (G.W.)
5   School of Mechanical and Mining Engineering, the University of Queensland, Brisbane, QLD 4072, Australia; k.hooman@uq.edu.au
*   Correspondence: guozhaoji@dlut.edu.cn or guozhao.ji@uqconnect.edu.au; Tel.: +86-0411-8470-6679; Fax: +86-0411-8470-6679

Received: 31 October 2018; Accepted: 21 November 2018; Published: 23 November 2018

Abstract: Cobalt oxide silica membranes were prepared and tested to separate small molecular gases, such as He (d_k = 2.6 Å) and H_2 (d_k = 2.89 Å), from other gases with larger kinetic diameters, such as CO_2 (d_k = 3.47 Å) and Ar (d_k = 3.41 Å). In view of the amorphous nature of silica membranes, pore sizes are generally distributed in the ultra-microporous range. However, it is difficult to determine the pore size of silica derived membranes by conventional characterization methods, such as N_2 physisorption-desorption or high-resolution electron microscopy. Therefore, this work endeavors to determine the pore size of the membranes based on transport phenomena and computer modelling. This was carried out by using the oscillator model and correlating with experimental results, such as gas permeance (i.e., normalized pressure flux), apparent activation energy for gas permeation. Based on the oscillator model, He and H_2 can diffuse through constrictions narrower than their gas kinetic diameters at high temperatures, and this was possibly due to the high kinetic energy promoted by the increase in external temperature. It was interesting to observe changes in transport phenomena for the cobalt oxide doped membranes exposed to H_2 at high temperatures up to 500 °C. This was attributed to the reduction of cobalt oxide, and this redox effect gave different apparent activation energy. The reduced membrane showed lower apparent activation energy and higher gas permeance than the oxidized membrane, due to the enlargement of pores. These results together with effective medium theory (EMT) suggest that the pore size distribution is changed and the peak of the distribution is slightly shifted to a larger value. Hence, this work showed for the first time that the oscillator model with EMT is a potential tool to determine the pore size of silica derived membranes from experimental gas permeation data.

Keywords: activation energy; pore size distribution; silica based membrane; effective medium theory; gas separation

1. Introduction

Microporous silica (SiO_2) membrane has attracted significant research attention for gas separation, including H_2 separation which is an energy carrier of great interest whilst also being a precursor for the production of chemical products [1–4]. The development of silica membranes was born out of a desire to control pore sizes at a molecular level. By tuning the pore size of silica membranes, the permeation of gas molecules is affected. As a result, gases with smaller kinetic diameters tend to permeate fast through the silica membranes whilst the permeation of the larger kinetic diameter gases are very slow or hindered. The initial development of silica membranes in the late 1980s and early 1990s was based on a sol-gel method using the conventional silica precursor tetraethyl orthosilicate (TEOS) [5–7]. A major development in silica membranes occurred with embedding metal oxides in the silica matrix. Since then, there have been a significant number of reports on metal oxide silica membranes, including those containing oxides of nickel [8,9], cobalt [10–14], niobium [15], palladium [16], zirconium [17], titanium [18] and aluminum [19]. A special feature of these membranes were high gas selectivities at high temperatures, which is of interest in several chemical engineering process applications. Subsequently, research groups reported on the use of binary mixed metal oxide in silica membranes, including those containing cobalt iron oxides [20,21], palladium cobalt oxide [22] and lanthanum cobalt oxide [23]. This allowed interesting pore size tailoring by reducing only a single metal oxide in the silica matrix, whilst the other one was left in the oxidized state. Nevertheless, based on the number of publications, cobalt oxide silica membranes remained the most popular metal oxide silica membranes, due to very high gas selectivity close to values of 1000 [24].

According to the molecular sieve mechanism, the most important parameter of a silica based membrane is the pore size. The determination of pore sizes in silica derived matrices is generally carried out by conventional characterization techniques, such as N_2 adsorption. This technique uses xerogels (i.e., dry gel) which is generally gelled in bulk. As bulk evaporation and gelation is very slow compared to thin film (30 nm thickness) deposition on substrates, it is known that the structures of xerogel and thin film are not equivalent. For instance, the properties of a deposited thin film may be quite different, due to non-equivalent gelation and drying conditions [25]. Hence, N_2 adsorption on bulk silica xerogel is generally used as a qualitative measure of thin film structures. Other more complex techniques have been applied to characterize xerogels and thin films. For instance, using Positron Annihilation Spectroscopy Duke and co-workers demonstrated that amorphous silica matrices have narrow tri-modal pore size distribution (PSD) at 2.5–4 Å, 6.7–7.8 Å and 12–14 Å [26]. Another technique to characterize the pore sizes of thin films deposited on substrate is molecular probing which uses gas molecules of different kinetic diameters to determine the average pore size, by plotting permeance versus the kinetic diameter of the gas tested.

A number of mass transfer studies suggested the gas permeation value was a result of membrane pore size and gas molecular size. In other words, if the pore size distribution (PSD) is known, the gas permeation data could be calculated from a proper model. The question in this work is whether it is possible to use a permeation model to derive the PSD, instead of the conventional approach of using permeation data of particular gases to determine the PSD. In a membrane gas permeation test, the measurable parameters are generally the gas permeance ($mol·m^{-2}·s^{-1}$) and the apparent activation ($kJ·mol^{-1}$) energy. Even though permeability ($mol·m^{-1}·s^{-1}$) is reported in some studies, it is not very reliable, due to the technical difficulty in accurately measuring the membrane thickness. Permeance may not be an ideal data for evaluating pore size, because it also depends on the membrane thickness. Fortunately, the apparent activation energy is not a function membrane thickness, and in principle it only depends on the pore size (pore diameter) and the size of gas molecule (molecular size).

If the gas permeation across a membrane follows an activated transport model, activation energy is a measure of the energy required by a gas molecule to permeate through a pore as follows:

$$E_a = \frac{d \ln(P/l)}{d(RT)^{-1}} \tag{1}$$

where P is permeability and l the membrane thickness. A number of studies reported apparent activation energies for different gases through the silica based membrane (Table 1). An interesting phenomenon is that H_2 and He generally exhibit positive apparent activation energy, but other gases present negative apparent activation energy. This current study will employ the correlation between pore size and apparent activation energy to evaluate the pore size distribution of silica based membrane by the measured apparent activation energy, and use this pore size distribution to explain the observed transport phenomena.

Table 1. Apparent activation energies of gas permeation across silica based membranes.

Apparent Activation Energy (kJ·mol^{-1})					Reference
H_2	He	CO_2	N_2	Ar	
16.4~17.1	16.2~17.1	-	-	-	[27]
2.2	2.8	−9.8	−5.4	-	[28]
15.4~19.4	-	-	8.5~11.9	-	[29]
12.8	20.7	−20	-	-	[24]
-	9.5	-	−5.0	-	[11]
-	13.6	-	-	-	[30]
6	-	-	-	-	[31]
4.91	4.70	−2.91	−1.41	-	[14]
10.1	-	−3.1	-	−1.9	[32]
14.1	7.7	-	12.3	-	[33]
~9	7.2	-	-	-	[34]

2. Theory of Mass Transfer Through Membrane

A general model of mass transfer is always expressed as a coefficient multiplied by a driving force as

$$J = k \times f \tag{2}$$

where J is the mass transfer flux, k is the mass transfer coefficient and f is the driving force. The driving force of the gas permeation is usually the chemical potential, which can be reduced to gradient of pressure, concentration, depending on the system. The mass transfer coefficient can be permeability, diffusivity or other term depending on the term used for driving force. For membrane processing of gases, the pressure difference and permeate flux are the quantities generally determined from experimental measurements, so the most common form in membrane mass transfer is

$$J = \left(\frac{P}{l}\right)\Delta p \tag{3}$$

where membrane thickness l is lumped together with permeability P into a term called permeance (P/l) which is a convenient form avoiding the difficulty in accurately measuring the thickness of silica thin films. Sometimes, silica thin films interpenetrate into the pores of porous of substrates, depending on the property of the sol and substrate. Hence, the membrane thickness is not always easy to determine.

2.1. Activated Transport

For ultra-microporous ($d_p < 5$ Å) material, the pore size of the silica thin-films is close to the Lennard Jones (L-J) diameter which results in overlap of the L-J potentials from the pore wall. In such cases, the L-J potential inside a pore could be several magnitudes higher than that of a flat surface. Therefore, the motion of gas molecules in a pore is not as free as in the bulky gas phase, instead it must be significantly affected by the potential fields. Pores of size close to the diameter of a gas molecule generally present a positive potential for the gas molecule. The potential becomes even higher with narrower pore size. To overcome the potential barrier inside a pore, the gas molecule needs activation energy to complete a successful penetration. Therefore, this type of mass transfer is called activated transport. Gas transport through silica based membranes generally follows activated transport [10,35,36]. The original expression of mass transfer across the membrane is given as [37–39]

$$J = -qD\frac{1}{RT}\frac{d\mu}{dz} \tag{4}$$

where J is the permeate flux, q the molar concentration of gas in the pore, D the diffusivity, R the gas constant, T the temperature, μ the chemical potential and z the space coordinate in the permeation direction.

Assuming equilibrium between the membrane adsorbate and the bulk gas phase, the following relationship for the chemical potential [40,41] is applicable

$$\mu = \mu_0 + RT \ln p \tag{5}$$

where p is the absolute pressure.

Substituting Equation (5) in Equation (4) gives

$$J = -D\frac{d\ln p}{d\ln q}\frac{dq}{dz} = -qD\frac{d\ln p}{dq}\frac{dq}{dz} = -D\Gamma\frac{dq}{dz} = -D\frac{1}{1-\theta_{ocp}}\frac{dq}{dz} \tag{6}$$

where $\Gamma = q(d\ln p/dq)$ is a thermodynamic factor, obtained as $\Gamma = 1/(1-\theta_{ocp})$ for a Langmuir isotherm. θ_{ocp} is the fractional occupancy of the adsorbate (the ratio of adsorbate concentration to the maximum adsorbate concentration). As the silica membrane is designed for separating hot gas (200–500 °C), the high temperature adsorption is very weak ($\theta_{ocp} \approx 0$), so that the thermodynamic factor is $\Gamma = 1/(1-\theta_{ocp}) \approx 1$, and the permeate flux is finally approximated as

$$J = -D\frac{dq}{dz} \tag{7}$$

The diffusivity D is a function of temperature, and is activated. The temperature dependence generally obeys the Arrhenius relation

$$D = D_0 \exp\left(-\frac{E_d}{RT}\right) \tag{8}$$

where D_0 is a pre-exponential coefficient, and E_d is an activation energy. At high temperatures, the adsorption is weak and the Langmuir isotherm approaches Henry's law [42]

$$q = Kp \tag{9}$$

K is the Henry's constant and a function of temperature, following the van't Hoff relation:

$$K = K_0 \exp\left(\frac{Q}{RT}\right) \tag{10}$$

where K_0 is a pre-exponential coefficient, and Q is the heat of adsorption.

Equations (7)–(10) can be combined as

$$J = -D_0 K_0 \exp\left(-\frac{E_d - Q}{RT}\right)\frac{dp}{dz} = -D_0 K_0 \exp\left(-\frac{E_a}{RT}\right)\frac{dp}{dz} \tag{11}$$

E_a is the apparent activation energy for activated transport, which has a relation to E_d and Q in Equation (12)

$$E_a = E_d - Q \tag{12}$$

Assuming a uniform pressure gradient, Equation (11) is simplified to

$$J = D_0 K_0 \exp\left(-\frac{E_a}{RT}\right) \frac{\Delta p}{l} \tag{13}$$

The permeance (P/l) is the coefficient between flux and pressure drop according to Equation (3), and is obtained as

$$\left(\frac{P}{l}\right) = \frac{D_0 K_0}{l} \exp\left(-\frac{E_a}{RT}\right) \tag{14}$$

the apparent activation energy is experimentally derived from the permeance measured at different temperatures. If Equation (14) is logarithmically transformed

$$\ln\left(\frac{P}{l}\right) = \ln\left(\frac{D_0 K_0}{l}\right) - \frac{E_a}{RT} \tag{15}$$

The slope of $\ln(P/l)$ to $-1/(RT)$ gives the apparent activation energy as indicated by Equation (1).

2.2. Potential in Cylindrical Pores

The motion of gas molecules in micro-pores is strongly affected by the potential inside the pore. The potential between two atoms based on the Lennard-Jones (L-J) potential is [43]

$$\phi(\rho) = 4\varepsilon \left[\left(\frac{\sigma}{\rho}\right)^{12} - \left(\frac{\sigma}{\rho}\right)^{6} \right] \tag{16}$$

where ϕ is the potential between two atoms, σ the L-J collision diameter, ε the L-J well depth, and ρ the distance between the centers of two atoms. For two different atoms, the L-J collision diameter and L-J well depth are estimated by the Lorentz-Berthelot mixing rules [44,45]

$$\sigma_{AB} = \frac{1}{2}(\sigma_A + \sigma_B) \tag{17}$$

$$\varepsilon_{AB} = \sqrt{\varepsilon_A \varepsilon_B} \tag{18}$$

In this work the pore is simplified as a cylinder consisting of L-J particles. It is important to clarify the definition of pore size beforehand (Figure 1). The diameter (d) of the pore is defined as the distance from an oxygen particle center to the opposite oxygen particle center. The inner diameter d_i (surface to surface) is more relevant in size sieving, but is not unambiguously defined for soft-sphere L-J particles. Therefore, d is used instead of d_i throughout this work and pore radius is $r_p = d/2$.

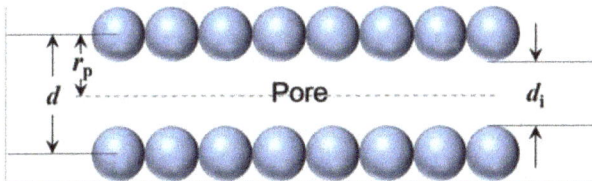

Figure 1. The pore of the silica membrane.

Equation (16) has provided the potential between two atoms. The potential in a pore is actually the summation from all the atoms which formed the pore. Calculating the potential inside a pore

145

could be conducted by integrating all the atom-atom potentials geometrically. By assuming the pore is cylindrical, with radius r (Figure 2), the potential distribution inside a pore is given by

$$
\begin{aligned}
\phi(r, r_p) &= 8\varepsilon\eta_d r_p \int_0^{\frac{L}{2}} \int_0^{2\pi} \left[\left(\frac{\sigma}{\rho}\right)^{12} - \left(\frac{\sigma}{\rho}\right)^6 \right] d\theta dz \\
&= 8\varepsilon\eta_d r_p \int_0^{\frac{L}{2}} \int_0^{2\pi} \left[\frac{\sigma^{12}}{\left(r_p^2 + r^2 - 2r_p r \cos\theta + z^2\right)^6} - \frac{\sigma^6}{\left(r_p^2 + r^2 - 2r_p r \cos\theta + z^2\right)^3} \right] d\theta dz
\end{aligned}
\tag{19}
$$

where r_p is the pore radius, L the pore length, η_d the surface atomic density on the pore wall and r the radial coordinate with a range of $r \in [0, r_p)$. The meaning of all the variables in Equation (19) are visually explained in Figure 2.

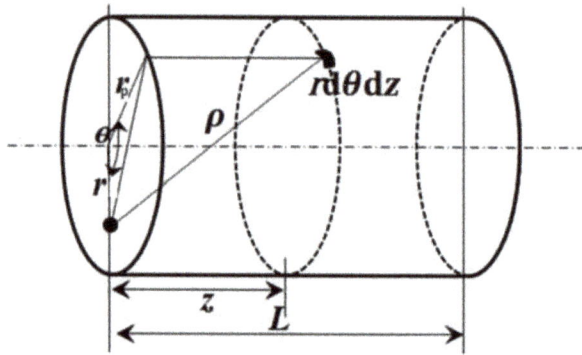

Figure 2. Geometrical simplification of a hydrogen molecule in a cylindrical pore.

An example of the H_2 potential distribution based on Equation (19) inside a silica pore with varied pore sizes is displayed in Figure 3. It is the nature of gases to follow thermodynamic laws and seek the lowest potential as displayed in Figure 4.

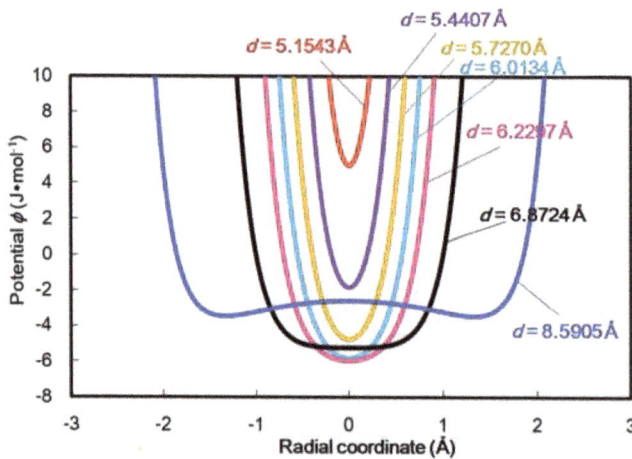

Figure 3. Hydrogen potential distribution inside cylindrical silica pores with different pore sizes.

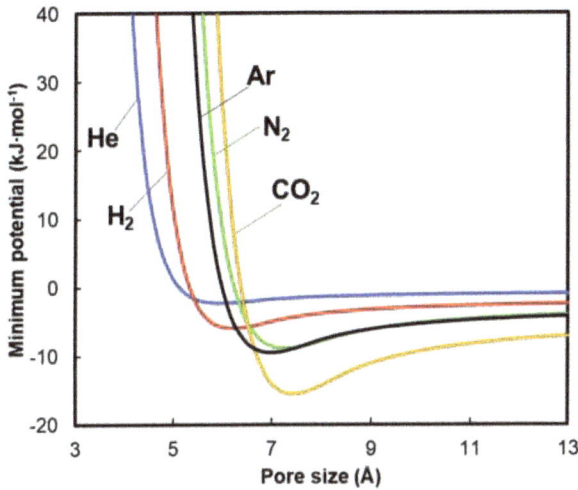

Figure 4. The minimum potential with respect to pore size, for different gases.

2.3. Pore Size Distribution (PSD) Estimation of A Membrane

Since a silica membrane is generally amorphous, there is a distribution of pore sizes and the pore size distribution of membrane differs from that in the sample powder for N_2 adsorption test. The first reason is attributed to the effect from the substrate, which may consequently affect the pore structure of the membrane. The second reason is that the membrane has several layers and each layer has a different thermal history which results in variation in pore structure [46].

The pore size distribution of sample powder is usually characterized by gas adsorption which is operated at a low temperature. Pores smaller than adsorbate kinetic diameter are hardly detected by adsorption, as the L-J potential in a small pore is too repulsive for adsorbate molecules to enter. However, gas permeation is operated at high temperatures, for which gas molecules travel with high kinetic energy and have the possibility to access pores smaller than their diameter. Since such pores also contribute to the total permeation, the pore size distribution can be extracted from permeation data.

To this end, we consider a Rayleigh distribution of pore sizes for the silica matrix [47–53]

$$f(r_p) = \frac{(r_p - r_o)}{(r_m - r_o)^2} \exp\left[-\frac{(r_p - r_o)^2}{2(r_m - r_o)^2}\right] \tag{20}$$

where $f(r_p)$ is the distribution function, r_p is the possible pore radius, r_o is the minimum pore radius and r_m is the modal pore radius. When there is a pair of (r_0, r_m), $f(r_p)$ is the probability density of finding pores of which the size is r_p. The oscillator model computes the permeate flow (mol·s^{-1}) of a pore with size r_p, such that

$$i = -\pi r_p^2 K(r_p) D(r_p) \nabla p \tag{21}$$

and the conductance for permeation through a pore with size r_p is defined as

$$\lambda(r_p) = \pi r_p^2 K(r_p) D(r_p) \tag{22}$$

The details of computing $D(r_p)$ and $K(r_p)$ are provided by Bhatia et al. [2,53,54]. The contribution of any pore size to the total bulk permeate flow depends on the pore size distribution. The total permeate bulk flow is the summation of the flow from all the pores. To calculate the total permeate flow, the effective medium theory (EMT) is generally applied [55–58]

$$\int_{r_{min}}^{r_{max}} f(r_p)\frac{\lambda(r_p) - \lambda_e}{\lambda(r_p) + \left(\frac{N}{2} - 1\right)\lambda_e}dr_p = 0 \tag{23}$$

where $f(r_p)$ is the pore size number distribution, N is the connectivity of pores, r_{min} and r_{max} are the minimum and maximum pore size. The solution λ_e of Equation (23) is the effective conductance of the entire membrane, which can be used to calculate the bulk permeate flux

$$J = -\frac{\varepsilon\lambda_e}{\tau_T\pi\langle r_p^2\rangle}\nabla p \tag{24}$$

where ε is the porosity, τ_T is the tortuosity and $\pi < r_p^2 >$ is the mean pore cross section area. For a disordered network, the tortuosity is normally given by [59,60]

$$\tau_T = \frac{3(N + 1)}{N - 1} \tag{25}$$

Equation (24) provides the bulk permeate flux. For a specific Rayleigh pore size distribution $f(r_o,r_m,r_p)$, the bulk permeate flux at different temperatures can be obtained from the model above. When permeate fluxes at different temperatures are available, the bulk apparent activation energy is then calculated by Equations (1) and (15). This bulk apparent activation energy shows one-to-one correspondence to the pore size distribution $f(r_o,r_m,r_p)$. By varying r_o and r_m of Equation (20), the calculated bulk apparent activation energy will change. The deviation of calculated apparent activation energy to the experimental apparent activation energy is defined as

$$\frac{|E_{a,osc} - E_{a,exp}|}{E_{a,exp}} = \frac{|E_{a,osc}^{He} - E_{a,exp}^{He}|}{E_{a,exp}^{He}} + \frac{|E_{a,osc}^{H_2} - E_{a,exp}^{H_2}|}{E_{a,exp}^{H_2}} \tag{26}$$

It is assumed that the value (r_o,r_m) which gives the activation energy closest to the experimental activation energy provides the best representation of the membrane pore size distribution.

3. Experiment

3.1. Sol-Gel Preparation and Membrane Coating

The cobalt oxide silica sol-gel method involved mixing several chemical precursors is reported elsewhere [11]. Briefly, cobalt nitrate hexahydrate ($Co(NO_3)_2 \cdot 6H_2O$, 98%, Sigma-Aldrich) was dissolved in 30 vol.% aqueous hydrogen peroxide (H_2O_2, 30 wt%) and ethanol (EtOH, 99%, AR grade). Then tetraethoxysilane (TEOS, 99%, Fluka) was added drop wise to form a final molar ratio of 255 ethanol:4 TEOS:1 $Co(NO_3)_2 \cdot 6H_2O$:9 H_2O_2:40 H_2O. The sol-gel was dip-coated on an alumina support purchased from the Energy Centre of the Netherlands (ECN). Each cobalt oxide silica layer was calcined up to 630 °C in air at a ramping rate of 1 °C·min^{-1} and with a dwell time of 4 h. A total of five layers were sequentially dip-coated on the outside surface of the alumina support and calcined to ensure defect-free membranes. The alumina support was composed of asymmetric layers. The top layer was derived from γ-alumina with an intrinsic pore size distribution around 4 nm, and was coated on a mechanically robust porous α-alumina substrate with the following dimensions: Length ~70 mm, external and internal diameters of 14 and 10 mm, respectively.

3.2. Gas Separation Measurement

The membrane module testing configuration is shown in Figure 5. A membrane tube was initially sealed by a set of Swagelok fittings, nut and graphite ferrule. After sealing, the membrane effective length was 48 mm. The membrane was finally assembled into a cylindrical module, tested and adjusted until a leak free membrane/seal interface was achieved. Once the membrane was leak free, it was tested for He and H_2 permeation from 100% of the pure feed gas. Then the mixed gases He/CO_2 and H_2/Ar were tested down to 30% He and 20% H_2 mixtures, respectively. The gas mixture testing included strong adsorbing gas (such as CO_2) and non-adsorbing gas (He) for the permeation for He/CO_2. The flow rate was measured by a bubble flow meter and the composition of the mixed gas was determined by Shimadzu gas chromatograph (GC).

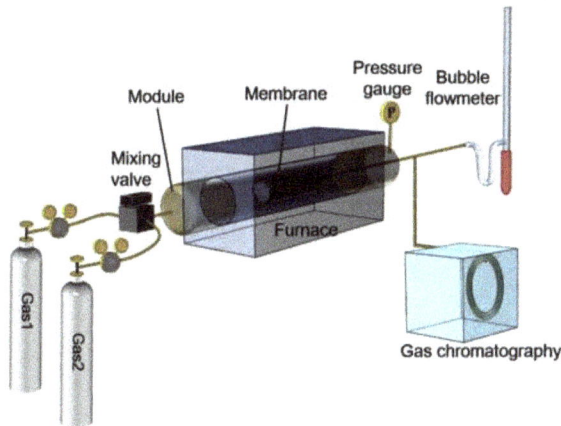

Figure 5. Diagram of the permeation/separation experimental set up.

4. Results and Discussion

In view of the asymmetric nature of the cobalt oxide silica membranes, gas permeation was also carried out on the substrate prior to dip coating with cobalt oxide silica sol-gel. This test was conducted to investigate the resistance level of the support alone. Figure 6 shows that the permeance of He and H_2 across the uncoated support was ~1 × 10^{-5} mol·m^{-2}·s^{-1}·Pa^{-1}, whilst for coated cobalt oxide silica support as a full membrane was ~2 × 10^{-8} mol·m^{-2}·s^{-1}·Pa^{-1}. If the resistance is defined as the reciprocal of permeance, the resistance in the substrate only accounts for ~0.2% of the total resistance which is much lower than the permeation experimental error of ±~8%. Therefore, the resistance of the substrate is ignored in the modelling in this work, and the transport of gases is assumed to be limited by cobalt oxide silica layers only.

Figure 7 shows the permeation test for 200 to 500 °C for He/CO_2 binary mixtures from 90% He (10% CO_2) down to 10% He (90% CO_2). The feed pressure was kept constant at 6 atm, and controlled by a back pressure valve. The permeate side was open to the atmosphere, to maintain pressure at 1 atm. More details about the mixed gas permeation could be found in Ji et al. [61]. The CO_2 single gas permeation proved to be too slow to be measured, and was considered below the measurable region of the bubble flow meter. It is interesting to observe in Figure 7 that the strong adsorbing gas of CO_2 did not greatly affect the permeation of He, which is a non-adsorbing gas under the tested conditions. The competitive adsorption effect is generally more prevalent in low temperature gas separation conditions, where the stronger adsorbing gas covers more surface and pores of the membrane, hindering the transport of the other gas in the membrane pores [62,63]. However, gas permeation was carried out at high temperatures, and the competitive adsorption effect barely affects the He permeation. It was observed that He permeate flow rate increased linearly as a function of the He feed fraction (or partial pressure), yielding a constant He permeance.

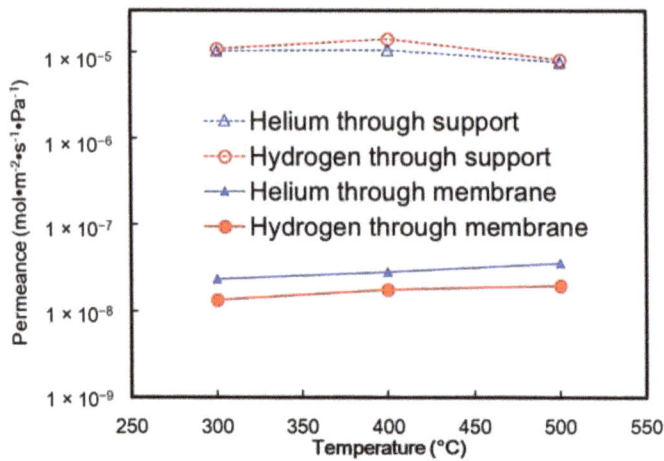

Figure 6. He and H_2 permeances through cobalt oxide silica coated substrate and uncoated substrate as a function of temperature.

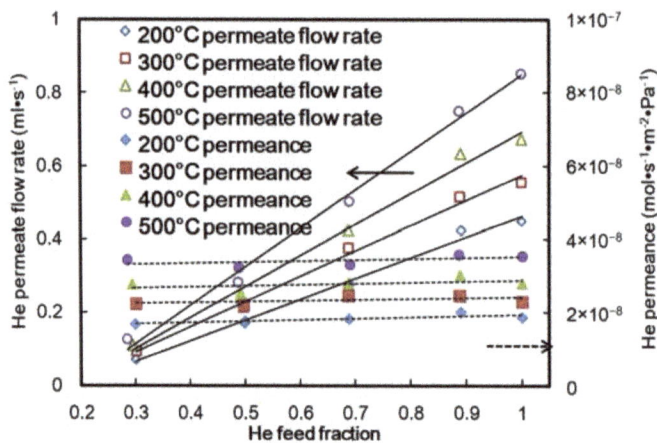

Figure 7. He permeation performance of cobalt oxide silica membrane.

Further work was carried out for H_2 and Ar mixtures. Due to the catalytic effect of the cobalt oxide, mixtures of H_2 and CO_2 were not considered to avoid the reverse reaction of the water gas shift reaction, which would result in the production of CO and water vapor [64]. This would generate a quaternary mixture of H_2, CO_2, CO and H_2O. Hence, Ar was used as a subrogated molecule to maintain a binary gas mixture of H_2/Ar instead of a multiple transient gas mixture. In addition, Ar ($d_k = 3.41$ Å) and CO_2 ($d_k = 3.47$ Å) [65,66] have similar kinetic diameters showing similar permeation performance through a silica membrane [67,68]. Single gas permeation was also carried for H_2 and Ar prior to the binary mixture testing. Again, single Ar permeation did not produce permeate flows, thus below the minimum detectable level similar to the case of CO_2 as mentioned above. After the H_2 single gas permeation test, H_2 and Ar mixture separation was measured for feed compositions from 90%, 70%, 50%, and 30% down to 20% H_2. The H_2 feed fraction was reduced step-wise after each day of testing [61].

The H_2 permeance rate is shown in Figure 8. Unlike the test for He, it is interesting to observe that the H_2 permeance is not constant. These unusual results suggest that H_2 is possibly reducing the cobalt oxide embedded in the silica matrix, thus leading to a different micropore structure. Based on Ballinger et al.'s study, the exposure to H_2 could reduce the metal oxide and widen the neck of pores [22]. To ensure a systematic study, each feed fraction was tested in a full single day, and this experiment was carried out in order from 100% H_2 down to 20% H_2. The results in Figure 8 implied that the reaction between membrane material and H_2 is a function of H_2 exposure time and H_2 concentration in the feed gas. This will be discussed in detail in the following section.

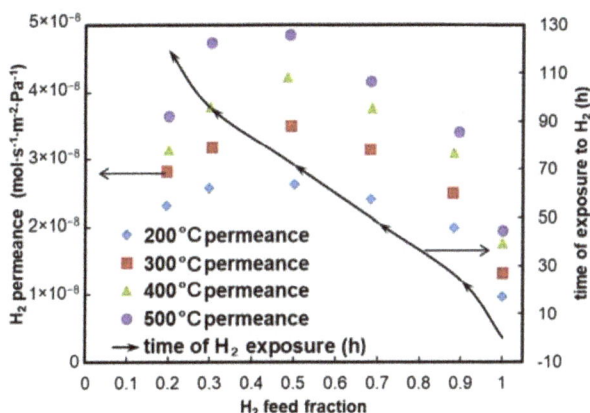

Figure 8. The permeation performance of H_2.

The separation of He/CO_2 shows that the presence of CO_2 did not significantly hinder the He permeance, though the flow rates increased with He feed partial pressure, which in turn caused a higher driving force or pressure difference. The apparent activation energy for He permeation has an average value of 6.3 kJ·mol^{-1} with 4.8% fluctuation for all the feed fractions.

However, it is interesting to observe that the H_2 permeance was higher for gas mixtures than for the pure gas. Miller et al. [69] and Ji et al. [61] reported that cobalt oxide embedded in silica undergoes reduction after exposure to H_2, resulting in pore size enlargement. As the single H_2 gas permeation was tested at the beginning in this series of experiments, the permeance was the lowest as the membrane was just exposed to H_2 for a short time and possibly in an oxidized state. After one day (24 h) of exposure to H_2, mixed gas separation was carried out from 90% H_2 and was observed that the membrane permeance increased by about two times in the reduced state than in the oxidized state, thus confirming the redox effect of H_2 on the membrane.

From Equation (15), the apparent activation energies before and after reduction can be obtained from the permeance at different temperatures. The apparent activation energy was reduced from 7.3 kJ·mol^{-1} (first day testing using pure H_2) to 5.2 kJ·mol^{-1} (5th day testing using 30% H_2). Activation energy is an energy barrier to overcome during permeation. Figures 3 and 4 have demonstrated that in the range of <5 Å the L-J potential which acts as a barrier increased when the pore size is reduced. As there is a correlation between the apparent activation energy and pore size [70–72], this implies that the change of pore size is responsible for the reduction of cobalt oxide.

Figure 9 shows that when r_o = 0.1402 nm and r_m = 0.2657 nm, the Rayleigh distribution gives the closest apparent activation energy to the experimental value for the oxidized membrane. For reduced membrane, the best fitting value is r_o = 0.1400 nm and r_m = 0.2673 nm. The pore size distribution was derived from the fitting to He and H_2 activation energies. If other gases could also permeate across this membrane, the pore size distribution estimated by fitting with permeation activation energies of more gases would be more reliable. But unfortunately other gases were impermeable through this membrane. The pore size distributions of these two states were displayed in Figure 10. The peak of inner diameter distribution for the oxidized membrane is d_i = 0.2514 nm, and the peak for the reduced membrane is d_i = 0.2546 nm. There are significant portion of pores below d_i < 0.3 nm that are below the detectable limit of a traditional method by adsorption of N_2, Ar or CO_2.

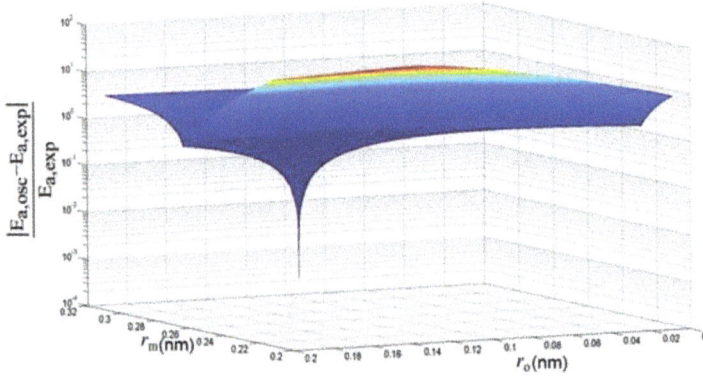

Figure 9. Screening of (r_o, r_m) to detect minimum deviation to the experimental activation energy.

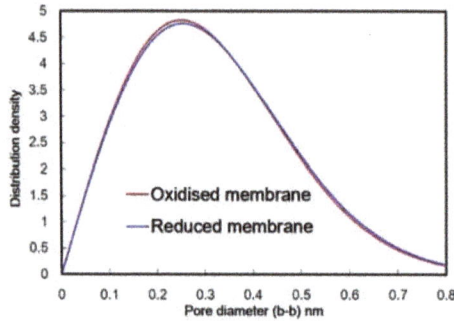

Figure 10. Calculated pore size distribution of silica membrane in oxidized state and reduced state.

With a pore size distribution, effective medium theory (EMT) could compute the effective transport conductance λ_e. Therefore, the bulk permeability from EMT is

$$P = \frac{\lambda_e}{\langle r_p^2 \rangle} \tag{27}$$

Based on the PSD estimated for the oxidized membrane and reduced membrane in Figure 10, the bulk permeability is calculated for He, H_2. The calculated permeability for He and H_2 was also compared with available experimental permeability as shown in Figure 11. The modelled permeability based on the pore size distribution above fits well with actual pore size distribution. This suggests that the estimated pore size distribution in Figure 10 could be a reasonable representation of the real pore size distribution.

The bulk permeabilities of Ar and CO_2 from these two pore size distribution are at least five orders of magnitude lower than for He and H_2. The distribution in Figure 10 shows that there is also a significant portion of pores larger than Ar and CO_2 kinetic diameter. However, it should be noted that a pathway for permeation consists of a serious of connected pores and only the narrowest part restricts the permeation, thus the permeability for Ar and CO_2 are much lower. This low permeability is why the permeation of Ar and CO_2 could not be measured by the bubble flow meter.

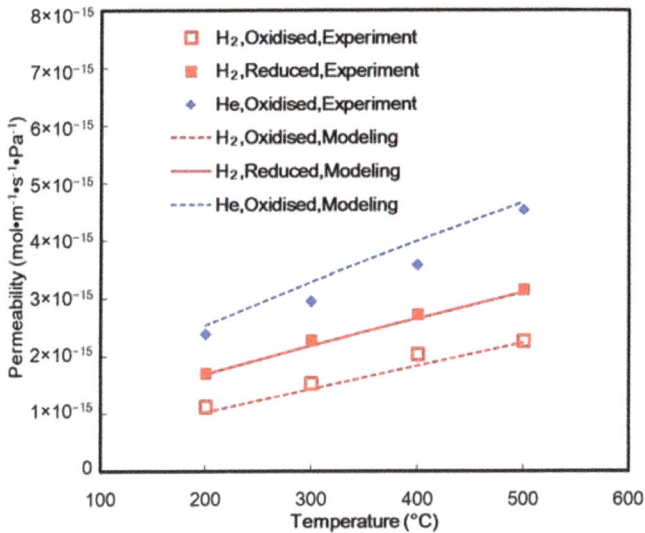

Figure 11. Comparison of permeability obtained from calculation and experiment.

5. Conclusion

The findings in this work supports the use of membrane permeation results to determine the apparent activation energy, and on coupling with the oscillator model permeation data can be used to estimate the pore size distribution. Therefore, better accuracy of pore sizes and pore size distribution can be achieved, which is otherwise not readily possible for xerogels and conventional characterization techniques. This study demonstrated that a Rayleigh distribution with r_o = 0.1402 nm and r_m = 0.2657 nm was a reasonable representation of the pore size distribution in a silica based membrane. Based on this estimated pore size distribution, the calculated permeability fitted well the experimental permeability. This method proved to be robust for calculating the pore size for cylindrical pore geometry and known material properties.

Author Contributions: G.J. designed and conducted the experiment, wrote the codes, analyzed the data, and wrote the manuscript; X.G. contributed in the modelling work; S.S. contributed in the membrane fabrication; S.K.B. provided the oscillator model and revised the manuscript; G.W. and K.H. contributed in supervision; J.C.D.d.C. contributed in supervision, membrane test and manuscript revision.

Funding: This research was funded by Australian Research Council via the Future Fellowship Program grant number FT130100405, National Natural Science Foundation of China grant number: 21878147, and Fundamental Research Funds for the Central Universities grant number: DUT18RC(3)036.

Acknowledgments: J.C.D.d.C. thanks the support from Australian Research Council via the Future Fellowship Program (FT130100405). X.G. thanks the financial support from National Natural Science Foundation of China (Grant number: 21878147). G.J. would appreciate the financial support from the Fundamental Research Funds for the Central Universities (Grant number: DUT18RC(3)036), and the technical support from Ling Yun HPC in Dalian University of Technology.

Conflicts of Interest: The authors declare no conflict of interest.

References

1. Ji, G.; Wang, G.; Hooman, K.; Bhatia, S.; Diniz da Costa, J.C. Scale-Up Design Analysis and Modelling of Cobalt Oxide Silica Membrane Module for Hydrogen Processing. *Processes* **2013**, *1*, 49–66. [CrossRef]
2. Ji, G.; George, A.; Skoulou, V.; Reed, G.; Millan, M.; Hooman, K.; Bhatia, S.K.; Diniz da Costa, J.C. Investigation and simulation of the transport of gas containing mercury in microporous silica membranes. *Chem. Eng. Sci.* **2018**, *190*, 286–296. [CrossRef]

3. Smart, S.; Lin, C.X.C.; Ding, L.; Thambimuthu, K.; Diniz da Costa, J.C. Ceramic membranes for gas processing in coal gasification. *Energy Environ. Sci.* **2010**, *3*, 268–278. [CrossRef]

4. Ji, G.; Yao, J.G.; Clough, P.T.; Diniz da Costa, J.C.; Anthony, E.J.; Fennell, P.S.; Wang, W.; Zhao, M. Enhanced hydrogen production from thermochemical processes. *Energy Environ. Sci.* **2018**, *11*, 2647–2672. [CrossRef]

5. De Vos, R.M.; Verweij, H. High-Selectivity, High-Flux Silica Membranes for Gas Separation. *Science* **1998**, *279*, 1710–1711. [CrossRef] [PubMed]

6. Raman, N.K.; Brinker, C.J. Organic "template" approach to molecular sieving silica membranes. *J. Membr. Sci.* **1995**, *105*, 273–279. [CrossRef]

7. Kusakabe, K.; Sakamoto, S.; Saie, T.; Morooka, S. Pore structure of silica membranes formed by a sol–gel technique using tetraethoxysilane and alkyltriethoxysilanes. *Sep. Purif. Technol.* **1999**, *16*, 139–146. [CrossRef]

8. Kanezashi, M.; Fujita, T.; Asaeda, M. Nickel-Doped Silica Membranes for Separation of Helium from Organic Gas Mixtures. *Sep. Sci. Technol.* **2005**, *40*, 225–238. [CrossRef]

9. Kanezashi, M.; Asaeda, M. Hydrogen permeation characteristics and stability of Ni-doped silica membranes in steam at high temperature. *J. Membr. Sci.* **2006**, *271*, 86–93. [CrossRef]

10. Battersby, S.; Duke, M.C.; Liu, S.; Rudolph, V.; Diniz da Costa, J.C. Metal doped silica membrane reactor: Operational effects of reaction and permeation for the water gas shift reaction. *J. Membr. Sci.* **2008**, *316*, 46–52. [CrossRef]

11. Uhlmann, D.; Liu, S.; Ladewig, B.P.; Diniz da Costa, J.C. Cobalt-doped silica membranes for gas separation. *J. Membr. Sci.* **2009**, *326*, 316–321. [CrossRef]

12. Uhlmann, D.; Smart, S.; Diniz da Costa, J.C. High temperature steam investigation of cobalt oxide silica membranes for gas separation. *Sep. Purif. Technol.* **2010**, *76*, 171–178. [CrossRef]

13. Uhlmann, D.; Smart, S.; Diniz da Costa, J.C. H_2S stability and separation performance of cobalt oxide silica membranes. *J. Membr. Sci.* **2011**, *380*, 48–54. [CrossRef]

14. Liu, L.; Wang, D.K.; Martens, D.L.; Smart, S.; Diniz da Costa, J.C. Influence of sol–gel conditioning on the cobalt phase and the hydrothermal stability of cobalt oxide silica membranes. *J. Membr. Sci.* **2015**, *475*, 425–432. [CrossRef]

15. Boffa, V.; Blank, D.H.A.; Ten Elshof, J.E. Hydrothermal stability of microporous silica and niobia–silica membranes. *J. Membr. Sci.* **2008**, *319*, 256–263. [CrossRef]

16. Kanezashi, M.; Fuchigami, D.; Yoshioka, T.; Tsuru, T. Control of Pd dispersion in sol–gel-derived amorphous silica membranes for hydrogen separation at high temperatures. *J. Membr. Sci.* **2013**, *439*, 78–86. [CrossRef]

17. Yoshida, K.; Hirano, Y.; Fujii, H.; Tsuru, T.; Asaeda, M. Hydrothermal Stability and Performance of Silica-Zirconia Membranes for Hydrogen Separation in Hydrothermal Conditions. *J. Chem. Eng. Jpn.* **2001**, *34*, 523–530. [CrossRef]

18. Gu, Y.; Oyama, S.T. Permeation properties and hydrothermal stability of silica–titania membranes supported on porous alumina substrates. *J. Membr. Sci.* **2009**, *345*, 267–275. [CrossRef]

19. Fotou, G.P.; Lin, Y.S.; Pratsinis, S.E. Hydrothermal stability of pure and modified microporous silica membranes. *J. Mater. Sci.* **1995**, *30*, 2803–2808. [CrossRef]

20. Darmawan, A.; Motuzas, J.; Smart, S.; Julbe, A.; Diniz da Costa, J.C. Binary iron cobalt oxide silica membrane for gas separation. *J. Membr. Sci.* **2015**, *474*, 32–38. [CrossRef]

21. Darmawan, A.; Motuzas, J.; Smart, S.; Julbe, A.; Diniz da Costa, J.C. Temperature dependent transition point of purity versus flux for gas separation in Fe/Co-silica membranes. *Sep. Purif. Technol.* **2015**, *151*, 284–291. [CrossRef]

22. Ballinger, B.; Motuzas, J.; Smart, S.; Diniz da Costa, J.C. Palladium cobalt binary doping of molecular sieving silica membranes. *J. Membr. Sci.* **2014**, *451*, 185–191. [CrossRef]

23. Ballinger, B.; Motuzas, J.; Miller, C.R.; Smart, S.; Diniz da Costa, J.C. Nanoscale assembly of lanthanum silica with dense and porous interfacial structures. *Sci. Rep.* **2015**, *5*, 8210. [CrossRef] [PubMed]

24. Battersby, S.; Tasaki, T.; Smart, S.; Ladewig, B.; Liu, S.; Duke, M.C.; Rudolph, V.; Diniz da Costa, J.C. Performance of cobalt silica membranes in gas mixture separation. *J. Membr. Sci.* **2009**, *329*, 91–98. [CrossRef]

25. Meixner, D.L.; Dyer, P.N. Characterization of the transport properties of microporous inorganic membranes. *J. Membr. Sci.* **1998**, *140*, 81–95. [CrossRef]

26. Duke, M.C.; Pas, S.J.; Hill, A.J.; Lin, Y.S.; Diniz da Costa, J.C. Exposing the Molecular Sieving Architecture of Amorphous Silica Using Positron Annihilation Spectroscopy. *Adv. Funct. Mater.* **2008**, *18*, 3818–3826. [CrossRef]

27. Diniz da Costa, J.C.; Lu, G.Q.; Rudolph, V.; Lin, Y.S. Novel molecular sieve silica (MSS) membranes: Characterisation and permeation of single-step and two-step sol–gel membranes. *J. Membr. Sci.* **2002**, *198*, 9–21. [CrossRef]

28. Gopalakrishnan, S.; Diniz da Costa, J.C. Hydrogen gas mixture separation by CVD silica membrane. *J. Membr. Sci.* **2008**, *323*, 144–147. [CrossRef]

29. Gopalakrishnan, S.; Yoshino, Y.; Nomura, M.; Nair, B.N.; Nakao, S.-I. A hybrid processing method for high performance hydrogen-selective silica membranes. *J. Membr. Sci.* **2007**, *297*, 5–9. [CrossRef]

30. Zivkovic, T. *Thin Supported Silica Membranes*; The University Twente: Enschede, The Netherlands, 2007.

31. Ha, H.Y.; Nam, S.W.; Lee, W.K. Chemical vapor deposition of hydrogen-permselective silica films on porous glass supports from tetraethylorthosilicate. *J. Membr. Sci.* **1993**, *85*, 279–290.

32. Ghasemzadeh, K.; Aghaeinejad-Meybodi, A.; Vaezi, M.J.; Gholizadeh, A.; Abdi, M.A.; Babaluo, A.A.; Haghighi, M.; Basile, A. Hydrogen production via silica membrane reactor during the methanol steam reforming process: Experimental study. *RSC Adv.* **2015**, *5*, 95823–95832. [CrossRef]

33. Kanezashi, M.; Sasaki, T.; Tawarayama, H.; Nagasawa, H.; Yoshioka, T.; Ito, K.; Tsuru, T. Experimental and Theoretical Study on Small Gas Permeation Properties through Amorphous Silica Membranes Fabricated at Different Temperatures. *J. Phys. Chem. C* **2014**, *118*, 20323–20331. [CrossRef]

34. Kanezashi, M.; Sasaki, T.; Tawarayama, H.; Yoshioka, T.; Tsuru, T. Hydrogen Permeation Properties and Hydrothermal Stability of Sol–Gel-Derived Amorphous Silica Membranes Fabricated at High Temperatures. *J. Am. Ceram. Soc.* **2013**, *96*, 2950–2957. [CrossRef]

35. Yoshioka, T.; Nakanishi, E.; Tsuru, T.; Asaeda, M. Experimental studies of gas permeation through microporous silica membranes. *AlChE J.* **2001**, *47*, 2052–2063. [CrossRef]

36. De Vos, R.M.; Maier, W.F.; Verweij, H. Hydrophobic silica membranes for gas separation. *J. Membr. Sci.* **1999**, *158*, 277–288. [CrossRef]

37. Krishna, R.; Baur, R. Modelling issues in zeolite based separation processes. *Sep. Purif. Technol.* **2003**, *33*, 213–254. [CrossRef]

38. Van den Bergh, J.; Ban, S.; Vlugt, T.J.H.; Kapteijn, F. Modeling the Loading Dependency of Diffusion in Zeolites: The Relevant Site Model Extended to Mixtures in DDR-Type Zeolite. *J. Phys. Chem. C* **2009**, *113*, 21856–21865. [CrossRef]

39. Ji, G.; Wang, G.; Hooman, K.; Bhatia, S.; Diniz da Costa, J.C. Simulation of binary gas separation through multi-tube molecular sieving membranes at high temperatures. *Chem. Eng. J.* **2013**, *218*, 394–404. [CrossRef]

40. Burggraaf, A.J. Single gas permeation of thin zeolite (MFI) membranes: Theory and analysis of experimental observations. *J. Membr. Sci.* **1999**, *155*, 45–65. [CrossRef]

41. Krishna, R. A unified approach to the modelling of intraparticle diffusion in adsorption processes. *Gas Sep. Purif.* **1993**, *7*, 91–104. [CrossRef]

42. Krishna, R.; Van Baten, J.M. A simplified procedure for estimation of mixture permeances from unary permeation data. *J. Membr. Sci.* **2011**, *367*, 204–210. [CrossRef]

43. Thornton, A.W.; Hilder, T.; Hill, A.J.; Hill, J.M. Predicting gas diffusion regime within pores of different size, shape and composition. *J. Membr. Sci.* **2009**, *336*, 101–108. [CrossRef]

44. Bonilla, M.R.; Bhatia, S.K. The low-density diffusion coefficient of soft-sphere fluids in nanopores: Accurate correlations from exact theory and criteria for applicability of the Knudsen model. *J. Membr. Sci.* **2011**, *382*, 339–349. [CrossRef]

45. Bhatia, S.K. Tractable molecular theory of transport of Lennard-Jones fluids in nanopores. *J. Chem. Phys.* **2004**, *120*, 4472–4485. [CrossRef] [PubMed]

46. Tomozawa, M. Chapter 3—Amorphous silica. In *Silicon-Based Material and Devices*; Nalwa, H.S., Ed.; Academic Press: Burlington, NJ, USA, 2001; pp. 127–154.

47. Lang, N.R.; Münster, S.; Metzner, C.; Krauss, P.; Schürmann, S.; Lange, J.; Aifantis, K.E.; Friedrich, O.; Fabry, B. Estimating the 3D Pore Size Distribution of Biopolymer Networks from Directionally Biased Data. *Biophys. J.* **2013**, *105*, 1967–1975. [CrossRef] [PubMed]

48. Metzner, C.; Krauss, P.; Fabry, B. Poresizes in random line networks. *arXiv* **2011**, arXiv:1110.1803v1.

49. Phirani, J.; Pitchumani, R.; Mohanty, K.K. Transport Properties of Hydrate Bearing Formations from Pore-Scale Modeling. In Proceedings of the SPE Annual Technical Conference and Exhibition, New Orleans, LA, USA, 4–7 October 2009.

50. Panfilov, M.; Panfilova, I.; Stepanyants, Y. Mechanisms of Particle Transport Acceleration in Porous Media. *Transp. Porous Media* **2008**, *74*, 49–71. [CrossRef]
51. Park, C.-Y.; Ihm, S.-K. New hypotheses for Mercury porosimetry with percolation approach. *AIChE J.* **1990**, *36*, 1641–1648. [CrossRef]
52. Mishra, B.K.; Sharma, M.M. Measurement of pore size distributions from capillary pressure curves. *AIChE J.* **1988**, *34*, 684–687. [CrossRef]
53. Bhatia, S.K. Modeling Pure Gas Permeation in Nanoporous Materials and Membranes. *Langmuir* **2010**, *26*, 8373–8385. [CrossRef] [PubMed]
54. Jepps, O.G.; Bhatia, S.K.; Searles, D.J. Wall Mediated Transport in Confined Spaces: Exact Theory for Low Density. *Phys. Rev. Lett.* **2003**, *91*, 126102. [CrossRef] [PubMed]
55. Zhang, L.; Seaton, N.A. Prediction of the effective diffusivity in pore networks close to a percolation threshold. *AIChE J.* **1992**, *38*, 1816–1824. [CrossRef]
56. Burganos, V.N.; Sotirchos, S.V. Diffusion in pore networks: Effective medium theory and smooth field approximation. *AIChE J.* **1987**, *33*, 1678–1689. [CrossRef]
57. Kirkpatrick, S. Percolation and Conduction. *Rev. Mod. Phys.* **1973**, *45*, 574–588. [CrossRef]
58. Landauer, R. The Electrical Resistance of Binary Metallic Mixtures. *J. Appl. Phys.* **1952**, *23*, 779–784. [CrossRef]
59. Bhatia, S.K. Directional autocorrelation and the diffusional tortuosity of capillary porous media. *J. Catal.* **1985**, *93*, 192–196. [CrossRef]
60. Bhatia, S.K. Stochastic theory of transport in inhomogeneous media. *Chem. Eng. Sci.* **1986**, *41*, 1311–1324. [CrossRef]
61. Ji, G.; Smart, S.; Bhatia, S.K.; Diniz da Costa, J.C. Improved pore connectivity by the reduction of cobalt oxide silica membranes. *Sep. Purif. Technol.* **2015**, *154*, 338–344. [CrossRef]
62. Bakker, W.J.W.; Kapteijn, F.; Poppe, J.; Moulijn, J.A. Permeation characteristics of a metal-supported silicalite-1 zeolite membrane. *J. Membr. Sci.* **1996**, *117*, 57–78. [CrossRef]
63. Kapteijn, F.; Bakker, W.J.W.; Van de Graaf, J.; Zheng, G.; Poppe, J.; Moulijn, J.A. Permeation and separation behaviour of a silicalite-1 membrane. *Catal. Today* **1995**, *25*, 213–218. [CrossRef]
64. Smart, S.; Vente, J.F.; Diniz da Costa, J.C. High temperature H_2/CO_2 separation using cobalt oxide silica membranes. *Int. J. Hydrog. Energy* **2012**, *37*, 12700–12707. [CrossRef]
65. Bhatia, S.K.; Nicholson, D. Some pitfalls in the use of the Knudsen equation in modelling diffusion in nanoporous materials. *Chem. Eng. Sci.* **2011**, *66*, 284–293. [CrossRef]
66. Gao, X.; Diniz da Costa, J.C.; Bhatia, S.K. Adsorption and transport of gases in a supported microporous silica membrane. *J. Membr. Sci.* **2014**, *460*, 46–61. [CrossRef]
67. Yacou, C.; Smart, S.; Diniz da Costa, J.C. Long term performance cobalt oxide silica membrane module for high temperature H_2 separation. *Energy. Environ. Sci.* **2012**, *5*, 5820–5832. [CrossRef]
68. Ji, G.; Wang, G.; Hooman, K.; Bhatia, S.; Diniz da Costa, J.C. The fluid dynamic effect on the driving force for a cobalt oxide silica membrane module at high temperatures. *Chem. Eng. Sci.* **2014**, *111*, 142–152. [CrossRef]
69. Miller, C.R.; Wang, D.K.; Smart, S.; Diniz da Costa, J.C. Reversible Redox Effect on Gas Permeation of Cobalt Doped Ethoxy Polysiloxane (ES40) Membranes. *Sci. Rep.* **2013**, *3*, 1–6. [CrossRef] [PubMed]
70. Xiao, J.; Wei, J. Diffusion mechanism of hydrocarbons in zeolites—I. Theory. *Chem. Eng. Sci.* **1992**, *47*, 1123–1141. [CrossRef]
71. Xiao, J.; Wei, J. Diffusion mechanism of hydrocarbons in zeolites—II. Analysis of experimental observations. *Chem. Eng. Sci.* **1992**, *47*, 1143–1159. [CrossRef]
72. Hacarlioglu, P.; Lee, D.; Gibbs, G.V.; Oyama, S.T. Activation energies for permeation of He and H2 through silica membranes: An ab initio calculation study. *J. Membr. Sci.* **2008**, *313*, 277–283. [CrossRef]

processes

MDPI

Article

Structure Manipulation of Carbon Aerogels by Managing Solution Concentration of Precursor and Its Application for CO_2 Capture

Pingping He [1,2], Xingchi Qian [1,2], Zhaoyang Fei [1,2], Qing Liu [1,3,]*, Zhuxiu Zhang [1], Xian Chen [1], Jihai Tang [1,2,4], Mifen Cui [1] and Xu Qiao [1,2,4,]*

1 College of Chemical Engineering, Nanjing Tech University, Nanjing 210009, China; shengrenping@njtech.edu.cn (P.H.); qianxc@njtech.edu.cn (X.Q.); zhaoyangfei@njtech.edu.cn (Z.F.); zhuxiu.zhang@njtech.edu.cn (Z.Z.); chenxian@njtech.edu.cn (X.C.); jhtang@njtech.edu.cn (J.T.); mfcui@njtech.edu.cn (M.C.)
2 State Key Laboratory of Materials-Oriented Chemical Engineering, Nanjing Tech University, Nanjing 210009, China
3 Department of Chemistry Centre for Catalysis Research and Innovation (CCRI), University of Ottawa, Ottawa, ON K1N 6N5, Canada
4 Jiangsu National Synergetic Innovation Center for Advanced Materials (SICAM), Nanjing 210009, China
* Correspondence: qing_liu@njtech.edu.cn (Q.L.); qct@njtech.edu.cn (X.Q.); Tel.: +86-25-8358-7168 (Q.L.); +86-25-8317-2298 (X.Q.)

Received: 2 February 2018; Accepted: 7 April 2018; Published: 12 April 2018

Abstract: A series of carbon aerogels were synthesized by polycondensation of resorcinol and formaldehyde, and their structure was adjusted by managing solution concentration of precursors. Carbon aerogels were characterized by X-ray diffraction (XRD), Raman, Fourier transform infrared spectroscopy (FTIR), N_2 adsorption/desorption and scanning electron microscope (SEM) technologies. It was found that the pore structure and morphology of carbon aerogels can be efficiently manipulated by managing solution concentration. The relative micropore volume of carbon aerogels, defined by V_{micro}/V_{tol}, first increased and then decreased with the increase of solution concentration, leading to the same trend of CO_2 adsorption capacity. Specifically, the CA-45 (the solution concentration of precursors is 45 wt%) sample had the highest CO_2 adsorption capacity (83.71 cm^3/g) and the highest selectivity of CO_2/N_2 (53) at 1 bar and 0 °C.

Keywords: carbon aerogels; concentration; structure manipulation; CO_2 capture

1. Introduction

Global climate and eco-environment changes are largely caused by elevated atmospheric CO_2 concentration mainly owing to the use of fossil fuels [1]. In order to solve these serious environmental problems, the implementation of carbon capture technologies have been proposed to control CO_2 emissions at the existing energy structure [2]. Research on CO_2 capture technologies has mainly been performed by using absorption [3], membrane separation [4], adsorption method [5], and so forth.

In absorption, elevated equipment size and corrosion rate, the large energy penalty caused by the regeneration of absorbents are challenging for absorption, especially aqueous amine solutions [6]. Monoethanolamine (MEA) was one of the earliest alkanolamines used for carbon capture, which has a high reaction rate, good absorption capacity. However the major drawbacks such as high energy penalties for regeneration, degradation in oxidizing environment, and corrosive effects limited its application. Membrane separation of CO_2 from flue gases depends on the difference in the diffusivity, solubility, absorption and adsorption abilities of different gases on different materials for separation. It was the best economical separation technique compared to other separation methods, when a high purity product is not desired [7]. The main limitation in case of membrane separation for carbon capture is need of very high selectivity to extract a relatively low concentration of CO_2 from flue gases. Thus, low selectivity is a huge challenge in commercializing this process. Carbon capture by sorbents is much more energy efficient as compared to aqueous amine solutions. Recently, many groups [8] focused on the study of synthetic methods and performance estimations on the new materials, such as metal-organic frameworks (MOFs), porous organic polymers (POPs), zeolites, activated carbons (ACs). Among these materials, porous carbons [9] have been widely researched, which are promising alternatives for CO_2 capture by virtue of their high specific surface areas, moderate heat of adsorption, low-cost preparation, relatively easy regeneration, and less sensitivity to the humidity than the other CO_2-philic materials.

Porous carbons derived from coal, petroleum and coconut shells have the uncontrollable surface chemistry and pore size due to uncertain structures of various precursors. Recently, carbon aerogels have received an increasing interest for their wide applicability as CO_2 adsorbents due to their special pore structure and variable surface properties [10]. Structure manipulation was the important method to enhance CO_2 adsorption performance. In addition, the amount of catalyst [11] and doped nitrogen [12] in the preparation process of carbon aerogels were crucial to pore volume and surface defect, thus affecting its CO_2 adsorption behavior. Post-modification of carbon aerogels [13] by adding amine was another effective measure to modulate its CO_2 adsorption performance. In addition, the preparation of new type carbon aerogel synthesized from bio-based nanocomposites [14] was the hot spot, showing a high CO_2 adsorption capacity.

Nowadays, a large number of reports have shown that structure design of the material is crucial to its performance [15,16]. It has been widely investigated to regulate the properties of the material simply and effectively. In the preparation process of the material, concentration as a simple yet effective factor for its structure and properties had been extensively studied [17,18]. It was reported by Zhang et al. [19] that the concentration of triphenylene-2,6,10-tricarboxylic acid (H_3TTCA) at the liquid–solid interface controlled self-assembling structure to fabricate a chicken-wire porous 2D network, which was confirmed by scanning tunneling microscopic (STM) measurements and density function theory (DFT) calculations. Han et al. [20] discussed the effect of concentration on the lyophilization-induced self-assembly of cellulose particles in aqueous suspensions. They found that cellulose particles self-organized into lamellar structured foam composed of aligned membrane layers with adjustable widths by regulating the concentration. Volpe et al. [21] explored that the concentrations of sodium caseinate (SC) and chitosan (CH) affected the structure and physical properties of the obtained blended films. It was found that the hydrophilic nature of films was reduced by increasing the ratio between CH and CS.

Carbon aerogels synthesized by using precursors of resorcinol and formaldehyde have been received much attention [22,23]. This paper demonstrates that the structure of carbon aerogel can be managed by solution concentration of its precursors, thus improving its CO_2 adsorption performance. Herein, we developed a series of structure-adjustable carbon aerogels by managing the concentration of precursors. The pore structure and surface morphologies of as-prepared sorbents were characterized by X-ray diffraction (XRD), Raman, Fourier transform infrared spectroscopy (FT-IR), N_2 adsorption/desorption and scanning electron microscope (SEM) technologies. In addition, adsorption capacity, isosteric heat of adsorption for CO_2, selectivity of CO_2/N_2, water-resistant performance and adsorption stability of carbon aerogels were investigated to exhibit the adsorption performance of the adsorbents.

2. Materials and Methods

All chemicals purchased by Aladdin in this study were of analytical grade and used as received without further purification, as follows: formaldehyde (37–40 wt% aqueous solution), resorcinol, deionized water, acetone and cetyltrimethyl ammonium bromide.

The preparation process of carbon aerogels was reported elsewhere with some modifications [24]. The molar ratio of cetyltrimethyl ammonium bromide (CTAB), formaldehyde (F) and resorcinol (R) is 1:125:250, and the solution concentration was controlled as X wt% managing by the amount of water.

The solution was under magnetic stirring in a glass vial. Then sealed and made sure that the solution underwent a sol-gel process at 85 °C for 72 h. Subsequently, the as-prepared organic aerogels were dried at room temperature for 36 h and then soaked in acetone for 72 h, replacing acetone once a day, finally dried in an oven at 100 °C at ambient pressure. Afterwards the as-obtained organic aerogels were pre-carbonized at 200 °C for 2 h under N_2 atmosphere [25]. Then, the pre-carbonized product was carbonized at 900 °C for 3 h with a heating rate of 5 °C/min under N_2 atmosphere to get the CA-X (X = 25, 35, 45, 55) sample.

The phase structure of the CA-X was characterized by powder X-ray diffraction (XRD) on a SarmtLab powder diffractometer using Ni-filtered Cu Kα radiation (λ = 0.15406 nm) at a setting of 40 kV and 100 mA. XRD patterns were recorded within the range 10~80° at a scan rate of 2°/min. Raman spectroscopy was measured with spectral resolution of 2 cm^{-1} in a scanning range of 100–4000 cm^{-1} on Labram HR800 apparatus (JY Horiba Corporation, Palaiseau, France). In addition, the He-Cd laser at 514 nm line was used as the excitation source. The surface morphologies of the samples were detected by scanning electron microscope (SEM) at an acceleration voltage of 15 kV on Hitachi S-4800 instrument (FEI, Hillsboro, OR, USA). The Fourier transform infrared (FTIR) spectra were collected on a Nicolit iS50 IR spectrometer (Thermo Nicolet Corporation, Madison, WI, USA) with a DTGS KBr detector (Thermo Nicolet Corporation, Madison, WI, USA) in the range of 4000 to 1000 cm^{-1} at room temperature. N_2-physisorption at −196 °C was performed on a BETSORP-II analyzer (MicrotracBEL, Osaka, Japan) to gain the textural properties of the materials. These samples were outgassed at 200 °C for 2 h prior to the adsorption measurements. The Brunauer–Emmett–Teller (BET) method was employed to determine the total surface area in the p/p₀ range between 0.05 and 0.20. The micropore volume was determined by the t-plot method. The mesopore volume and size distribution were calculated from the adsorption branch of the isotherm by the Barrett–Joyner–Halenda (BJH) method.

Static CO_2 adsorption experiments were measured by a BELSORP-II adsorption apparatus (MicrotracBEL, Osaka, Japan) at different temperature, which was controlled by a constant temperature water tank. The adsorption isotherms were fitted with the Langmuir model [25] as shown in Equation (1), and isotherm parameters were listed in the Tables S1–S4 at the supporting information.

$$q = q_c \frac{k_c p_t}{1 + k_c p_t} \tag{1}$$

where p is the pressure, and q is the adsorption capacity. In addition, q_c and k_c are the Langmuir model parameters with the subscripts c denoting the channels.

The isosteric heat of adsorption for CO_2 over the samples was calculated from the result of three adsorption isotherms at 0, 12.5 and 25 °C by using the Clausius–Clapeyron equation [25], which was shown in Equation (2). Specifically, the isosteric heat of adsorption was determined by evaluating the slope of the plots of ln(P) versus 1/T at the same adsorbed amount, where P and T are respectively the absolute pressure and temperature.

$$\ln \frac{P_2}{P_1} = \frac{Q}{R}\left(\frac{1}{T_1} - \frac{1}{T_2}\right) \tag{2}$$

The selectivity of CO_2/N_2 (15%/85% in volume) was calculated by using the ideal adsorption solution theory (IAST), which was described particularly in the Appendix A [26].

The test of CO_2 breakthrough was measured with a packed-bed column (length = 10.0 cm, inner diameter = 1.0 cm) connected to a QGA mass spectrometer (Hiden, Warrington, UK) at the presence of 15 vol% CO_2 with N_2 gas. The complete removal of adsorbed species from the adsorbent was achieved through thermal activation at 200 °C under a purge flow of N_2 gas. Besides, in order to determine the effect of water vapor on CO_2 adsorption, the N_2 gas passed through a water saturator (30 °C) located in a temperature-controlled water bath.

In addition, the regeneration experiments were also carried out with the BELSORP-II adsorption apparatus. These samples were saturated with CO_2 up to 1 bar at 0 °C. In addition the recovered adsorbents were degassed at 100 °C under vacuum for 30 min prior to each measurement.

3. Results and Discussion

3.1. Material Characterizations

3.1.1. XRD Analysis

As presented in Figure 1, XRD analysis was carried out to characterize the phase structure of carbon aerogels. The characteristic diffraction peaks around $2\theta = 23°$ and $43°$ were assigned to the (002) and (101) reflections, which indicated that all carbon aerogels showed short-range-ordered amorphous carbon materials as a partly graphitized carbon according to the reported literatures [27,28]. It indicated that the crystal form of the carbon aerogel was hardly affected by the change of solution concentration.

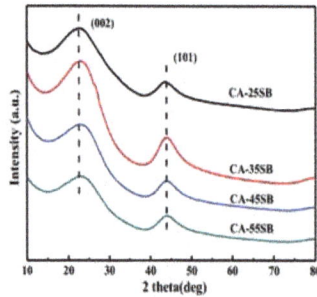

Figure 1. XRD patterns of CA-X samples.

3.1.2. Raman Analysis

Raman spectra of the carbon aerogels were employed to further observe the structure and surface defects of carbon aerogels as shown in Figure 2. The Raman spectra of carbon materials were similar to that of the graphite structure. The prominent Raman G-band near 1590 cm^{-1} was related to the E_{2g} active modes, which reflected the sp2 type hybridization. In addition, the strong and rather broad D-band, called the defect band, at ca. 1350 cm^{-1} attributed to a A_{1g} mode, which assigned to the vibration of carbon atoms with dangling bonds [29]. As expected, the structure of amorphous carbon material as a partly graphitized carbon was formed, which was compatible with the result of XRD analysis. In addition, the relatively adjacent I_D/I_G ratio, showing the surface defect of the material, indicated that the change of solution concentration had a little effect on its surface defects.

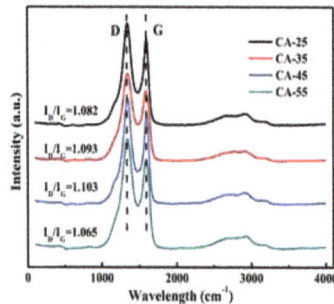

Figure 2. Raman patterns of CA-X samples.

3.1.3. FTIR Analysis

The FTIR spectra of carbon aerogels were exhibited in Figure 3, which were used to detect the surface functional groups of these materials. As shown in the picture, there were no obvious characteristic peak of organic functional groups after the calcination. The typical weak bands at 1570 cm^{-1} and 1350 cm^{-1} stem from the absorption peak of carbon in the skeleton over carbon aerogels. And the dominant absorption in the 2800–3000 cm^{-1} region was an indicative sign of the C–H symmetric and asymmetric stretching of CH$_2$ and CH$_3$ groups [29,30]. These results forcefully showed that the change of solution concentration can't lead to the variation of surface functional groups.

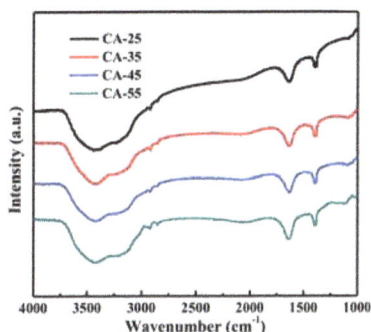

Figure 3. FT-IR spectra of CA-X samples.

3.1.4. SEM Analysis

SEM images of these materials were depicted in Figure 4, which were used to investigate the morphologies of carbon aerogels. As shown in the below pictures, it can be seen that the polymerized particle of materials exhibited the special coral shape, which was the typical shape of carbon aerogels. Besides, it can be found that the solution concentration appreciably influenced the size and morphology of carbon aerogel [31]. The size of particle was within the confines of dozens of nanometer caused by the difference of cross-linking strength among clusters, which caused a large number of pores produced between the clusters. Moreover, we can observe that the size of particle became smaller and the degree of cross-linking among clusters was further strengthened as the increase of solution concentration.

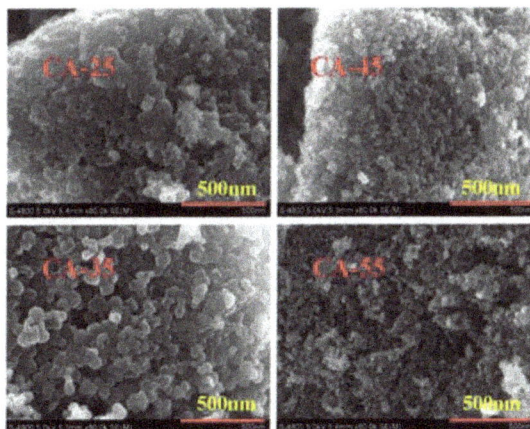

Figure 4. SEM images of CA-X samples.

3.1.5. N_2 Adsorption/Desorption Analysis

N_2 adsorption/desorption isotherms and BJH pore distribution curves of carbon aerogels were exhibited in Figure 5. As shown in Figure 5a, all materials displayed the typical IV-type isotherm with obvious type H2 hysteresis loop at P/P_0 range of 0.8~0.9, indicating the existence of slit pore structures inside the materials [25,32]. It can be indicated that the mesoporous structure of the material was formed by the agglomeration of the nanoparticle building blocks, which was previously confirmed by the SEM observation.

In addition, the textural parameters of samples were listed in Table 1. From the below table, it can be concluded that the CA-45 sample had a specific surface area of 848 m^2/g and total pore volume of 0.95 cm^3/g. To be specific, the surface area and the ratio of V_{micro}/V_{tol} firstly enlarged with the increase of the solution concentration, which was probably as the result of the comparatively large pore disappearing that occurred upon the reduction of the water. It was confirmed that the surface area and the ratio of V_{micro}/V_{tol} decreased as the solution concentration further rising, which was possibly that hypo-water usage was bad for the formation of micropore causing by dispersion of water. Furthermore, the BJH pore distribution exhibited the opposite rule due to the same reason.

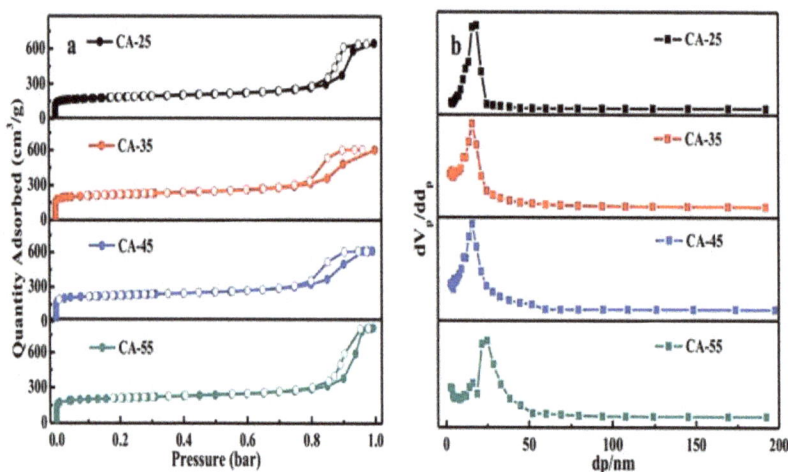

Figure 5. (a) N_2 adsorption/desorption isotherms and (b) pore size distributions of CA-X samples.

Table 1. Textural structure parameters of CA-X samples.

Sample	S_{BET} (m^2/g)	V_p (cm^3/g)			dp [2] (nm)	V_{micro}/V_{tol}
		Micropore [1]	Mesopore [2]	Total		
CA-25	638	0.20	0.79	0.99	7.37	0.20
CA-35	813	0.25	0.67	0.92	4.52	0.27
CA-45	848	0.27	0.68	0.95	4.46	0.28
CA-55	784	0.24	1.00	1.24	6.34	0.19

[1] Calculated by the t-plot method; [2] Calculated by the BJH method.

3.2. CO_2 Adsorption

3.2.1. CO_2 Adsorption Capacity

Figure 6 displays the CO_2 adsorption isotherms for carbon aerogels at 0, 12.5 and 25 °C, which showed the static adsorption of the materials at different pressure. As shown in Figure 6, we can find that the adsorption isotherms of CO_2 at 12.5 and 25 °C for carbon aerogels follow the same law compared to adsorption isotherm at 0 °C. These adsorption isotherms can be well confirmed by the parameters of Langmuir model. Then these parameters were listed in Tables S1–S4 at the supporting information. It was worth noting that the CO_2 uptake over carbon aerogels was greatly enhanced as the pressure increasing. CO_2 uptake shows the opposite result along with the temperature increasing. With the increase of the solution concentration, the CO_2 uptake firstly increased from 68.5 cm^3/g (CA-25 sample) to 83.7 cm^3/g (CA-45 sample), then the CO_2 uptake over CA-55 sample sharply decreased to 62.7 cm^3/g. On the basis of the aforementioned results, especially pore structure, CA-45 sample exhibited the highest surface area and ratio of V_{micro}/V_{tol}, which was beneficial to CO_2 capture. The adsorption capacities of carbon aerogels had a good correspondence with their surface area, pore volume and pore size. It was worth noting that the CO_2 adsorption capacity of CA-45 sample was the highest under the test pressure. The increasing trend of CO_2 uptake over carbon aerogels was more obvious at lower relative pressure, which was related to the existence of micropores over the adsorbent. As is well-known that micropores can enhance the contact possibility between CO_2 and the pore walls. While the existence of abundant mesopores will provide low-resistant pathways for CO_2 through the porous material, which was beneficial to enhance the adsorption performance of the material at higher relative pressure. So the proper value of V_{micro}/V_{tol}, which was defined to weight the value of micropores and mesopores, can improve the CO_2 adsorption performance.

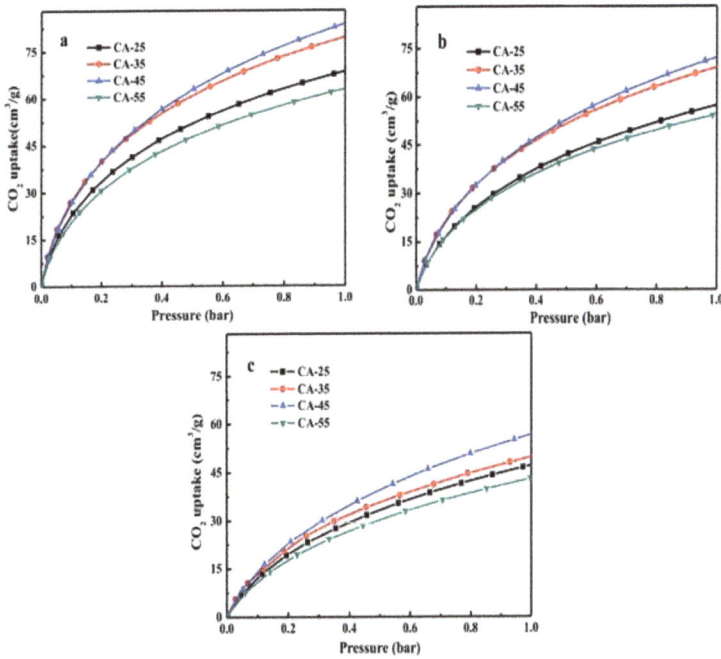

Figure 6. Adsorption isotherms of CO_2 at (**a**) 0; (**b**) 12.5 and (**c**) 25 °C over CA-X samples.

3.2.2. Isosteric Heat of Adsorption for CO_2

The estimated isosteric heat of adsorption for CO_2 over carbon aerogels was listed at the Table 2. In order to understand the adsorbate–adsorbent interaction, the isosteric heat of adsorption (Q_{st}) was calculated by using the Clausius–Clapeyron equation from the adsorption isotherms collected at 0, 12.5 and 25 °C. As displayed in Table 2, the values of adsorption heat were near 25 kJ/mol, indicating that CO_2 adsorption over carbon aerogels was mainly based on physical adsorption caused by the channels function. This result was confirmed by the outcomes of Raman which indicated that the change of solution concentration had a little effect on the surface defects. Besides, the low Q_{st} value was consistent with FTIR analysis which exhibited that changes of concentration can't lead to the variation of surface functional groups. In addition, the relatively low isosteric heat of adsorption for CO_2 over the samples had an advantage of low renewable energy consumption.

Table 2. CO_2 adsorption capacities of samples under different pressure and operating temperature.

Sample	CO$_2$ Uptake (cm^3/g)							Q_{st} ^{3} (kJ/mol)
	0 °C		12.5 °C		25 °C		50 °C	
	0.15 bar ^{1}	1 bar ^{1}	0.15 bar ^{1}	1 bar ^{1}	0.15 bar ^{1}	1 bar ^{1}	0.15 bar (Humid) ^{2}	
CA-25	28.7	68.5	22.2	57.3	16.3	47.3	13.2 (12.1)	25.1
CA-35	34.1	79.5	27.5	68.7	17.6	49.6	15.1 (14.3)	24.8
CA-45	34.2	83.7	27.6	71.8	19.1	56.5	18.5 (16.2)	24.3
CA-55	26.2	62.7	22.1	54.4	14.5	43.3	13.1 (11.9)	25.6

^{1} Determined by CO_2 adsorption isotherms; ^{2} Determined by CO_2 breakthrough curves; ^{3} Calculated by CO_2 adsorption isotherms using the Equation (2).

3.2.3. CO_2 Adsorption Selectivity

In order to further understand the relationship between structure and CO_2 performance, the adsorption isotherms of N_2 and CO_2 were measured at 0 °C as shown in Figure S1. All carbon aerogels showed the high uptake of CO_2, and N_2 uptake was barely adsorbed at 0 °C. For example, the N_2 uptake on CA-45 sample was just 13.8 cm^3/g at 0 °C and 1 bar, which was much lower than the uptake of CO_2 (83.7 cm^3/g). Similar results were also observed on other carbon aerogels. The adsorption selectivity of CO_2/N_2 was calculated by IAST which has been widely used to predict adsorption selectivity of gas mixtures. In the calculation, the ratio of CO_2/N_2 in volume was 15%/85%, which is the typical component of flue gases. Fitting parameters of Langmuir model were listed in Tables S1–S4 at the Supporting Information, and the IAST selectivity results were shown in Figure 7. We can see that the IAST selectivity of CO_2/N_2 over carbon aerogels increased with the pressure rising, which was mainly caused by the effect of active adsorption sites on micro- and mesopores. To be specific, the CA-45 sample kept relatively high adsorption selectivity compared to other samples because of the highest relative micropore volume (determined by V_{micro}/V_{tol}) and surface area. At a relatively low pressure, the adsorption potential of the mesopore was much lower than that of the micropore. When the pore size of the micropore was close to the size of adsorbate molecules (CO_2 at 0.33 nm and N_2 at 0.364 nm), which caused that CO_2 molecules were easier to enter micropores compared to N_2. The selectivity of CO_2/N_2 on CA-55 sample was 48 at 1 bar, which was the lowest among all samples because that the presence of abundant mesopores provided more adsorbed space to the N_2.

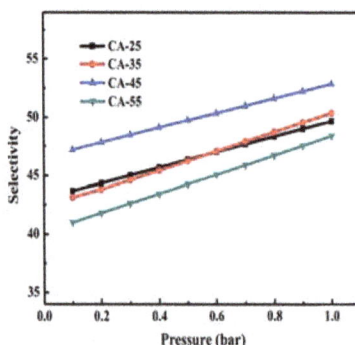

Figure 7. IAST selectivity of CO_2/N_2 (15%/85% in volume) on the CA-X samples at 0 °C.

3.2.4. Water-Resistant Experiment

In order to test the adsorption performance of the materials in flue gases and considering that the flue gases were still warm after CO_2 scrubbing (~50 °C), the CO_2 breakthrough curves of all samples were measured at 50 °C in the presence of 15 vol% CO_2 as shown in Figure 8a. The adsorption capacities of carbon aerogels were listed in Table 2. In the table, we can see that the CO_2 adsorption capacity of CA-45 sample was 18.5 cm^3/g, respectively. As the flue gases always contain water vapor, it was important to evaluate the effect of moisture on CO_2 adsorption performance over the current adsorbents [33]. All samples were also selected to research the effect of water vapor (13.74%) on CO_2 uptake at 50 °C in the CO_2/N_2 (15%/85% in volume) mixture to investigate the water-resistant performance of these materials, which was shown in the Figure 8b. Typical CO_2 breakthrough curves in the presence of dry and humid gas feed showed that the water vapor had a negative effect on the CO_2 adsorption for carbon aerogels. As discussed previously, there was no obvious variation of the adsorption capacity and adsorption rate in the presence of water vapor compared to adsorption behavior on dry gas feed. This is probably attributed to the appropriate relative micropore volume adjusted by managing solution concentration, which was contributed to the fast pathways for CO_2 through the porous network at the presence of water vapor.

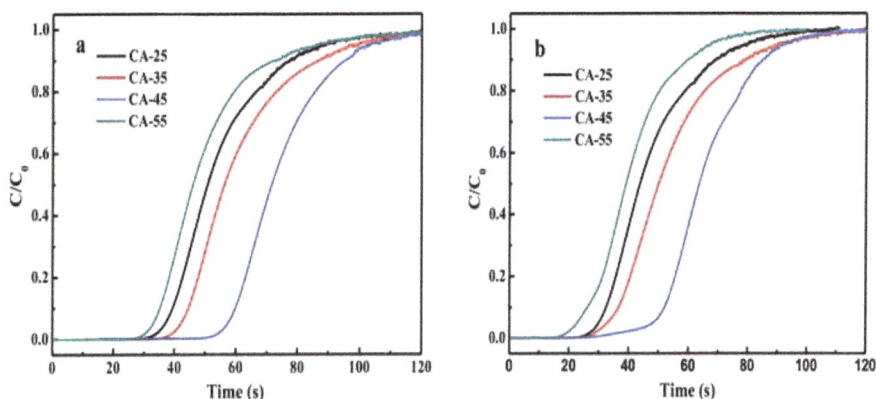

Figure 8. CO_2 breakthrough curves of CA-X samples under (**a**) dry and (**b**) humid conditions.

3.2.5. Adsorbent Stability

Taking account into the practical industrial application, the recyclability of adsorbent is of great significance. Figure 9 summarizes the CO_2 adsorption capacity at 0 °C and recycle times for the entire sorbents. After each adsorption cycle, quite mild conditions (vacuum, 100 °C and 30 min) were executed for the regeneration. No obvious loss of CO_2 adsorption capacity (less than 1%) on samples took place after several cycles. These sorbents were very stable and exhibited outstanding CO_2 adsorption capacities for 6 cycles, which suggests that carbon aerogel is a stable and promising adsorbent for CO_2 capture [34]. Excellent recyclability of carbon aerogels should be ascribed to the proper relative micropore volume, which provides a relative loose pathway for CO_2 adsorption.

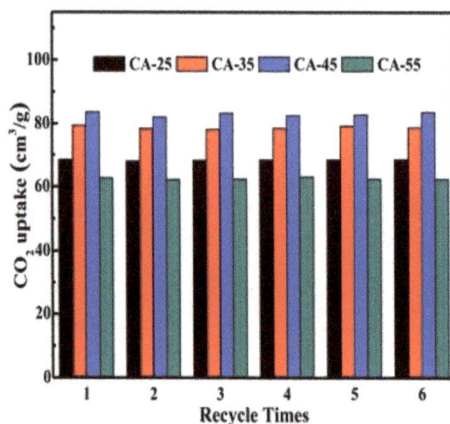

Figure 9. Adsorption isotherms of CO_2 over CA-X samples at 1 bar and 0 °C.

4. Conclusions

In this research, we prepared a series of carbon aerogels synthesized by polycondensation of resorcinol and formaldehyde by managing solution concentration. It was found that the pore structure and morphology of carbon aerogels were influenced by the solution concentration of their precursors. The relative micropore volume (determined by the ratio of V_{micro}/V_{tol}) of carbon aerogels was effectively regulated. With the solution concentration increasing, we interestingly found that the particle size decreased and the relative micropore volume was first increasing and then decreasing. The CA-45 sample had the largest relative micropore volume, which shows the optimum adsorption capacity of CO_2 and selectivity of CO_2/N_2. At the same time, the phase structure, surface defects and functional groups of carbon aerogels have no obvious difference with the change of solution concentration. The CA-45 sample exhibited the highest CO_2 adsorption capacity up to 83.71 cm^3/g and highest selectivity of CO_2/N_2 (15%/85% in volume) (53) at 1 bar, respectively. Moreover, all samples can be completely regenerated under mild conditions, and little loss of CO_2 adsorption capacity was detected after six cycles, showing excellent adsorbent stability.

Supplementary Materials: The following are available online at http://www.mdpi.com/2227-9717/6/4/35/s1, Figure S1: Adsorption isotherms of CO_2 and N_2 over all samples at 0 °C, Tables S1–S4: Fitting parameters derived from isotherms of all samples.

Acknowledgments: This work was supported by National Natural Science Foundation of China (Grant Nos. 21606130, 21306089), the State Key Laboratory of Materials-Oriented Chemical Engineering (Grant ZK201610, ZK201703), and the Priority Academic Program Development of Jiangsu Higher Education Institutions (PAPD).

Author Contributions: Q.L., X.Q. and P.H. conceived and designed these experiments; P.H. and X.Q. performed the experiments and analyzed the data; P.H., X.Q. and Q.L. wrote the paper; Z.F., Z.Z., X.C., J.T., and M.C. supervised the process of experiments and paper.

Conflicts of Interest: The authors declare no conflict of interest.

Appendix A

The well-known IAST has been extensively reviewed in literature [35] and hence only the working equations will be given here. The theory assumes that the adsorbed phase is an ideal solution of the adsorbed components and the reduced spreading pressure (π_i^*) of all the components in the mixture in their standard states is equal to the reduced spreading pressure of the adsorbed mixture (π^*). Thus,

$$\pi_1^* = \pi_1^* = \cdots = \pi_n^* = \pi^* \tag{A1}$$

The reduced spreading pressure of each component is computed from Gibb's adsorption isotherm as follows:

$$\pi_i^* = \frac{\pi_i A}{RT} \int_0^{P_i^0} \frac{q_i}{P_i} dP \tag{A2}$$

Bring Equation (A3) into Equation (1), we can obtain the DL-IAST model. When the gas mixtures achieve balance, the spreading pressures of component i and j are equal. Thus,

$$q_{ci} \ln(1 + \frac{k_{ci} y_i P_t}{x_i}) - q_{cj} \ln[1 + \frac{k_{cj}(1 - y_i)P_t}{1 - x_i}] = 0 \tag{A3}$$

where q_c, and k_c, are the Langmiur model parameters with the subscripts c denoting the channels, respectively. P_t is the system pressure, and q is the adsorption amount. What's more, x_i and y_i are the molar fractions of component i in the adsorbed and bulk phases.

The ideal adsorption solution theory (IAST) has been reported for predicting binary gas mixture adsorption in solid adsorbent. The selectivity of x_i over x_j has been defined according to Equation (A4). Thus,

$$S = \frac{(x_i/y_i)}{(x_j/y_j)} \tag{A4}$$

The adsorbed-phase mole fraction of the different components are related to those of the gas phase by the Raoult's law for ideal solutions, analogous to vapor-liquid systems:

$$Py_i = P_i^0(\pi^*)x_i \tag{A5}$$

With the constraint that

$$y_i + y_j = 1 \quad x_i + x_j = 1 \tag{A6}$$

So, the selectivity of CO_2/N_2 can be calculated as long as the y_i is known.

References

1. Haszeldine, R.S. Carbon Capture and Storage: How Green Can Black Be? *Science* **2009**, *325*, 1647–1652. [CrossRef] [PubMed]
2. Boot-Handford, M.E.; Abanades, J.C.; Anthony, E.J.; Blunt, M.J.; Brandani, S.; Mac Dowell, N.; Fernandez, J.R.; Ferrari, M.C.; Gross, R.; Hallett, J.P.; et al. Carbon capture and storage update. *Energy Environ. Sci.* **2014**, *7*, 130–189. [CrossRef]
3. Leimbrink, M.; Nikoleit, K.G.; Spitzer, R.; Salmon, S.; Bucholz, T.; Gorak, A.; Skiborowski, M. Enzymatic reactive absorption of CO_2 in MDEA by means of an innovative biocatalyst delivery system. *Chem. Eng. J.* **2018**, *334*, 1195–1205. [CrossRef]
4. Lu, J.G.; Ge, H.; Chen, Y.; Ken, R.T.; Xu, Y.; Zhao, Y.X.; Zhao, X.; Qian, H. CO_2 capture using a functional protic ionic liquid by membrane absorption. *J. Energy Inst.* **2017**, *90*, 933–940. [CrossRef]
5. Puthiaraj, P.; Lee, Y.R.; Ahn, W.S. Microporous amine-functionalized aromatic polymers and their carbonized products for CO_2 adsorption. *Chem. Eng. J.* **2017**, *319*, 65–74. [CrossRef]
6. Pires, J.C.M.; Martins, F.G.; Alvim-Ferraz, M.C.M.; Simoes, M. Recent developments on carbon capture and storage: An overview. *Chem. Eng. Res. Des.* **2011**, *89*, 1446–1460. [CrossRef]
7. Merkel, T.C.; Lin, H.Q.; Wei, X.T.; Baker, R. Power plant post-combustion carbon dioxide capture: An opportunity for membranes. *J. Membr. Sci.* **2010**, *359*, 126–139. [CrossRef]
8. Younas, M.; Sohail, M.; Kong, L.L.; Bashir, M.J.; Sethupathi, S. Feasibility of CO_2 adsorption by solid adsorbents: A review on low-temperature systems. *Int. J. Environ. Sci. Technol.* **2016**, *13*, 1839–1860. [CrossRef]
9. Rashidi, N.A.; Yusup, S. An overview of activated carbons utilization for the post-combustion carbon dioxide capture. *J. CO₂ Util.* **2016**, *13*, 1–16. [CrossRef]
10. Wu, X.B.; Wu, D.C.; Fu, R.W. Studies on the adsorption of reactive brilliant red X-3B dye on organic and carbon aerogels. *J. Hazard. Mater.* **2007**, *147*, 1028–1036. [CrossRef] [PubMed]

11. Liu, Q.; He, P.P.; Qian, X.C.; Fei, Z.Y.; Zhang, Z.X.; Chen, X.; Tang, J.H.; Cui, M.F.; Qiao, X. Carbon Aerogels Synthesizd with Cetyltrimethyl Ammonium Bromide (CTAB) as a Catalyst and its Application for CO$_2$ Capture. *Z. Anorg. Allg. Chem.* **2017**, *644*, 155–160. [CrossRef]
12. Jeon, D.H.; Min, B.G.; Oh, J.G.; Nah, C.; Park, S.J. Influence of Nitrogen moieties on CO$_2$ capture of Carbon Aerogel. *Carbon Lett.* **2015**, *16*, 57–61. [CrossRef]
13. Kong, Y.; Jiang, G.D.; Wu, Y.; Cui, S.; Shen, X.D. Amine hybrid aerogel for high-efficiency CO$_2$ capture: Effect of amine loading and CO$_2$ concentration. *Chem. Eng. J.* **2016**, *306*, 362–368. [CrossRef]
14. Alhwaige, A.A.; Ishida, H.; Qutubuddin, S. Carbon Aerogels with Excellent CO$_2$ Adsorption Capacity Synthesized from Clay-Reinforced Biobased Chitosan-Polybenzoxazine Nanocomposites. *ACS Sustain. Chem. Eng.* **2016**, *4*, 1286–1295. [CrossRef]
15. Li, Z.N.; Gadipelli, S.; Yang, Y.C.; Guo, Z.X. Design of 3D Graphene-Oxide Spheres and Their Derived Hierarchical Porous Structures for High Performance Supercapacitors. *Small* **2017**, *13*. [CrossRef] [PubMed]
16. Zhang, C.; He, Y.W.; Mu, P.; Wang, X.; He, Q.; Chen, Y.; Zeng, J.H.; Wang, F.; Xu, Y.H.; Jiang, J.X. Toward High Performance Thiophene-Containing Conjugated Microporous Polymer Anodes for Lithium-Ion Batteries through Structure Design. *Adv. Funct. Mater.* **2018**, *28*. [CrossRef]
17. Ahmadi, H.; Lotfollahi-Yaghin, M.A. Stress concentration due to in-plane bending (IPB) loads in ring-stiffened tubular KT-joints of offshore structures: Parametric study and design formulation. *Appl. Ocean Res.* **2015**, *51*, 54–66. [CrossRef]
18. Oh, T.H.; Oh, S.K.; Kim, H.; Lee, K.; Lee, J.M. Conceptual Design of an Energy-Efficient Process for Separating Aromatic Compounds from Naphtha with a High Concentration of Aromatic Compounds Using 4-Methyl-N-butylpyridinium Tetrafluoroborate Ionic Liquid. *Ind. Eng. Chem. Res.* **2017**, *56*, 7273–7284. [CrossRef]
19. Zhang, S.Q.; Zhang, J.Y.; Deng, K.; Xie, J.L.; Duan, W.B.; Zeng, Q.D. Solution concentration controlled self-assembling structure with host-guest recognition at the liquid-solid interface. *Phys. Chem. Chem. Phys.* **2015**, *17*, 24462–24467. [CrossRef] [PubMed]
20. Han, J.Q.; Zhou, C.J.; Wu, Y.Q.; Liu, F.Y.; Wu, Q.L. Self-Assembling Behavior of Cellulose Nanoparticles during Freeze-Drying: Effect of Suspension Concentration, Particle Size, Crystal Structure, and Surface Charge. *Biomacromolecules* **2013**, *14*, 1529–1540. [CrossRef] [PubMed]
21. Volpe, S.; Cavella, S.; Masi, P.; Torrieri, E. Effect of solid concentration on structure and properties of chitosan-caseinate blend films. *Food Packag. Shelf Life* **2017**, *13*, 76–84. [CrossRef]
22. Aegerter, M.A.; Leventis, N.; Koebel, M.M. *Aerogels Handbook*; Springer: New York, NY, USA, 2011; pp. 215–233.
23. Far, H.M.; Donthula, S.; Taghvaee, T.; Saeed, A.M.; Garr, Z.; Sotiriou-Leventis, C.; Leventis, N. Air-oxidation of phenolic resin aerogels: Backbone reorganization, formation of ring-fused pyrylium cations, and the effect on microporous carbons with enhanced surface areas. *RSC Adv.* **2017**, *7*, 51104–51120. [CrossRef]
24. Li, C.F.; Yang, X.Q.; Zhang, G.Q. Mesopore-dominant activated carbon aerogels with high surface area for electric double-layer capacitor application. *Mater. Lett.* **2015**, *161*, 538–541. [CrossRef]
25. Kou, J.H.; Sun, L.B. Nitrogen-Doped Porous Carbons Derived from Carbonization of a Nitrogen-Containing Polymer: Efficient Adsorbents for Selective CO$_2$ Capture. *Ind. Eng. Chem. Res.* **2016**, *55*, 10916–10925. [CrossRef]
26. Liu, Q.; He, P.P.; Qian, X.C.; Fei, Z.Y.; Zhang, Z.X.; Chen, X.; Tang, J.H.; Cui, M.F.; Qiao, X.; Shi, Y. Enhanced CO$_2$ Adsorption Performance on Hierarchical Porous ZSM-5 Zeolite. *Energy Fuels* **2017**, *31*, 13933–13941. [CrossRef]
27. Jiang, S.F.; Zhang, Z.A.; Qu, Y.H.; Wang, X.W.; Li, Q.; Lai, Y.Q.; Li, J. Activated carbon aerogels with high bimodal porosity for lithium/sulfur batteries. *J. Solid State Electrochem.* **2014**, *18*, 545–551. [CrossRef]
28. Lee, Y.J.; Park, H.W.; Park, S.; Song, I.K. Electrochemical properties of Mn-doped activated carbon aerogel as electrode material for supercapacitor. *Curr. Appl. Phys.* **2012**, *12*, 233–237. [CrossRef]
29. Patole, A.S.; Patole, S.P.; Kang, H.; Yoo, J.B.; Kim, T.H.; Ahn, J.H. A facile approach to the fabrication of graphene/polystyrene nanocomposite by in situ microemulsion polymerization. *J. Colloid Interface Sci.* **2010**, *350*, 530–537. [CrossRef] [PubMed]
30. Patole, A.S.; Patole, S.P.; Jung, S.Y.; Yoo, J.B.; An, J.H.; Kim, T.H. Self assembled graphene/carbon nanotube/polystyrene hybrid nanocomposite by in situ microemulsion polymerization. *Eur. Polym. J.* **2012**, *48*, 252–259. [CrossRef]

31. Nallusamy, S. Synthesis and Characterization of Carbon Black-Halloysite Nanotube Hybrid Composites Using XRD and SEM. *J. Nano Res.* **2017**, *45*, 208–217. [CrossRef]
32. Zhuo, H.; Hu, Y.J.; Tong, X.; Zhong, L.X.; Peng, X.W.; Sun, R.C. Sustainable hierarchical porous carbon aerogel from cellulose for high-performance supercapacitor and CO_2 capture. *Ind. Crop. Prod.* **2016**, *87*, 229–235. [CrossRef]
33. Sayari, A.; Liu, Q.; Mishra, P. Enhanced Adsorption Efficiency through Materials Design for Direct Air Capture over Supported Polyethylenimine. *Chemsuschem* **2016**, *9*, 2796–2803. [CrossRef] [PubMed]
34. Qian, Y.; Delgado, J.D.L.P.; Veneman, R.; Brilman, D.W.F. Stability of a Benzyl Amine Based CO_2 Capture Adsorbent in View of Regeneration Strategies. *Ind. Eng. Chem. Res.* **2017**, *56*, 3259–3269. [CrossRef]
35. Jiang, W.J.; Yin, Y.; Liu, X.Q.; Yin, X.Q.; Shi, Y.Q.; Sun, L.B. Fabrication of supported cuprous sites at low temperatures: An efficient, controllable strategy using vapor-induced reduction. *J. Am. Chem. Soc.* **2013**, *135*, 8137–8140. [CrossRef] [PubMed]

processes

MDPI

Article

A Facile Synthesis of Hexagonal Spinel λ-MnO₂ Ion-Sieves for Highly Selective Li⁺ Adsorption

Fan Yang, Sichong Chen, Chentao Shi, Feng Xue * , Xiaoxian Zhang, Shengui Ju * and Weihong Xing

College of Chemical Engineering, Nanjing Tech University, Nanjing 210009, China; yangfan930416@njtech.edu.cn (F.Y.); chensc@njtech.edu.cn (S.C.); shichentao@njtech.edu.cn (C.S.); 2637561934@njtech.edu.cn (X.Z.); xingwh@njtech.edu.cn (W.X.)
* Correspondence: xuefeng@njtech.edu.cn (F.X.); jushengui@njtech.edu.cn (S.J.); Tel.: +86-25-8358-7182 (F.X.)

Received: 28 April 2018; Accepted: 14 May 2018; Published: 17 May 2018

Abstract: Ion-sieves are a class of green adsorbent for extraction Li⁺ from salt lakes. Here, we propose a facile synthesis of hexagonal spinel $LiMn_2O_4$ (LMO) precursor under mild condition which was first prepared via a modified one-pot reduction hydrothermal method using $KMnO_4$ and ethanol. Subsequently, the stable spinel structured λ-MnO_2 (HMO) were prepared by acidification of LMO. The as-prepared HMO shows a unique hexagonal shape and can be used for rapid adsorption-desorption process for Li⁺ adsorption. It was found that Li⁺ adsorption capacity of HMO was 24.7 mg·g⁻¹ in Li⁺ solution and the HMO also has a stable structure with manganese dissolution loss ratio of 3.9% during desorption process. Moreover, the lithium selectivity (α_{Mg}^{Li}) reaches to 1.35×10^3 in brine and the distribution coefficients (K_d) of Li⁺ is much greater than that of Mg^{2+}. The results implied that HMO can be used in extract lithium from brine or seawater containing high ratio of magnesium and lithium.

Keywords: $LiMn_2O_4$; λ-MnO_2; ion-sieve; hydrothermal reaction; adsorption

1. Introduction

Lithium and its compounds—known as "industrial monosodium glutamate" [1]—are widely used in significant fields such as batteries, ceramics, glass, alloy, lubricants, refrigerants and the nuclear industry [2,3]. The lithium reserves in China are the world's second-largest, which are primarily distributed in the salt lakes of Qinghai and Tibet [4]. However, the ratio of magnesium to lithium in the salty brine is extremely high, making it difficult to extract and recover lithium using conventional separation technologies [5,6]. Compared with precipitation and solvent extraction methods, ion-sieve adsorption has many technical merits, such as excellent selectivity and relatively low cost [7,8], which is considered to be the most promising environmentally benign technology for extracting lithium from salt lakes [9,10].

Manganese series spinel ion-sieves are widely used in lithium ion adsorption, which primarily includes λ-MnO_2, $MnO_2 \cdot 0.3H_2O$ and $MnO_2 \cdot 0.5H_2O$, after removal of lithium by acidification from precursors $LiMn_2O_4$ [11], $Li_4Mn_5O_{12}$ [12,13] and $Li_{1.6}Mn_{1.6}O_4$ [14–16], respectively. $LiMn_2O_4$ (LMO) is commonly used adsorbent precursor[8], which is fabricated through embedding the target Li⁺ in the Mn-O chemical skeleton to construct composite $Li_xMn_yO_z$. After extracting Li⁺ by acidification without damages in the structure, of λ-MnO_2 (HMO) with regular vacancy [17]. The cubic spinel structures and adsorption-desorption relationship of HMO and LMO is shown in Figure 1. Oxygen atoms(O), Mn^{3+}/Mn^{4+} and lithium atoms (Li) occupy 32 e, 16 d and 8 a of the Wyckoff site, respectively [18]. Then, lithium at the 8a position is extracted by hydrogen because of ion exchange process which can adsorption Li⁺ subsequently.

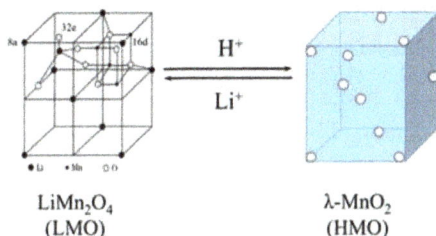

Figure 1. Illustrated microstructures of $LiMn_2O_4$ and λ-MnO_2.

In general, existing methods of preparing LMO can be boiled down to two categories, the solid-phase and the liquid-phase. Yuan et al. [19] utilized Li_2CO_3 and $MnCO_3$ (Li/Mn molar ratio was 0.5) as raw materials and calcined the mixture at 800 °C in air for 5 h to obtain the LMO. Park et al. [20] prepared spinel LMO by a simple spray mixed $Li(NO)_3$ and $Mn(NO)_3$ pyrolysis at 700 °C, with the deficiencies of inhomogeneity and large particle sizes (1 μm). The above-mentioned solid-phase LMO preparation methods often require high energy consumption and involve multiple steps. Besides, they always result in yielding large size particles because of agglomeration which decrease its contact area with solution for Li^+ extraction. The liquid-phase method (also known as soft-chemical process) for fabrication of LMO usually exhibits high purity, excellent crystal integrity and good dispersion. Tang et al. [21] prepared a nano-chain LMO using a sol-gel method. $LiNO_3$ and $Mn(NO_3)_2$ were stirred with the assistance of the starch at 110 °C for 1.5 h, followed by heating at 250 °C for 3 h and a thermal treatment at 700 °C for 3 h. Xiao et al. [22] prepared the ultrafine LMO powder by mixing $Mn(NO_3)_2$ with ammonia to produce precipitate, then they impregnated the precipitate with $LiOH \cdot H_2O$ and calcined the mixture at 830 °C for 8 h. Zhang et al. [23] prepared cubic phase LMO via a hydrothermal method by reacting $Mn(NO_3)_2$ with LiOH and H_2O_2 at 110 °C for 8 h. Despite the liquid-phase method being well investigated and developed, simplifying LMO synthetic process and improving the adsorbing ability and selectivity are still challenging. The existing methods mainly use $LiOH \cdot H_2O$ solution or acidic salts as raw materials. To our best knowledge, neutral synthetic routes through one-pot hydrothermal reaction to produce LMO and the corresponded HMO are rarely reported. Besides, HMO synthesized from high-valence manganese always shows higher adsorption capacity and selectivity than that of HMO synthesized from low-valence manganese, which is beneficial for lithium extraction from brine with high Li^+/Mg^{2+} ratio.

In this study, a series of LMO was prepared by a facile one-pot hydrothermal method using ethanol as reductant, $KMnO_4$ and $LiCl \cdot H_2O$ as precursors. We first optimized several synthetic parameters (i.e., $LiCl \cdot H_2O$ concentration, mass of $KMnO_4$, volume of ethanol, reaction time and reaction temperature) in preparing of LMO. Then we prepared the stable HMO by acidification treatment of LMO. The crystallization phase, morphology characteristic and chemical phase of as-prepared ion-sieves were systematically investigated. The Li^+ adsorption performance of HMO was studied and relevant adsorption kinetic model and adsorption isotherm were fitted. Finally, Li^+ extraction capacity and selectivity in brine containing high ratio of Mg^{2+} and Li^+ were studied.

2. Experimental

2.1. Preparation of LMO and HMO Ion Sieve

All chemicals used in this work are AR reagents unless otherwise noted. The detailed synthetic parameters are given in Table 1. Briefly, a certain amount of $LiCl \cdot H_2O$ and $KMnO_4$ were added to 75 mL deionized water. Then, ethanol was dropwise added into the mixed homogeneous solution. The final solution was obtained with the addition of deionized water to 150 mL. Next, the solution was transferred into a polytetrafluoroethylene (PTFE)-lined stainless-steel autoclave, heated at the specified temperatures (130–180 °C) for the specified time and cooled naturally to room temperature. The black precipitate was collected, filtered, washed completely and then dried at 80 °C for 12 h to obtain the as-prepared LiMn$_2$O$_4$ (LMO). Subsequently, the obtained LMO was added in hydrochloric acid solution (0.1 mol·L^{-1}) at 20 °C for 24 h until the lithium were completely extracted. The resulting precipitate was filtered, washed completely and dried at 80 °C for 12 h to obtain the λ-MnO$_2$ (HMO).

Table 1. Experimental parameters of synthesis LiMn$_2$O$_4$ (LMO) at different schemes.

Experiment Group	LiCl·H$_2$O (mol·L^{-1})	KMnO$_4$ (g)	Ethanol (V, %)	React. Time (h)	React. Temp. (°C)
1	a	3	7.5	12	160
2	11	b	7.5	12	160
3	11	3	c	12	160
4	11	3	7.5	d	160
5	11	3	7.5	12	e

Note: a = 4, 7, 11; b = 5, 7, 9; c = 2.5, 7.5, 8.75; d = 8, 10, 12; e = 130, 160, 180.

2.2. Characterization

The phase composition of the samples was characterized by X-ray powder diffraction (XRD, Mini Flex600, Rigaku Coporation, Tokyo, Japan with monochromatized Cu Kα radiation (λ = 1.54056 Å), operating at 40 kV and 15 mA, with a scanning rate of 20°/min from 10° to 80°. The concentration of each ion was measured by Inductively Coupled Plasma (ICP, Optima 7000DV, Perkin Elmer, Waltham, MA, USA), which was used to examine adsorption/desorption activity of the samples. The morphology of the samples was examined by scanning electron microscopy (SEM, S-4800, Hitachi, Tokyo, Japan) while morphology and crystal lattice were obtained by high resolution transmission electron microscopy (HRTEM, Libra120, Carl Zeiss AG, Jena, Germany). The chemical phase of manganese in the sample was analyzed by X-ray photoelectron spectroscopy (XPS, EscaLab 250Xi, Thermo Fisher, Shang Hai, China), with AlKα radiation (h$_v$ = 1103 eV), C1s of 20.05 eV to calibration.

2.3. Adsorption Behavior

2.3.1. Adsorption Capacity Test at Different pH Value

The lithium ion adsorption behavior test was measured by stirring (200 rpm) 0.1 g HMO in 500 mL LiCl·H$_2$O solution (pH value: 4, 5, 6, 7, 8, 9, 10 and 11, respectively), adjusted by a buffer solution composed of 0.1 mol·L^{-1} NH$_4$Cl and 0.1 mol·L^{-1} HCl and 0.1 mol·L^{-1} NH$_4$OH) with a uniform initial concentration of lithium ions (50 mg·L^{-1}) at 18 °C for 12 h.

The adsorption capacity is calculated by Equation (1).

$$Q_t = C_0 - C_t \times V/W \tag{1}$$

where C_0 is the initial concentration of metal ions (mg·L^{-1}); C_t is the concentration of metal ions at time t (mg·L^{-1}); V is the volume of solution (L); and W is the weight of HMO ion sieve (g).

2.3.2. Static Kinetic Test

The lithium ion adsorption behavior test was measured by stirring (200 rpm) 0.1 g HMO in 500 mL LiCl·H$_2$O solution (Ph = 10, adjusted by a buffer solution composed of 0.1 mol·L^{-1} NH$_4$Cl and 0.1 mol·L^{-1} NH$_4$OH) with a uniform initial concentration of lithium ions (50 mg·L^{-1}) at 18 °C for 12 h.

The data of the HMO adsorption capacity was fitted by a simplified Crank's single-hole diffusion model to obtain an efficient film coefficient (D_e) by Equation (2) [24,25].

$$\frac{Q_t}{Q_\infty} = 1 - \frac{6}{\pi^2} \times exp\left(-\frac{\pi^2 \times D_e \times t}{r^2}\right) \tag{2}$$

where Q_∞ is the adsorption capacity at the final time $(mg \cdot L^{-1})$; D_e is the diffusion coefficient $(cm^2 \cdot s^{-1})$; and r is the particle size of the adsorbent (cm).

The pseudo-first-order kinetic model (Equation (3)) and the pseudo-second-order kinetic model (Equation (4)) were used to simulate the saturated adsorption curve, aimed to confirm the kinetic constant of the adsorption process.

$$lg(Q_e - Q_t) = lgQ_e - \left(\frac{K_1}{2.303}\right) \times t \tag{3}$$

$$\frac{t}{Q_t} = \frac{1}{K_2} \times \frac{1}{Q_e^2} + \frac{1}{Q_e} \times t \tag{4}$$

where Q_e is the adsorption capacity when it reaches the adsorption equilibrium $(mg \cdot L^{-1})$; Q_t is the adsorption capacity calculated with Equation (1); K_1 is the adsorption rate constant of the pseudo-first-order kinetic model; and K_2 is the adsorption rate constant of pseudo-second-order kinetic model.

2.3.3. Adsorption Isotherm Test

The lithium ion adsorption behavior test was measured on HMO (0.04, 0.075, 0.11, 0.15 and 0.19 g) in 500 mL initial concentrations (10, 20, 30, 40 and 50 mg·L^{-1} LiCl·H$_2$O solution) were added to five flasks respectively, (Ph = 10, adjusted by a buffer solution composed of 0.1 mol·L^{-1} NH$_4$Cl and 0.1 mol·L^{-1} NH$_4$OH). The flasks were shaken on a shaker at 200 rpm at 18 °C for 12 h.

The adsorption isotherm curve is fitted according to the following isotherm models:
Langmuir isotherm model:

$$Q_{e1} = \frac{Q_m \times K_L \times C_e}{1 + K_L \times C_e} \tag{5}$$

Freundlich isotherm model:

$$Q_{e2} = K_F \times C_e^{1/n} \tag{6}$$

where Q_m is the theoretically calculated maximum adsorption capacity; K_L is the Langmuir constant; K_F is the Freundlish constant; and n is an empirical constant.

2.4. Selective Adsorption Behavior

The selectivity of lithium ions compared with other coexisting ions in brine was adjusted pH value to 10 by 0.1 mol·L^{-1} NH$_4$OH, carried out by stirring (200 rpm) 0.1 g ion sieve in 20 mL saline brine at 20 °C for 72 h. The adsorption capacity of metal ion at equilibrium (Q_e), distribution coefficient (K_d), separation factor (α_{Me}^{Li}) and concentration factor (C_F) are calculated according to the following equations:

$$K_d = C_{0,Me} - C_{e,Me} \times V/(C_{e,Me} \times W) \tag{7}$$

$$\alpha_{Me}^{Li} = K_{d,Li}/K_{d,Me} \quad (Me = K^+, Ca^{2+}, Na^+, Mg^{2+}, Li^+) \tag{8}$$

$$C_F = Q_{e,Me}/C_{0,Me} \quad (Me = K^+, Ca^{2+}, Na^+, Mg^{2+}, Li^+) \tag{9}$$

where $C_{0, Me}$ is the initial concentrate of ions in brine $(mg \cdot L^{-1})$; $C_{e, Me}$ is the final concentrate of ions in brine after adsorption $(mg \cdot L^{-1})$; V is the volume of solution (L); W is the weight of the HMO ion sieve (g); $Q_{e, Me}$ is the saturated adsorption capacity of ions in brine $(mg \cdot g^{-1})$.

2.5. Desorption Behavior

LMO was renamed LMO-1 after the Li$^+$ adsorption of the HMO. The curve of the Li$^+$ extraction and manganese dissolution was carried out by stirring (200 rpm) 0.1 g LMO-1 in 500 mL hydrochloric acid solution (0.04 mol·L^{-1}) for 24 h at 20 °C. The extraction ratio of lithium and the dissolution loss ratio of manganese were calculated using Equation (10).

$$R_{Me} = \frac{C_{t,Me} \times V}{W_{Me}} \times 100\% \quad (Me = Mn^{2+}, Li^+) \tag{10}$$

where R_{Me} is the extraction ratio of lithium or dissolution loss ratio of manganese; $C_{t,Me}$ is the element concentration of different times; V is the solution volume and W_{Me} is the weight of Me in the LMO-1. The influence of hydrochloric acid concentration was studied by stirring (200 rpm) 0.05 g LMO-1 in 100 mL hydrochloric acid solution (0.02–0.1 mol·L^{-1}) for 12 h at 20 °C.

3. Results and Discussion

3.1. Optimization of Synthesis Parameters

The XRD patterns of the products obtained under different conditions are shown in Figure 2. Figure 2a shows intermediate γ-MnOOH (JCPDS cards no. 50-0009) was produced at low Li$^+$ concentration. With the increase of Li$^+$ (>11 mol·L^{-1}), the target LMO was produced and intermediate γ-MnOOH was disappeared. Figure 2b indicates that with the increase of the amount of KMnO$_4$, the LMO lattice structure becomes stable gradually but when the amount of KMnO$_4$ was above 9 g, the impurity (∇) was generated. Figure 2c shows that using lower ethanol volume in the synthesis process resulted in the formation of intermediate Li$_4$Mn$_{14}$O$_{27}$·xH$_2$O (JCPDS cards no. 41-1379). When the volume fraction increases above 8.75%, the impurity peak (•) was observed. Figure 2d showed that the intermediate Li$_4$Mn$_{14}$O$_{27}$·xH$_2$O and γ-MnOOH were first formed within a short reaction time and LMO could be obtained after 12-h reaction. Figure 2e reflects the effect of reaction temperature on the LMO. Li$_4$Mn$_{14}$O$_{27}$·xH$_2$O and γ-MnOOH were produced at the lower temperature and LMO could be synthesized when the reaction temperature over 160 °C. Thus, we found the optimal LMO could be obtained at Li$^+$ concentration of 11 mol·L^{-1}, hydrothermal reaction at 160 °C for 12 h, ethanol volume fraction of 7.5%, using 3 g of KMnO$_4$. We speculate the synthesis is followed by the mechanism illustrated in Figure 3. In LiCl·H$_2$O solution, KMnO$_4$ is firstly reduced by ethanol and the intermediates Li$_4$Mn$_{14}$O$_{27}$·xH$_2$O and γ-MnOOH are formed. Then γ-MnOOH is oxidized by KMnO$_4$ and Li$_4$Mn$_{14}$O$_{27}$·xH$_2$O is furthered reduced by ethanol simultaneously. Finally, the lithium ion enters the Mn-O framework to form cubic LMO with the increase of lithium concentration.

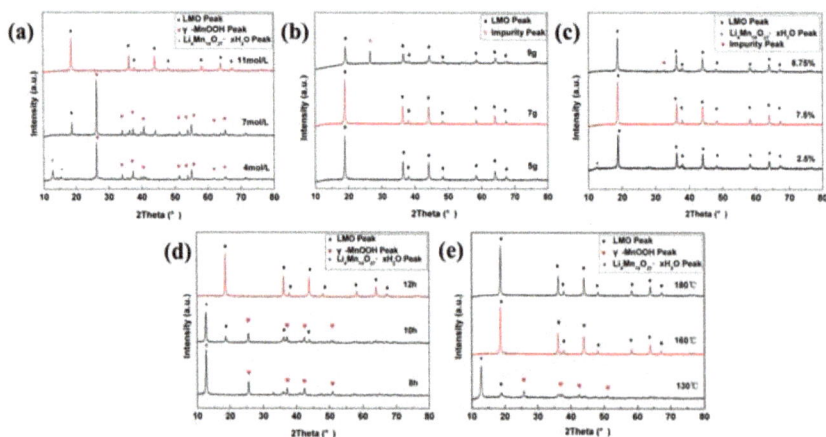

Figure 2. X-Ray diffraction (XRD) patterns of resultant under different preparation conditions: (**a**) the concentration of Li$^+$; (**b**) the amount of KMnO$_4$; (**c**) the volume ratio of ethanol; (**d**) reaction time; (**e**) reaction temperature.

$$KMnO_4^- + CH_3CH_2OH + LiCl \longrightarrow \gamma - MnOOH + Li_4Mn_{14}O_{27} \cdot xH_2O + CH_3COOH + KCl \quad (a)$$

$$\left.\begin{array}{l} \gamma - MnOOH + KMnO_4 + LiCl \\ Li_4Mn_{14}O_{27} \cdot xH_2O + CH_3CH_2OH \end{array}\right\} \longrightarrow LiMn_2O_4 \quad (b)$$

$$\underset{(LMO)}{LiMn_2O_4} \underset{Li^+}{\overset{H^+}{\rightleftharpoons}} \underset{(HMO)}{\lambda - MnO_2} \quad (c)$$

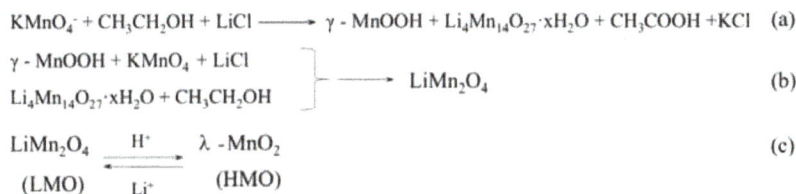

Figure 3. Synthetic mechanisms: (**a,b**) synthesis of LMO; (**c**) absorption-desorption mechanism of λ-MnO$_2$ (HMO) and LMO.

3.2. Ion-Sieves Characterization

Figure 4 shows the XRD patterns of the LMO, HMO and the sample after adsorption process (noted as LMO-1). The diffraction peak of LMO corresponds to a cubic spinel HMO structure [space group: Fd3m (JCPDS 35-0782)], with the lattice constants is 8.23 Å. It should be noted that the XRD patterns of HMO and LMO-1 are similar with the diffraction patterns of LMO, with lattice constants of 8.01 Å and 8.23 Å, respectively, indicating that the Li$^+$ is free to access the structure and the Mn-O lattice remains stable during the adsorption and desorption process. It is found that the diffraction peak of HMO shifts to a higher diffraction angle than that of LMO, which can be explained by the mechanism showed in Figure 3c. During the Li$^+$ desorption process, H$^+$ in the solution replaces the original position of Li$^+$ in the LMO the ionic radius of H$^+$ is smaller than Li$^+$, leading to cell shrinkage, which is also reported in literature [26]. The characteristic diffraction peaks of LMO-1 are still sharp and only the intensities decreased compared with the LMO, indicating that the HMO can be used for efficient adsorption of Li$^+$.

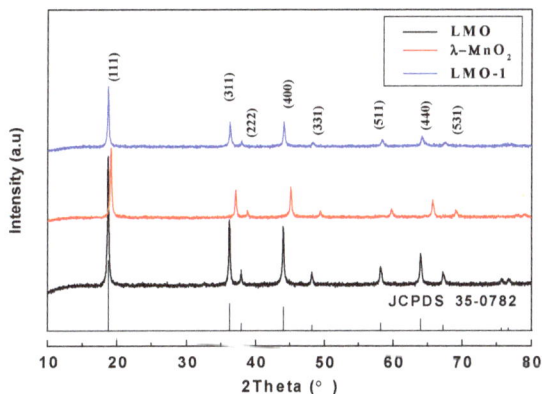

Figure 4. XRD patterns of optimized LMO, HMO and LMO-1.

Figure 5 describes the XPS spectra of LMO and HMO. As showed in Figure 5a, the spectra of the Mn3s orbit shows the binding energy difference of the two peaks was 5.15 eV (ΔE = 5.15 eV), indicating that the valences of Mn in LMO are +3 and +4. The binding energy of Mn^{3+} peak was 641.33 eV and Mn^{4+} peaks were 643.76 eV and 642.66 eV, which were obtained by means of peak-differentiation-imitating analysis at the Mn2p3/2 orbit (Figure 5b). The results are in line with a previous report [27]. The average valence of Mn in LMO (+3.65) could be calculated (Table 2), which is higher than the theoretical valence (+3.5), indicating that proportion of Mn^{3+} in LMO is lower than theoretical. Thus, it can be deduced that LMO has a more stable crystal structure. Figure 5c is the XPS spectra of HMO in the Mn3s orbit and the binding energy difference of the two peaks is 4.78 eV, indicating that the manganese valence in HMO is +4. Furthermore, this is also proven by the peak of HMO in the Mn2p3/2 orbital (Figure 5d).

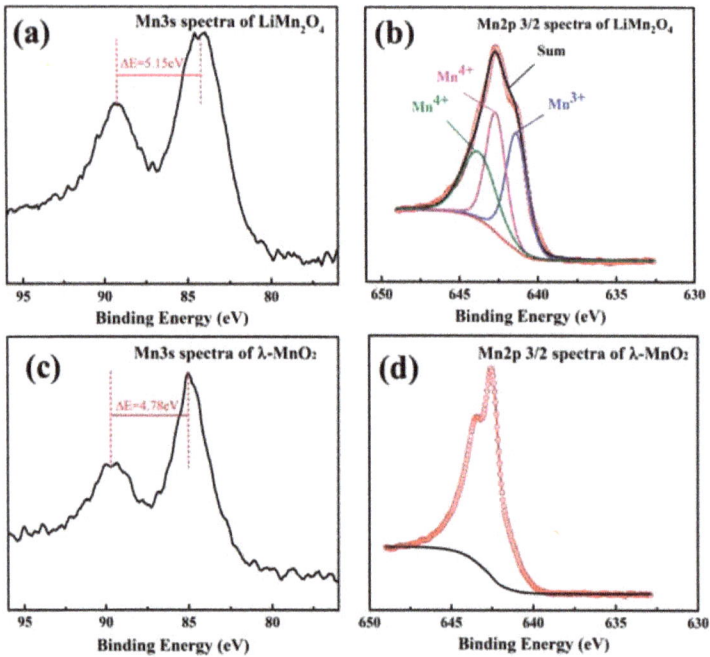

Figure 5. X-ray photoelectron spectroscopy (XPS) Mn3s and Mn2p spectra of LMO and HMO.

Table 2. Average valances of Mn element in LMO and HMO.

Sample	Binding Energy (eV)	Chemical State	Peak Area	Average Valences
LMO	643.76	Mn2p$_{3/2}$ Mn^{4+}	38,795.51	
	642.66	Mn2p$_{3/2}$ Mn^{4+}	40,557.36	+3.65
	641.33	Mn2p$_{3/2}$ Mn^{3+}	42,725.95	
HMO	—	Mn2p$_{3/2}$ Mn^{4+}	—	+4

Figure 6 shows the morphology of LMO, HMO and the ion-sieve after Li$^+$ adsorption (LMO-1). The LMO presents regular hexagonal shape with the thickness of 110 nm and the lateral size of ~300–400 nm (Figure 6a). It is apparent that the LMO (Figure 6b) have a smooth surface without agglomeration, while HMO and LMO-1 appear to have a small crack on the surface (Figure 6c). We speculate that it is attributed to the manganese loss after acid treatment that results in partial collapse of the crystal. However, HMO and LMO-1 can still remain their intact hexagonal structure and it is consistent with the XRD results in Figure 4.

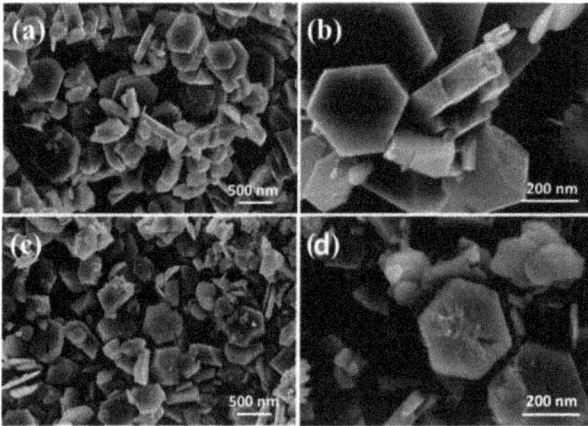

Figure 6. Scanning electron microscopy (SEM) images of LMO (**a,b**), HMO (**c**) and LMO-1 (**d**).

Figure 7 shows the HRTEM images of LMO and HMO. Both LMO and HMO were observed as a non-agglomerated particle with a regular hexagonal morphology (Figure 7a,d). The lattice spacing are 0.478 nm and 0.477 nm, respectively, as shown in Figure 7b,f, which agrees with the (111) crystal plane of the XRD pattern in Figure 4. The selected area electron diffraction (SAED) patterns of LMO and HMO can be seen in Figure 7c,g, the dot matrix confirms their cubic single-crystal structures.

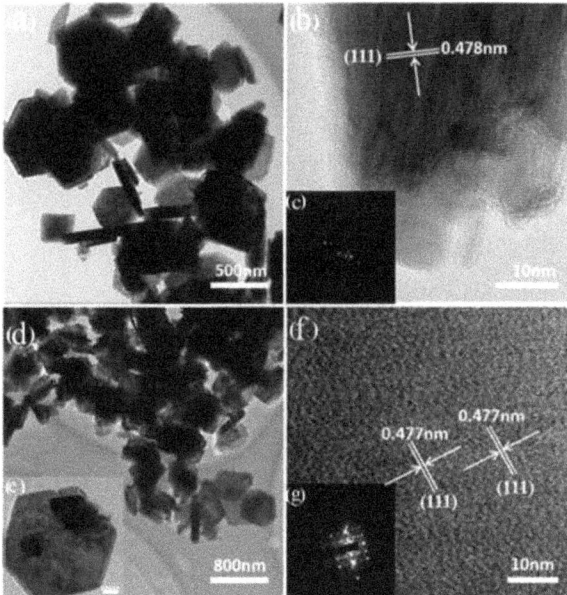

Figure 7. High resolution transmission electron microscopy (HRTEM) images and selected area electron diffraction (SAED) patterns of the LMO (**a–c**) and HMO (**d–g**).

3.3. Adsorption Behavior of the HMO

3.3.1. Effect of pH Value on Adsorption Capacity

Figure 8 describes the pH value effect on the Li⁺ adsorption process. The adsorption capacity of HMO was very low in acidic condition. The adsorption capacity of HMO increased sharply and then reached the maximum with the pH value increase in solution, which indicated that alkaline adsorption environment favored the adsorption of HMO. The adsorption-desorption mechanism of LMO can be explained by Figures 1 and 3c. Adsorption Li⁺ at alkaline condition is beneficial to the formation of LMO and desorption of Li⁺ at the acid condition is beneficial to the formation of HMO. The mathematic relationship [12] between adsorption capacity (Q_e) and pH could be described by the equation $Q_e = f (C_e, pH)$. The Q_e (the amount of Li⁺ insertion) increases by the increase of pH. When the pH was greater than 10, the adsorption capacity of HMO hardly increase with the increase of pH value. We speculated that the reduction of Mn^{4+} was accelerated under the strong alkaline condition, so the adsorption of the ion sieve was inhibited. Therefore, when pH > 10, the adsorption capacity Q_e tends to be stable. The similar phenomenon was also found by other reseachers [28].

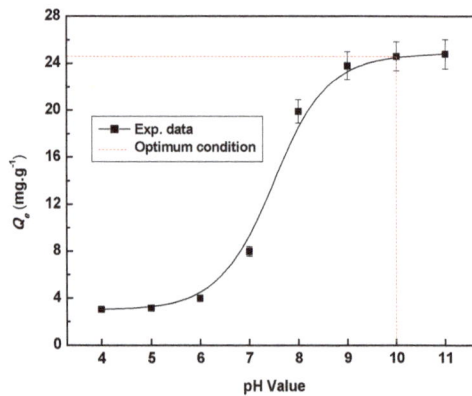

Figure 8. The Q_e—pH value curve with adsorption of 0.1 g HMO in 50 mg·L⁻¹ Li⁺ at 18 °C.

3.3.2. Static Adsorption Test

Figure 9 shows that the adsorption process occurs primarily in the rapid adsorption stage and the exchange of Li⁺ into the spinel lattice dominates the adsorption flat stage. Table 3 compares synthesis method and adsorption capacity of λ-MnO₂ in this paper with those of other paper. Solid-phase [19,20] method is often reacted with high energy consumption. It is apparent that hydrothermal method usually uses strong alkaline LiOH [29] or acidic manganese salt [23] as raw material with the disadvantage of corroding equipment. In this study λ-MnO₂ was obtained by the one-pot hydrothermal method under neutral and mild condition. The adsorption capacity is 24.7 mg·g⁻¹, 64.4% of the theoretical adsorption capacity $Q_{th} = \frac{M_{Li}}{M_{\lambda-MnO_2}} = \frac{6.94 \times 1000}{180.94} = 38.3$ mg·g⁻¹; $\frac{Q}{Q_{th}} = \frac{24.7}{38.3} = 64.4\%$, which is higher than the 49.2% of the theoretical adsorption capacity reported in the paper [23]; and 61.9% of the theoretical adsorption capacity in the paper. The Crank's model was used to predict the adsorption rate of Li⁺. The model fitted well with the experimental data. The efficient film coefficient (D_e) were calculated by Equation (2) as 1.35×10^{-5} cm²·s⁻¹. The correlation coefficient (R^2) was 0.9971. The coefficient of mass transfer (k) can be obtained by efficient film coefficient and physical property of adsorption system. In all, D_e derived from fitting calculation provides a vital parameter of feed height in adsorption tower design [30].

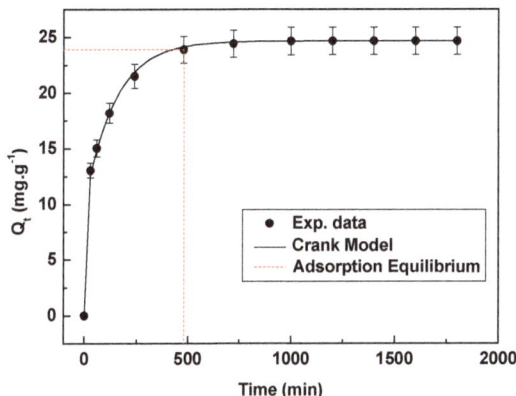

Figure 9. Fitting result of adsorption data by Crank model using $50 \, \text{mg·L}^{-1}$ Li^+ on $0.1 \, \text{g}$ HMO at $18 \, °C$.

Table 3. Similar method of adsorption capacity comparison.

Ion Sieve	Raw Materials	Method	*Temp.* (°C)	*t* (h)	Crystal Morphology	Q (mg·g^{-1})	$\frac{Q}{Q_{th}}$ (%)	Ref.
λ-MnO$_2$	Mn(NO$_3$)$_2$, LiOH, H$_2$O$_2$	hydrothermal	110	10	Nanowire	23.7	61.9	[29]
λ-MnO$_2$	MnSO$_4$, (NH$_4$)$_2$S$_2$O$_8$	hydrothermal	150 650	12 6	Nanowire	16.9	49.2	[23]
λ-MnO$_2$	LiNO$_3$, Mn(NO$_3$)$_2$	solid-phase	700	1	Sphere	-		[20]
λ-MnO$_2$	Li$_2$CO$_3$, MnCO$_3$	solid-phase	800	5	-	-		[19]
λ-MnO$_2$	LiCl KMnO$_4$ ethanol	hydrothermal	160	12	Hexagonal	24.7	64.4	This work

3.3.3. Adsorption Kinetic Test

Figure 10 shows the linear fitting of the pseudo-first-order kinetic model and the pseudo-second-order kinetic model. Table 4 compares the fitted kinetic data of the two models at same temperatures. Under the same test conditions, the two models both predicted the adsorption capacity and the correlation coefficient (R^2) of the pseudo-second-order kinetics equation is much larger than the pseudo-first-order kinetic equation ($R^2 = 0.7678$). These data reveal that the adsorption behavior of the HMO ion sieve conforms to the pseudo-second-order kinetics model and the adsorption process is primarily chemical adsorption [31].

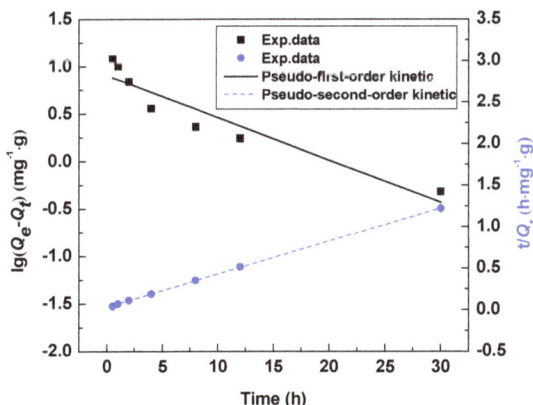

Figure 10. Pseudo-first-order and pseudo-second-order kinetic curves Li^+ adsorption by HMO at $18 \, °C$.

Table 4. Dynamic parameters of lithium adsorption.

Temperature	Pseudo-First-Order Kinetic Model			Pseudo-Second-Order Kinetic Model		
	K_1	Q_{e1}	R^2	K_2	Q_{e2}	R^2
18 °C	0.115	8.41	0.7678	0.0687	25.3	0.9998

3.3.4. Adsorption Isotherm of Li+ on HMO

The adsorption constants and the correction factors were obtained by Langmuir and Freundlich equations fittings. Table 5 lists the various parameter values for both models. Figure 11 show the fitting effect of the two models. The Langmuir isotherm model ($R^2 = 0.9999$) fitting was much better than that of the Freundlich isotherm model ($R^2 = 0.9918$) compared with the experimental data. This result indicates that the HMO has homogeneous adsorption sites.

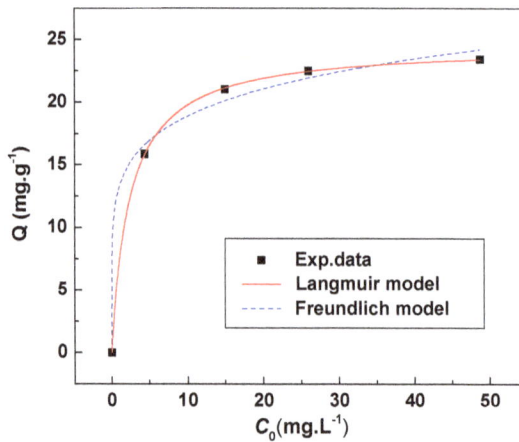

Figure 11. Langmuir and Freundlich isotherms of Li+ adsorption by HMO at 18 °C.

Table 5. Adsorption isotherm constants of Li+ on HMO.

Temperature	Langmuir Model			Freundlich Model		
	K_L	Q_m	R^2	K_F	n	R^2
18 °C	0.415	24.6	0.9999	13.2	6.38	0.9918

3.4. Absorption Selectivity of HMO

Table 6 shows the HMO ion sieve adsorption selectivity for Li+ compared with other coexisting metal ions in brine, including Na+, K+, Ca^{2+} and Mg^{2+}. According to Table 6, the adsorption capacity of HMO in brine is 6.26 mg·g^{-1}, which is lower than the value of that in the pure Li+ solution. We speculated that the acidic environment (pH = 5.64) is not conducive to the free insertion of lithium ions in λ-MnO$_2$. The distribution coefficients (K_d) are in the order of Li+ > Ca^{2+} > K+ > Na+ > Mg^{2+}, indicating high selectivity for Li+, compared with other metal ions. The ion sieve showed excellent ion selectivity, especially for Mg^{2+}, whose separation factor (α_{Mg}^{Li}) is 1.35×10^3. This solves the problem of separating Li+ and Mg^{2+} in brine with a high ratio of magnesium to lithium. Na+, K+, Ca^{2+}, Mg^{2+} in solution do not have competitive effect with Li+ during ion sieve adsorption process since the concentration factor (C_F) of Li+ is higher than other ions.

Table 6. Adsorption selectivity data of metal ions on HMO in brine.

Metal Ion	C_0 (mg·L^{-1})	C_e (mg·L^{-1})	C_F (L·g^{-1} × 10^{-3})	Q_e (mg·g^{-1})	K_d (mL·g^{-1})	ff$_{Me}^{Li}$
Li$^+$	319.3	288.0	19.6	6.26	19.6	1.00
Na$^+$	1810.0	1804.6	0.591	1.07	0.592	36.7
K$^+$	815.8	812.6	0.793	0.647	0.796	27.3
Ca^{2+}	121.8	120.2	2.63	0.320	2.63	8.16
Mg^{2+}	119,600.0	119,590.4	0.0161	1.93	0.0161	1.35×10^3

Experiment conditions: T = 18 °C, pH = 5.64, V = 20.0 mL, W = 0.100 g.

3.5. Desorption Behavior of LMO-1

Figure 12 shows the desorption curve of Li$^+$ or Mn^{2+} after adsorption. It was observed that the extraction of Li$^+$ and Mn^{2+} occurred rapidly at the beginning of the desorption process, almost reaching the maximum extraction rate at 20 min, then slightly rose up to 22.0 mg·g^{-1} and 38.9 mg·g^{-1}, respectively. The maximum extraction rate of R_{Li}^+ is 98.7%. The dissolution of Mn^{2+} (R_{Mn}^{2+}) is only 3.9%, which was calculated by Equation (10). This phenomenon may benefit from the unique layered structure. Integrity shape without defect with a larger surface can sufficiently contact with Li$^+$ of solution and maintain adsorption stability, accelerating the absorption and desorption process.

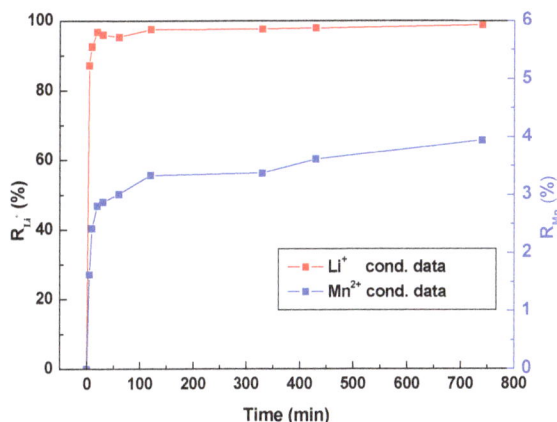

Figure 12. Desorption and Dissolution loss behavior of LMO-1.

4. Conclusions

A series of LMO was successfully prepared via a facile one-pot hydrothermal method and we optimized synthetic conditions as well. The HMO ion-sieve has unique hexagonal spinel structure with the thickness of 110 nm and lateral size of 300–400 nm. XRD patterns of LMO and HMO confirm their high crystallization degree. The average valence of Mn in LMO is +3.65, which higher than that in theory (+3.5). The adsorption capacity of HMO is 24.7 mg·g^{-1} in Li$^+$ solution and the dissolution of Mn^{2+} is only 3.9%. The adsorption equilibrium isotherms data are well fitted with Langmuir model. Moreover, the distribution coefficients (K_d) of HMO is much larger between Li$^+$ and Mg^{2+} and the separation factor (α_{Mg}^{Li}) was 1.35×10^3. Therefore, our HMO ion-sieve shows great potential to extract lithium in brine or seawater under high magnesium ratio conditions.

Author Contributions: F.Y. and S.J. conceived and designed the experiments; F.Y. and S.C. performed the experiments; X.Z. and C.S. analyzed the data; S.J. contributed reagents/materials/analysis tools; F.Y., F.X., S.J. and W.X. wrote the paper.

Acknowledgments: This work was financially supported by the National Natural Science Foundation of China (U1407122), Innovation Project of Jiangsu Province (SJZZ16_0136).

Conflicts of Interest: The authors declare no conflict of interest.

References

1. Yang, S.; Huimin, R.; Shen, J.; Gao, C. Preparation methods and analyses of structural performance of spinel-type lithium manganese oxide ion sieves. *Chem. Ind. Eng. Prog.* **2015**, *6*, 1690–1699.
2. Özgür, C. Preparation and characterization of LiMn$_2$O$_4$ ion-sieve with high Li$^+$ adsorption rate by ultrasonic spray pyrolysis. *Solid State Ion.* **2010**, *181*, 1425–1428. [CrossRef]
3. Xiao, J.L.; Sun, S.Y.; Song, X.; Li, P.; Yu, J.G. Lithium ion recovery from brine using granulated polyacry λ–MnO$_2$ ion-sieve. *Chem. Eng. J.* **2015**, *279*, 659–666. [CrossRef]
4. Kesler, S.E.; Gruber, P.W.; Medina, P.A.; Keoleian, G.A.; Everson, M.P.; Wallington, T.J. Global lithium resources: Relative importance of pegmatite, brine and other deposits. *Ore Geol. Rev.* **2012**, *48*, 55–69. [CrossRef]
5. Feng, G.; Ping, Z.M.; Zhen, N.; Hua, L.J.; Sheng, S.P. Brine Lithium Resource in the Salt Lake and Advances in Its Exploitation. *Acta Geosci. Sin.* **2011**, *32*, 483–492.
6. Xiao, J.L.; Sun, S.Y.; Wang, J.; Li, P.; Yu, J.G. Synthesis and Adsorption Properties of Li$_{1.6}$Mn$_{1.6}$O$_4$ Spinel. *Ind. Eng. Chem. Res.* **2013**, *52*, 11967–11973. [CrossRef]
7. Zandevakili, S.; Ranjbar, M.; Ehteshamzadeh, M. Improvement of lithium adsorption capacity by optimising the parameters affecting synthesised ion sieves. *Micro Nano Lett.* **2015**, *10*, 58–63. [CrossRef]
8. Li, L.; Qu, W.; Liu, F.; Zhao, T.; Zhang, X.; Chen, R.; Wu, F. Surface modification of spinel λ-MnO$_2$ and its lithium adsorption properties from spent lithium ion batteries. *Appl. Surf. Sci.* **2014**, *315*, 59–65. [CrossRef]
9. Xiao, G.; Tong, K.; Zhou, L.; Xiao, J.; Sun, S.; Li, P.; Yu, J. Adsorption and Desorption Behavior of Lithium Ion in Spherical PVC–MnO$_2$ Ion Sieve. *Ind. Eng. Chem. Res.* **2012**, *51*, 10921–10929. [CrossRef]
10. Zhu, G.; Wang, P.; Qi, P.; Gao, C. Adsorption and desorption properties of Li$^+$ on PVC-H$_{1.6}$ Mn$_{1.6}$O$_4$ lithium ion-sieve membrane. *Chem. Eng. J.* **2014**, *235*, 340–348. [CrossRef]
11. Wang, C.; Zhai, Y.; Wang, X.; Zeng, M. Preparation and characterization of lithium λ-MnO2 ion-sieves. *Front. Chem. Sci. Eng.* **2014**, *8*, 471–477. [CrossRef]
12. Xiao, J.; Nie, X.; Sun, S.; Song, X.; Ping, L.; Yu, J. Lithium ion adsorption–desorption properties on spinel Li$_4$Mn$_5$O$_{12}$ and pH-dependent ion-exchange model. *Adv. Powder Technol.* **2015**, *26*, 589–594. [CrossRef]
13. Singh, I.B.; Singh, A. A facile low-temperature synthesis of Li4Mn5O12 nanorods. *Colloid Polym. Sci.* **2017**, *295*, 689–693. [CrossRef]
14. Liu, L.; Zhang, H.; Zhang, Y.; Cao, D.; Zhao, X. Lithium extraction from seawater by manganese oxide ion sieve MnO$_2$·0.5H$_2$O. *Colloids Surf. A.* **2015**, *468*, 280–284. [CrossRef]
15. Park, M.J.; Nisola, G.M.; Beltran, A.B.; Torrejos, R.E.C.; Seo, J.G.; Lee, S.; Kim, H.; Chung, W. Recyclable composite nanofiber adsorbent for Li$^+$ recovery from seawater desalination retentate. *Chem. Eng. J.* **2014**, *254*, 73–81. [CrossRef]
16. Sorour, M.H.; El-Rafei, A.M.; Hani, H.A. Synthesis and characterization of electrospun aluminum doped Li$_{1.6}$Mn$_{1.6}$O$_4$ spinel. *Ceram. Int.* **2016**, *42*, 4911–4917. [CrossRef]
17. Sun, D.; Meng, M.; Yin, Y.; Zhu, Y.; Li, H.; Yan, Y. Highly selective, regenerated ion-sieve microfiltration porous membrane for targeted separation of Li$^+$. *J. Porous Mater.* **2016**, *23*, 1–9. [CrossRef]
18. Yu, Q.; Sasaki, K. In situ X-ray diffraction investigation of the evolution of a nanocrystalline lithium-ion sieve from biogenic manganese oxide. *Hydrometall* **2014**, *150*, 253–258. [CrossRef]
19. Yuan, J.S.; Yin, H.B.; Ji, Z.Y.; Deng, H.N. Effective Recycling Performance of Li$^+$ Extraction from Spinel-Type LiMn$_2$O$_4$ with Persulfate. *Ind. Eng. Chem. Res.* **2014**, *53*, 9889–9896. [CrossRef]
20. Park, H.K.; Rah, H.; Dong, J.K.; Chun, U.; Kim, S.G. Confined growth of lithium manganese oxide nanoparticles. *J. Sol-Gel Sci. Technol.* **2013**, *67*, 464–472. [CrossRef]
21. Tang, W.; Tian, S.; Liu, L.L.; Li, L.; Zhang, H.P.; Yue, Y.B.; Bai, Y.; Wu, Y.P.; Zhu, K. Nanochain LiMn$_2$O$_4$ as ultra-fast cathode material for aqueous rechargeable lithium batteries. *Electrochem. Commun.* **2011**, *13*, 205–208. [CrossRef]
22. Xiao, G.P. Granulation to LiMn$_2$O$_4$ Ion-Sieve and Its Lithium Adsorption Property. *Chin. J. Inorg. Chem.* **2010**, *26*, 435–439.
23. Zhang, Q.H.; Li, S.P.; Sun, S.Y.; Yin, X.S.; Yu, J.G. LiMn$_2$O$_4$ spinel direct synthesis and lithium ion selective adsorption. *Chem. Eng. Sci.* **2010**, *65*, 169–173. [CrossRef]

24. Crank, J. *The Mathematics of Diffusion*; Oxford University Press: Oxford, UK, 1977; Volume 7, pp. 1–10.

25. Xue, F.; Xu, Y.; Lu, S.; Ju, S.; Xing, W. Adsorption of Cefocelis Hydrochloride on Macroporous Resin: Kinetics, Equilibrium, and Thermodynamic Studies. *J. Chem. Eng. Data* **2016**, *61*, 2179–2185. [CrossRef]

26. Tian, L.; Ma, W.; Han, M. Adsorption behavior of Li+ onto nano-lithium ion sieve from hybrid magnesium/lithium manganese oxide. *Chem. Eng. J.* **2010**, *156*, 134–140. [CrossRef]

27. Sun, S.Y.; Song, X.; Zhang, Q.H.; Wang, J.; Yu, J.G. Lithium extraction/insertion process on cubic Li-Mn-O precursors with different Li/Mn ratio and morphology. *Adsorption* **2011**, *17*, 881–887. [CrossRef]

28. Ooi, K.; Miyai, Y.; Sakakihara, J. Mechanism of lithium (Li^+) insertion in spinel-type manganese oxide. Redox and ion-exchange reactions. *Langmuir* **1991**, *7*, 1167–1171. [CrossRef]

29. Zhang, Q.; Sun, S.; Li, S.; Jiang, H.; Yu, J. Adsorption of lithium ions on novel nanocrystal MnO_2. *Chem. Eng. Sci.* **2007**, *62*, 4869–4874. [CrossRef]

30. Chen, L.; Lin, J.W.; Yang, C.L. Absorption of NO_2 in a packed tower with Na_2SO_3 aqueous solution. *Environ. Prog. Sustain. Energy* **2002**, *21*, 225–230. [CrossRef]

31. Tian, M.J.; Liao, F.; Ke, Q.F.; Guo, Y.J.; Guo, Y.P. Synergetic effect of titanium dioxide ultralong nanofibers and activated carbon fibers on adsorption and photodegradation of toluene. *Chem. Eng. J.* **2017**, *328*, 962–976. [CrossRef]

processes

MDPI

Article

Seepage and Damage Evolution Characteristics of Gas-Bearing Coal under Different Cyclic Loading–Unloading Stress Paths

Qingmiao Li [1,2], Yunpei Liang [1,2,*] and Quanle Zou [1,2,*]

1   State Key Laboratory of Coal Mine Disaster Dynamics and Control, Chongqing University, Chongqing 400044, China; 20152001009@cqu.edu.cn
2   College of Resources and Environment Science, Chongqing University, Chongqing 400044, China
*   Correspondence: liangyunpei@cqu.edu.cn (Y.L.); quanlezou2016@cqu.edu.cn (Q.Z.); Tel.: +86-135-9460-8029 (Y.L.); +86-177-8337-2719 (Q.Z.)

Received: 19 September 2018; Accepted: 12 October 2018; Published: 15 October 2018

Abstract: The mechanical properties and seepage characteristics of gas-bearing coal evolve with changes in the loading pattern, which could reveal the evolution of permeability in a protected coal seam and allow gas extraction engineering work to be designed by using the effect of mining multiple protective seams. Tests on gas seepage in raw coal under three paths (stepped-cyclic, stepped-increasing-cyclic, and crossed-cyclic loading and unloading) were carried out with a seepage tester under triaxial stress conditions. The permeability was subjected to the dual influence of stress and damage accumulation. After being subjected to stress unloading and loading, the permeability of coal samples gradually decreased and the permeability did not increase before the stress exceeded the yield stage of the coal samples. The mining-enhanced permeability of the coal samples in the loading stage showed a three-phase increase with the growth of stress and the number of cycles and exhibited an N-shaped increase under the stepped-cyclic loading while it linearly increased under the other two paths in the unloading stage. With the increase of peak stress and the accumulation of damage in coal samples, the sensitivity of the permeability of coal samples to stress gradually declined. The relationship between the damage variable and the number of cycles conformed to the Boltzmann function.

Keywords: multiple protective seams; cyclic loads; recovery rate of permeability; stress sensitivity coefficient; loading–unloading response ratio; damage variable

1. Introduction

Coal seams in China are characterized by having a low permeability and a high gas content, which is an essential reason for the occurrence of coal and gas outbursts [1–6]. Permeability-enhancing measures must be taken and the gas content must be below the critical values specified by the related provisions before the mining of the coal seam that has an outburst risk [7–12]. The mining of multiple protective seams in coal seam groups is considered to be one of the more effective regional measures for preventing and controlling gas outbursts in protected seams [13–16]. In this process, the coal in a protected seam is subjected to cyclic loading–unloading effects, which significantly change the mechanical properties and seepage characteristics of the coal. Related research results show that gas has a significant influence on the mechanical properties and energy dissipation characteristics of coal [17–24]. Moreover, the mechanical properties and seepage characteristics of coal under different loading–unloading paths greatly differ from those under the conventional loading pattern. Therefore, it is necessary to investigate the mechanical properties and seepage characteristics of gas-containing coal under different loading–unloading paths. Various achievements have been made regarding the previously mentioned topics. An experimental investigation on the anisotropic permeability of coal under cyclic loading and unloading conditions was conducted and it was found that cyclic loading and unloading can induce an irreversible reduction of the permeability, which tends to be diminished with the increasing loading/unloading cycles [25]. The seepage properties, acoustic emission characteristics, and energy dissipation of coal under the tiered cyclic loading were also relatively revealed [26]. To analyze the spatial evolution characteristics of AE events and the failure process of coal samples, the single-link cluster method was used. Based on the microcrack density criterion, Zhang et al. proposed that the failure process of a coal sample involves the transformation from small-scale damage to large-scale damage, which results in changes in the spatial correlation length [27]. Afterward, multilevel cyclic loading tests on cylindrical coal specimens were performed by Yang et al. to investigate the fatigue failure of coal under uniaxial stress [28]. Furthermore, the relationship between electromagnetic radiation and the dissipated energy of coal during the cyclic loading process was elaborated by Song et al. [29]. Overall, the mechanical properties and seepage characteristics of coal have been shown to be closely related to the loading pattern. Investigating the mechanical properties and seepage characteristics of gas-containing coal under cyclic loading–unloading paths will provide more theoretical and practical guidance toward revealing the evolution of permeability in a protected coal seam and to design gas extraction engineering works under the effect of mining multiple protective seams. In addition, the damage evolution characteristics of gas-bearing coal under different cyclic loading–unloading stress paths have not been fully investigated.

In order to study the mechanical properties and seepage characteristics of gas-containing coal under complex mining stress, three different cyclic loading–unloading paths were designed. By analyzing stress–strain and stress–permeability curves, the deformation and seepage characteristics of gas-containing coal under different cyclic loading–unloading paths were revealed. By virtue of the theory of the loading–unloading response ratio (LURR), the damage evolution characteristics of coal under different cyclic loading–unloading paths were investigated. By applying a mining-enhanced permeability and a stress sensitivity coefficient of permeability, the characteristics of seepage evolution in gas-containing coal under different cyclic loading–unloading paths were quantified. The research results are expected to provide theoretical support to further reveal the mechanism of a permeability increase in coal seams under multiple protective seams.

2. Experimental Equipment and Test Scheme

2.1. Experimental Equipment

The experiment was carried out by using a triaxial seepage experiment device for heat–fluid–solid coupling in gas-containing coal made at the Chongqing University, China. The experimental device can be used in gas seepage experiments in coal under the effects of different stress regimes (including confining pressure and axial stress) and gas pressures. The experimental device and constituents of the experimental cavity are displayed in Figure 1.

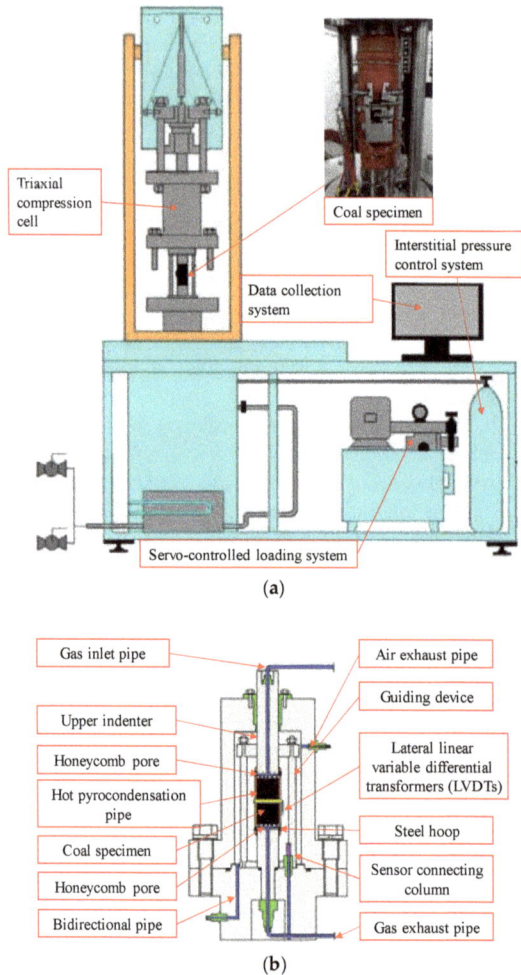

Figure 1. Triaxial seepage experimental device for the measurement of the heat–fluid–solid coupling property of gas-containing coal and the experimental chamber: (**a**) The whole experimental device and (**b**) the experimental chamber.

2.2. Coal Sample Preparation

The coal samples used for the experiment were taken from the 3-1 coal seam of Yuanzhuang Coal Mine, Anhui Province, China, which is shown in Figure 2. The 3-1 coal seam is located in the Permian Lower Stone Box Group, which is shown in Figure 2. The coal type is low metamorphic gas coal. The average thickness is 4.2 m and the gas content is 3.45 m^3/t. The main characteristics related to the coal's quality are shown in Table 1. The coal field where the mine is located is affected by the Huaxia tectonic system and the construction line is dominated by North-northeast (NNE), which is shown in Figure 3. These coal samples were cored, cut, and ground to form standard cylindrical coal samples that were 50 mm in diameter and 100 mm in length.

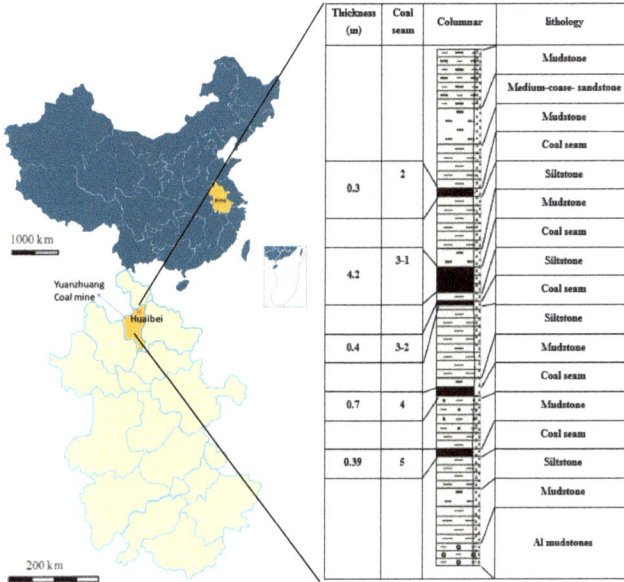

Figure 2. Location and geologic column of the Yuan Zhuang coal mine.

Table 1. The main characteristics related to the coal's quality.

Item	Fixed Carbon	Ash Content	Sulfur Content	Phosphorus Content	Volatile Content	Calorific Value	Bulk Density
3-1 Coal seam	60%	6%	0.40%	0.0004%	32%	27.86 MJ/kg	1350 Kg/m³

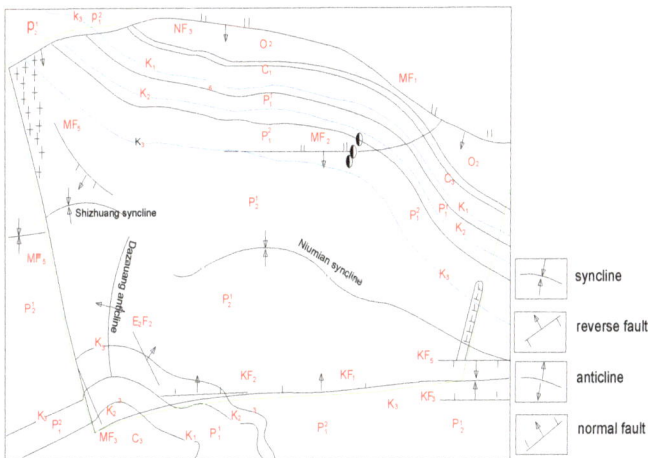

Figure 3. Geological structure outline map of the Yuan Zhuang coal mine.

2.3. Experimental Scheme

The authors carried out a similar simulation experiment involving the real-time monitoring of the stress state of a protected seam on the mining conditions of a coal seam group. The simulation experimental model is shown in Figure 4. During the experiment, under the conditions used to successively mine coal seams #5 and #6, the stress state of the coal seam #4 was monitored in real-time

(Figure 5). Before coal seam #5 was mined for 80 m, the coal masses at the monitored point of the coal seam #4 were subjected to a significant concentration of stress. Furthermore, during the mining of coal seam #6, the coal masses at the monitored point of coal seam $4 underwent exposure to multiple stresses and stress relief.

Figure 4. Similarity model for real-time monitoring of the stress state of the protected seam during the mining of multiple protective seams.

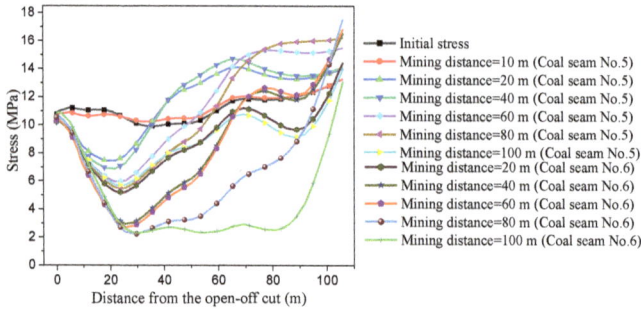

Figure 5. The stress state in the protected seam after the mining of protective seams.

Based on the previously mentioned analysis, three simplified cyclic loading–unloading stress paths were designed, which is shown in Figure 6. The experimental steps were as follows: the axial stress and confining pressures were synchronously loaded to 2 MPa at a rate of 0.05 MPa/s. Then, gas was constantly injected under a pressure of 1 MPa with the gas concentration consisting of 99.99%. After holding for 24 h in this state, the coal samples reached adsorption saturation. If the confining pressure was unchanged, the axial stress continued to be loaded or unloaded at a rate of 0.05 MPa/s until the samples were damaged.

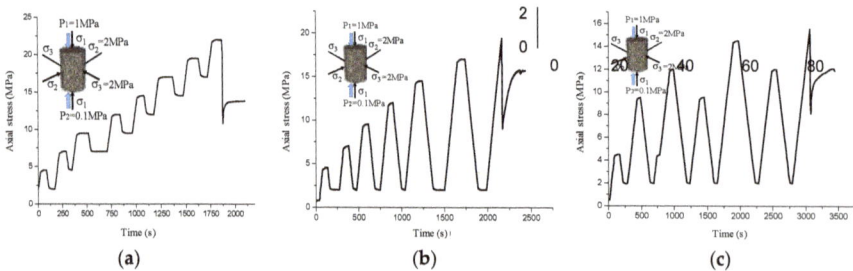

Figure 6. Three cyclic loading–unloading stress paths. (**a**) Stress path 1: stepped-loading and unloading. (**b**) Stress path 2: stepped-increasing-loading and unloading. (**c**) Stress path 3: crossed-cyclic-loading and unloading.

3. Methods and Data

3.1. Permeability Calculation

It is presumed that the gas permeation process in raw coal is an isothermal process and this gas is considered to be an ideal gas. According to Darcy's law, the permeability of coal is given by Reference [30].

$$K = \frac{2q_v P_a \mu L}{A(P_1^2 - P_2^2)} \tag{1}$$

where K is the permeability (m^2), q_v is the seepage velocity (m^3/s) of gas in coal masses, P_a is the atmospheric pressure (Pa), A is the cross-sectional area (m^2) of the specimen, L is the length (m) of the specimens, P_1 is the gas pressure at the air inlet (Pa), P_2 is the gas pressure at the air outlet (MPa), and μ is the gas viscosity (Pa·s).

During testing, the axial and lateral strains (ε_1 and ε_2) on the raw coal were monitored in real-time. Through the formula $\varepsilon_v = \varepsilon_1 + 2\varepsilon_2$, the volumetric strain (ε_v) in the raw coal was calculated. Based on the drawing, the axial stress–axial strain curve (σ_1–ε_1), axial stress–lateral strain curve (σ_1–ε_2), axial stress–volumetric strain curve (σ_1–ε_v), and permeability–axial strain curve (K–ε_1) of raw coal under different cyclic loading–unloading paths were obtained.

3.2. Permeability Ratio

Under cyclic loading–unloading paths 2 and 3, the ratio of the permeability at the end of unloading in each cycle to the permeability at the initial load is defined as the absolute recovery rate of permeability, which can be expressed in Formula (2) in percentage terms.

$$\chi_a = K_i / K_1 \times 100\% \tag{2}$$

where χ_a and K_i refer to the absolute recovery rates of the permeability and the permeability (mD) when the axial stress applied to coal samples was unloaded to 2 MPa during the ith cycle, respectively. Additionally, K_1 denotes the permeability (mD) of coal samples when the axial stress was loaded from 2 MPa during the first loading–unloading cycle.

The ratio of the permeability of coal samples after the stress unloading during each loading–unloading cycle to that in the loading process during this cycle is defined as the relative recovery rate of the permeability, which can be expressed as Formula (3) in percentage terms.

$$\chi_r = K_{i+1} / K_i \times 100\% \tag{3}$$

where χ_r and K_{i+1} represent the relative recovery rates of the permeability and the permeability (mD) when the axial stress was unloaded to 2 MPa during the $(i + 1)$th loading–unloading cycle, respectively.

3.3. Mining-Enhanced Permeability

A method was proposed by Xie et al. [31] to reflect the permeability increasing effect of coal seams by considering the contribution of volumetric change of a coal mass to its permeability. The mining-enhanced permeability (χ_p) is defined as the variation in permeability under the change per unit volume of coal masses.

$$\chi_p = \frac{dk}{d\varepsilon_v} \tag{4}$$

where k and ε_v represent the permeability (mD) of coal masses and the volumetric strain of damaged and fractured coal masses, respectively. The mining-enhanced permeability describes the permeability increase due to the fracturing of coal masses under the impact of mining and can allow the quantitative evaluation of the effect of permeability by increasing measures in coal seams.

3.4. Stress Sensitivity Coefficient of Permeability

To express the evolution of permeability, the dimensionless permeability (*DP*) was introduced, which is defined below [32].

$$DP = \frac{k}{k_0} \tag{5}$$

where k and k_0 refer to the permeability (mD) under different effective stresses and that under the initial effective stress, respectively.

The regression analysis showed that, during the loading and unloading of coal samples, the dimensionless permeability has a negative exponential relationship with the effective stress effect.

$$DP = \frac{k}{k_0} = be^{-\alpha_e \sigma_e} \tag{6}$$

where α_e, σ_e, and b separately denote the stress sensitivity coefficient (MPa^{-1}) of the permeability, the effective stress (MPa), and the dimensionless coefficient influenced by the initial permeability. Additionally, the stress sensitivity coefficient of the permeability of coal samples can be defined by the equation below.

$$\alpha_e = -\frac{\Delta k}{k} \cdot \frac{1}{\Delta \sigma_e} = -\frac{1}{k} \cdot \frac{dk}{d\sigma_e} \tag{7}$$

where Δk and $\Delta \sigma_e$ refer to the variation in the permeability of coal samples and the variation in the effective stress, respectively.

3.5. Damage Variable of Coal

The LURR is a parameter that was proposed by Yin Xiangchu to quantify damage when relating the extent of damage in a material to change in a physical quantity [33]. The commonly used axial stress and strain properties are separately considered as the load variable and the corresponding response variable, which are taken as the parameters for describing the LURR. The LURR Y is defined by the equation below.

$$Y = \frac{X_+}{X_-} \tag{8}$$

$$X = \lim_{\Delta P \to 0} \frac{\Delta R}{\Delta P} \tag{9}$$

where X_+ and X_- separately refer to the responses during loading and unloading, respectively. ΔP and ΔR denote the increments corresponding to the stress load variable (P) and the strain response variable (R), respectively, which can be acquired according to the stress–strain curve under the effect of experimental cyclic loading. When the load (P) is low, rocks are in an elastic stage and a linear or quasi-linear relationship arises between P and R with $X_+ = X_-$ and LURR $Y = 1$. When the load gradually increases, the damage increases, which shows that $X_+ > X_-$ and, correspondingly, $Y > 1$. Moreover, with increased damage, Y also rises. When media are close to becoming damaged, Y reaches a maximum. Therefore, the damage to coal samples during cyclic loading can be characterized by using LURR Y.

The relationship between LURR Y_E and damage variable (D) based on the fact that the fracture limit of materials conforms to a Weibull distribution on a mesoscopic scale was established by Zhang et al. [34].

$$Y_E = \frac{1}{m\left(\varepsilon_F^m - \varepsilon^m\right)} = \frac{1}{1 + m_{Wei}\ln(1 - D(\varepsilon))} \tag{10}$$

where m_{Wei} represents the Weibull index, ε denotes the strain ($\varepsilon = \left(\frac{1}{m}\right)^{\frac{1}{m}}$), and ε_F refers to the strain at the breaking point where the correspondence to damage is denoted by $D_F (D_F = 1 - e^{-\frac{1}{m}})$.

By transforming Formula (11), we get the equation below.

$$D = 1 - e^{\frac{1-Y_E}{m_{Wei}Y_E}} \tag{11}$$

Formula (11) displays the relationship between D and LURR Y. By utilizing the LURR Y calculated using Formula (11) and curve fitting, the parameters of the stress–strain curve were transformed to D to acquire the relationship between D and cyclic loading–unloading stresses.

4. Results

4.1. Characteristics of Deformation–Permeability of Gas-Containing Coal under Stress Path 1

The experimental results under stress path 1 are shown in Figure 7. Figure 7b presents the partially enlarged detail of the permeability–axial strain curve. As shown in the figure, during the stepped-cyclic loading–unloading experiment, the change in permeability of coal samples matched that of the axial stress–strain curve. The change in permeability of raw coal generally appeared as follows: with an increase in the number of cycles of loading and unloading, the permeability of the coal samples was significantly reduced during compaction, slowly declined in the elastic stage, and slowly rose in the yield stage. When coal samples were damaged, the permeability of the coal samples increased to a significant extent. The test results are described from two aspects below.

First, the stress–strain relationship met the following rules: hysteresis loops were seen in the axial stress–strain curve reflecting the plasticity of the coal samples. With an increasing axial stress, the area of these hysteresis loops gradually decreased. The radial strain was low before reaching the yield stage and coal samples were mainly subjected to axial deformation, which indicated that the axial strain played a dominant role while radial deformation exhibited little influence.

Second, the permeability–strain relationship had the following results: as cyclic loads were applied to the coal samples, their permeability generally declined with increasing stress. A significant hysteresis loop appeared during the first cycle and hysteresis loops gradually reduced in magnitude during subsequent cycles. This implied that, with growing stress, the plasticity of the coal samples reduced while their elasticity increased. The influence of stress loading and unloading on the permeability of coal samples gradually diminished.

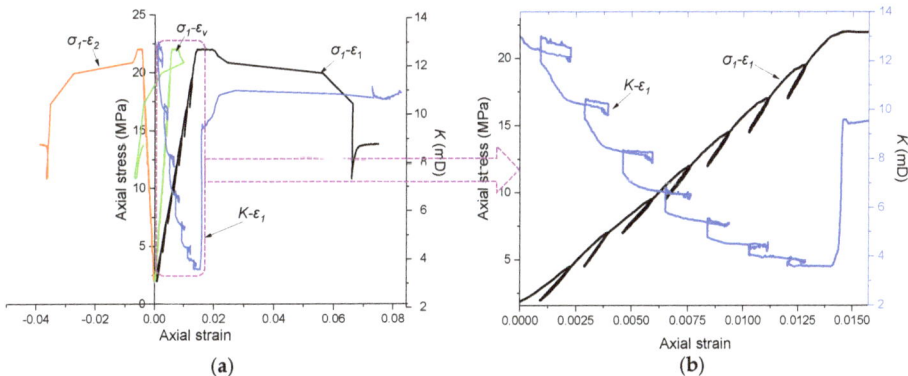

Figure 7. The relationships among the stress, strain, and permeability of raw coal under stress path 1. (**a**) Stress–strain and permeability–strain curves. (**b**) Partial enlargement of the axial stress–strain curve and the permeability–axial strain curve.

Additionally, under stress path 1, with an increasing number of cycles, the area enclosed by unloading and loading curves in the stress–strain space gradually declined and the elastic moduli gradually increased coinciding during the unloading and loading stages. In this case, the permeability of the coal samples in the loading stage gradually approximated that in the unloading stage and the hysteresis effect diminished. This indicated that, as cyclic loads were applied to the coal samples, their elasticity increased and the fracture deformation could be recovered to the greatest extent.

4.2. Characteristics of the Deformation–Permeability Curve of Gas-Containing Coal under Stress Path 2

Figure 8 shows the relationship curve between the stress–strain and the permeability of raw coal under the effect of the stepped-increasing peak load in stress path 2. As shown in the figure, the envelope line between the axial peak stress and axial strain of raw coal under stress path 2 was similar to that under uniaxial loading, which shows that the coal samples showed a favorable mechanical memory performance. Similar to stress path 1, the radial deformation of raw coal was less significant than its axial deformation. With an increase in the number of cycles, the area of the hysteresis loops in the axial stress–strain curve gradually increased.

Figure 9 shows the axial stress–axial strain and permeability–axial strain curves of raw coal in each cycle. As shown in the figure, with increasing axial stress, the permeability of raw coal declined. During the first two cycles, the permeability of raw coal in the unloading stage was lower than that in the loading stage. The reason for this was that, during the first two cycles, coal samples were compacted and, therefore, pores were smaller and subjected to irrecoverable deformation, which causes the permeability of the coal samples to fail to recover in the unloading stage after decreasing initially. From the third cycle to the failure stage, the permeability of the coal samples in the unloading stage was always greater than that in the loading stage during each cycle. The reason for this was due to the peak stress during each loading cycle damaging the coal samples, which resulted in increased fracturing and, thus, the permeability increased in the unloading stage. Eventually, the permeability rose dramatically in the yield stage and surpassed the initial permeability of the coal samples.

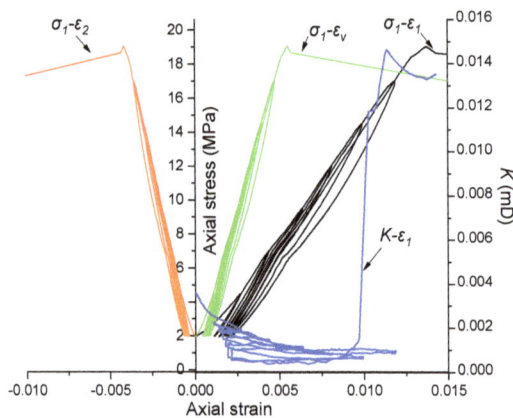

Figure 8. The relationships among the stress, strain, and permeability of raw coal under stress path 2.

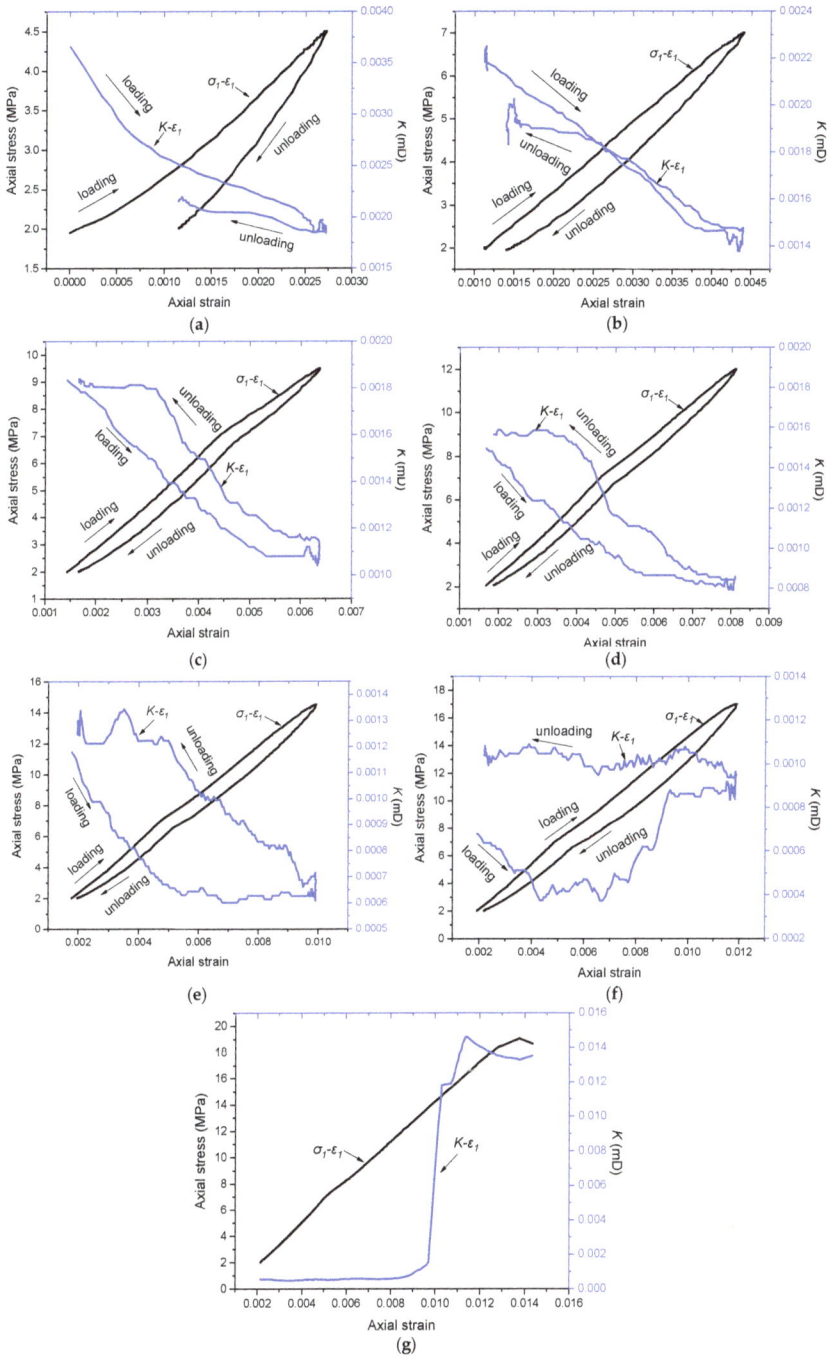

Figure 9. The axial stress–axial strain and permeability–axial strain curves of raw coal during each cycle under stress path 2. (**a**) The first cycle. (**b**) The second cycle. (**c**) The third cycle. (**d**) The fourth cycle. (**e**) The fifth cycle. (**f**) The sixth cycle. (**g**) The seventh cycle.

4.3. Characteristics of the Deformation–Permeability Curve of Gas-Containing Coal under Stress Path 3

The experimental results under stress path 3 are shown in Figure 10. Figure 10b shows the partially enlarged details of the permeability–axial strain curve. It can be seen from the figure that, under the effect of the crossed-cyclic loading–unloading path (stress path 3), the stress at the fourth cycle was the same as that at the second cycle and it was the same as that in the third and sixth cycles. However, the stress curves at the same stress level did not coincide and strain accumulation was found subjected to the previous high stress level.

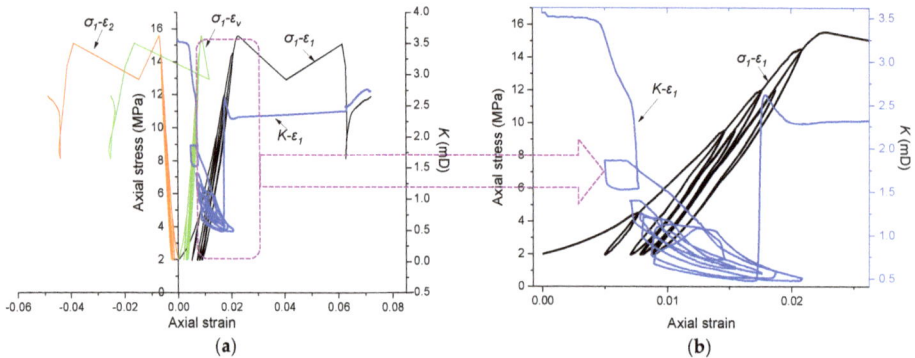

Figure 10. The relationships among stress, strain, and permeability of raw coal under stress path 3. (**a**) Stress–strain and permeability–strain curves. (**b**) Partial enlargement.

Figure 11 shows the axial stress–axial strain and permeability–axial strain curves of raw coal in each cycle. As shown in the figure, during the first two cycles, the permeability in the unloading process was far lower than that in the loading process, which indicates that coal samples were still in a compaction and closure stage. In this case, the pores failed to recover completely after being compressed. In contrast, during the third cycle, the permeability of coal samples in the unloading process approximated this in the loading process, which implies that the coal was in an elastic deformation stage. During the fourth cycle, the stress level was the same as that during the second cycle. However, during unloading, the stress–strain hysteresis loop was significantly smaller than during the second cycle. The permeability of coal samples in the unloading stage was larger than that in the loading stage, which indicates that the coal samples had entered a yielding stage. During this stage, the cumulative damage caused by cyclic loading and unloading constantly increased and the pores expanded after unloading, which resulted in increased permeability of the coal samples. During the fifth cycle, the stress further increased while the permeability further increased after unloading. During the sixth cycle, the stress applied to the coal samples was the same as that during the third cycle while the permeability of the coal samples was greater in the unloading stage, which indicates that damage accumulated over time in the post-yield state.

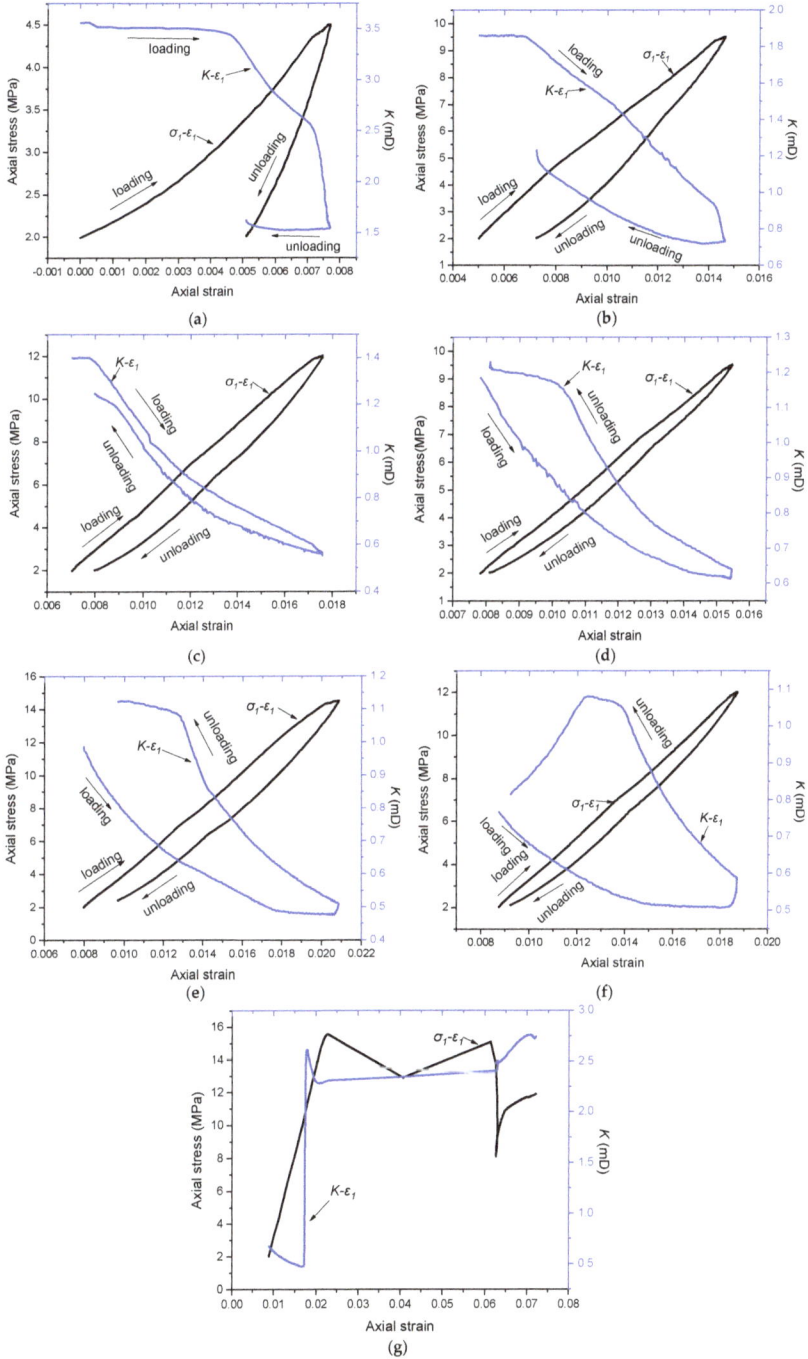

Figure 11. The axial stress–axial strain and permeability–axial strain curves of raw coal during each cycle under stress path 3. (**a**) The first cycle. (**b**) The second cycle. (**c**) The third cycle. (**d**) The fourth cycle. (**e**) The fifth cycle. (**f**) The sixth cycle. (**g**) The seventh cycle.

5. Discussion

5.1. Effects of the Loading–Unloading Processes on the Permeability of Gas-Containing Coal

For a given stress level, the permeability when the peak stress was applied during the first cycle is defined as Kf. Similarly, Ks refers to the permeability during the ascending stage of stress application at the same stress level as Kf while the permeability in the stress valley is defined as Kg. Furthermore, Ks/Kf and Kg/Kf denote the recovery rates of permeability in the ascent and valley stages of stress relative to those under peak stress. The previously mentioned permeability levels of raw coal under the effect of stress path 1 are displayed in Figure 12.

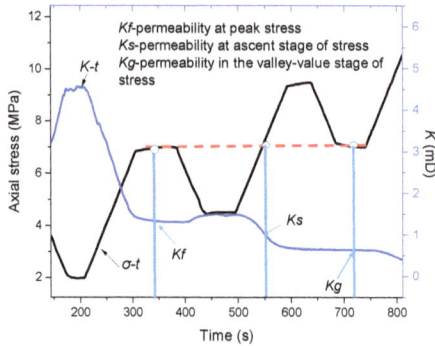

Figure 12. The definitions of permeability indices under stress path 1.

Figure 13 shows the changes in permeability of raw coal and its recovery rate at different stress levels under stress path 1. It can be seen from Figure 13 that, as the cyclic loads were applied and axial stress increased, the permeability decreased. Ks/Kf and Kg/Kf declined at first and then increased with increasing axial stress. At the same stress level, the permeability of coal samples after being subjected to unloading–loading decreased while their relative recovery rates fell at first and then rose with growing axial stress. The pore structure of coal samples before entering the yield stage showed plasticity under the effect of changing stress and, therefore, the loading–unloading effects were able to decrease the permeability of such coal samples.

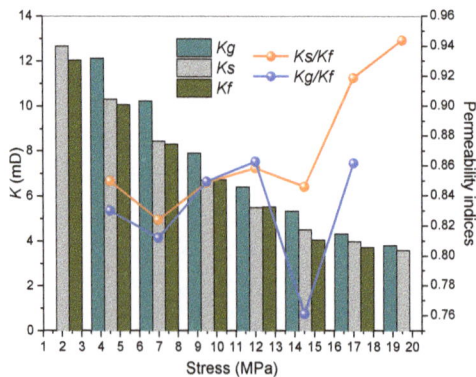

Figure 13. Changes in permeability of raw coal and its recovery rate at different stress levels under stress path 1.

5.2. Changes in the Relative and Absolute Recovery Rates of Permeability under Different Cyclic Loading–Unloading Paths

Figure 14 shows the changes in the relative and absolute recovery rates of the permeability of raw coal under stress paths 2 and 3. It can be seen from the figure that the evolution of the permeability of coal samples under stress paths 2 and 3 was similar. As cyclic loads were applied, the permeability in the loading and unloading processes declined and the permeability after loading showed a more rapid rate of reduction than that in the unloading stage. Under stress path 2, the relative recovery rate exhibited a three-phase characteristic—a rapid, then slow, and finally rapid increase—and, in contrast, the absolute recovery rate linearly decreased. Under stress path 3, the relative recovery rate rapidly, slowly, and steadily rose (a three-phase increase) while the rate of reduction of the absolute recovery was lower than that under stress path 2. Under stress path 2, the pores in coal samples, as gas seepage passages, were gradually compressed with the growth in the number of cycles. In the first two cycles, the permeability of the coal samples after stress relief failed to recover to the level before stress loading. However, the relative recovery rate during the second cycle was far larger than that during the first cycle, which implies that the recovery rate of pores from deformation increased. From the third to the fifth cycles, the increment of the recovery rate of permeability was low, which indicates that the seepage pores in coal samples were further compressed under the effect of cyclic loading and damage to the coal samples due to increasing peak stress made an insignificant contribution to changes in the permeability. During the sixth cycle, the relative recovery rate rose dramatically, which suggests that the coal samples entered the yield deformation stage. The stress cycle-induced damage to coal samples resulted in an increase in their permeability. Under stress path 3, from the first to the third cycles, the relative recovery rate of the permeability rose rapidly. Due to the stress level during the fourth cycle being the same as that during the second cycle, the stress-induced damage to coal samples was alleviated and the rate of increase in the recovery of the permeability decreased. Similarly, the stress during the sixth cycle was equivalent to that during the third cycle. As a result, the recovery rate of the permeability was basically the same as that during the fifth cycle and the change in the permeability was near-zero. Through the previously mentioned analysis, it can be seen that cyclic loading and unloading caused certain levels of damage to the coal samples, which shows a negative influence on the permeability of coal samples. Cyclic loading and unloading failed to increase the permeability before the stress exceeded the yield stress of the coal samples.

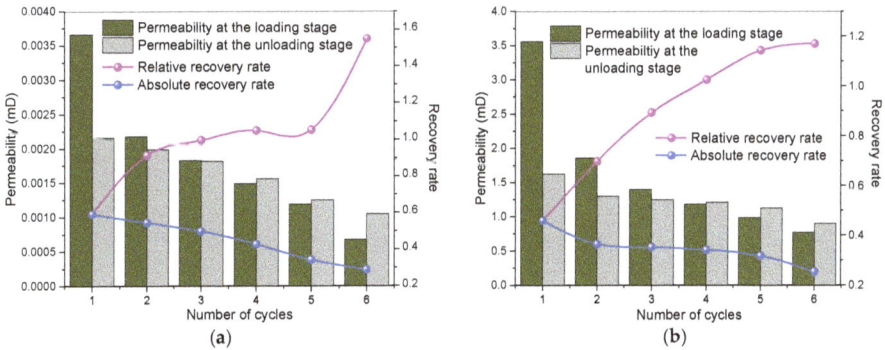

Figure 14. Changes in the absolute and relative recovery rates of the permeability of raw coal under cyclic loading–unloading paths 2 and 3. (**a**) Stress path 2. (**b**) Stress path 3.

5.3. Evolution of Mining-Enhanced Permeability of Gas-Containing Coal under Different Cyclic Loading–Unloading Paths

Under the effect of mining-induced disturbance, the equilibrium state of in situ stress on deep underground coal and rock masses was disturbed and the stress state continuously changed.

The structures of the coal masses also underwent continuous change with fracture initiation, fracture propagation, cracking, and rupture of coal masses. Mining-induced damage, fractures, and rupture of coal masses caused fundamental changes in the permeability of such coal masses, which also affected the extraction of gas. The permeability of coal masses reflects the number of pores in and connectivity of the material's structure and the change in the permeability of coal mass is closely related to volumetric changes in the coal mass.

The change law of the mining-enhanced permeability of coal mass under three paths was calculated by using Formula (4), as shown in Figure 15. It can be seen from the figure that, under stress path 1, the permeability of the coal samples declined with increasing axial stress so that the calculated mining-enhanced permeability in the loading was negative and gradually rose. This indicates that loading during cyclic loading increased the permeability of the samples. Furthermore, during loading, the mining-enhanced permeability fell at first, then rapidly rose, and increased slowly with a rise in stress. In contrast, the mining-enhanced permeability in the unloading stage first increased, then decreased, and, finally, rose to a positive value with increasing axial stress, which implies that the growth in volumetric strain during unloading resulted in the increase, reduction and growth in permeability of the coal samples. The increase in volume was not the only cause of the increase in permeability. Under stress path 2, with an increasing number of cycles, a three-phase change was seen in the mining-enhanced permeability in the loading stage, which is a rapid increase in the first two cycles. Then there was slow growth from the third to the sixth cycles and a rapid increase during the seventh cycle. This indicates that, with an increasing number of cycles, the mining-enhanced permeability rises and the increment of permeability under per unit volumetric strain is different. The mining-enhanced permeability and the number of cycles increased linearly during unloading. Under stress path 3, the change in mining-enhanced permeability under loading during different cycles was similar to that under stress path 2. The difference was that the increments of mining-enhanced permeability during the fourth and sixth cycles were lower than those during the corresponding previous cycles due to the stress levels during the fourth and sixth cycles being the same as those during the second and third cycles. In a similar way, the mining-enhanced permeability in the unloading also deviated from the predicted linear increase.

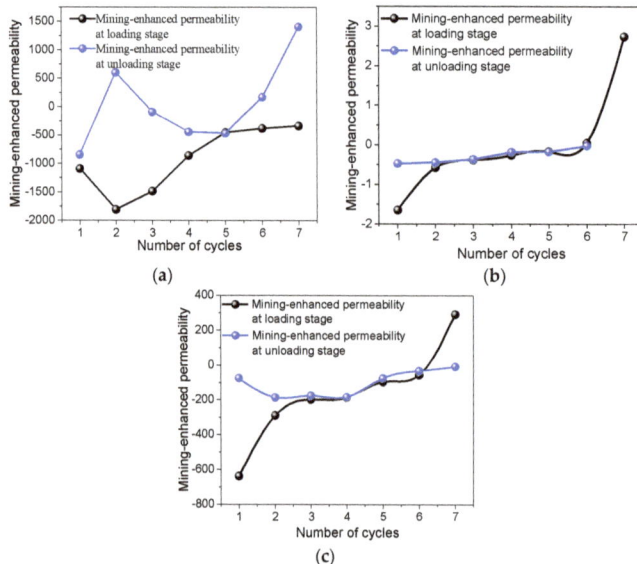

Figure 15. Changes in the mining-enhanced permeability of coal samples under different cyclic loading–unloading paths. (**a**) Stress path 1. (**b**) Stress path 2. (**c**) Stress path 3.

5.4. Stress Sensitivity of the Permeability of Gas-Containing Coal under Different Cyclic Loading–Unloading Paths

The large value of α_e in Formula (7) implies that the permeability of coal samples is more sensitive to changes in the effective stress and the permeability of coal samples changes more significantly under changes in the effective stresses when the amplitude of variation is kept constant. On the contrary, a small value of α_e indicates that the permeability of coal samples is less sensitive to changes in the effective stress and the permeability of coal samples changes less significantly with changes in the effective stress.

The stress sensitivity coefficients of the permeability of coal samples under three stress paths were calculated by using the above formula, as shown in Figure 16. It can be seen from the figure that, under stress path 1, the stress sensitivity coefficient of the permeability of coal samples in the loading stage varied with the growth in effective stress in the following manner. With an increase in the effective stress, the stress sensitivity coefficient of the permeability of coal samples rose at first and then reduced and, finally, it fluctuated to some extent. However, the stress sensitivity coefficient of the permeability of coal samples during unloading showed the opposite trend, which indicates that, during stepped-cyclic loading, when the stress applied on the coal samples was less than 10 MPa, the level of permeability was sensitive to stress but not to stress unloading. This occurs because the pores in the coal samples were subjected to plastic deformation and failed to recover after stress unloading. When the stress was greater than 10 MPa, the pores in the coal samples shrank under load and recovered upon unloading to some extent, which showed that the permeability became less sensitive to stress loading but more sensitive to stress unloading. Under stress path 2, the stress sensitivity coefficient of permeability under load declined with an increasing number of cycles and this can be divided into three phases: rapid reduction during the first two cycles, a slow decrease from the third to the fifth cycle, and rapid reduction during the last two cycles. This indicates that, with an increase in peak stress and the accumulation of damage, the sensitivity of the permeability to stress gradually decreased. The stress sensitivity coefficient of the permeability during unloading slowly increased and then linearly decreased with an increasing number of cycles, which implied that, as stress was applied, the influence of stress unloading on the permeability of coal samples was strengthened at first and then weakened. Under stress path 3, the change in the stress sensitivity coefficient of permeability with the number of cycles was similar to that under stress path 2. The difference was that the stress levels during the fourth and sixth cycles were, respectively, the same as those during the second and third cycles so that the stress sensitivity coefficients of permeability were, respectively, equivalent to those during the third and fifth cycles. The change in the stress sensitivity coefficient of permeability shows that the increase in effective stress caused the reduction in the stress sensitivity coefficient of permeability. The growth in the number of cyclic loading and unloading events also resulted in the decrease of the stress sensitivity coefficient of the permeability of such coal samples.

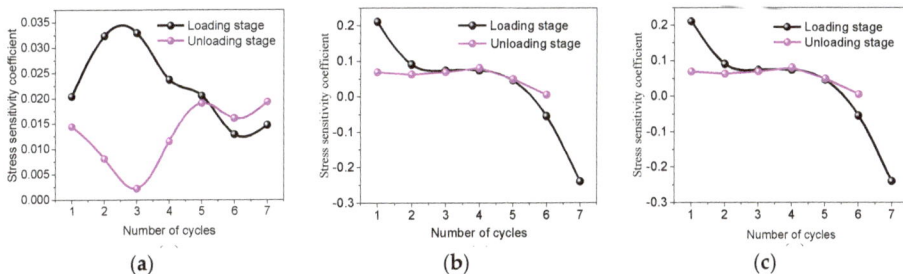

Figure 16. The changes in the stress sensitivity coefficient of the permeability of coal samples under different cyclic loading–unloading paths. (**a**) Stress path 1. (**b**) Stress path 2. (**c**) Stress path 3.

5.5. Damage Evolution Characteristics of Gas-Containing Coal under Different Cyclic Loading–Unloading Paths

Figure 17 shows the change in LURR Y and D of raw coal with stress under different cyclic loading–unloading paths. It can be seen from Figure 16 that, under stress path 1, with increasing axial stress, the LURR Y rose slowly at first, then rapidly increased, and. finally, increased in a quasi-linear manner. However, the curve relating to the damage suffered by coal samples exhibited a three-phase characteristic: a slow increase, rapid growth, and finally, a slow increase. It can be seen from the figure that the elastic modulus during loading was lower than that during unloading, which implies that damage accumulated in coal samples under load. The LURR gradually increased. When the stress reached a certain level, the elastic modulus under loading was further decreased and the LURR rose suddenly, which indicated that coal samples were about to be damaged and, therefore, they entered a yielding stage. This is consistent with the previously mentioned result in which the recovery rate of the permeability of coal samples rose dramatically. The change in D with the number of cycles obtained through Formula (11) was fitted by using the Boltzmann formula and the goodness of fit reached 0.95. The Boltzmann formula can be used as an empirical formula for damage prediction. Under the effects of stress paths 2 and 3, the elastic modulus under load remained unchanged with an increasing number of cycles in the initial stage while it suddenly reduced in the later stages of testing. In contrast, the elastic modulus remained unchanged during unloading. In terms of LURR Y, it rose slowly at the start and suddenly increased later. This indicated that coal samples underwent yield and failure, which agrees with the observed evolution of the recovery rate of the permeability mentioned above. Correspondingly, D, calculated through the Y value, also showed a similar trend and can be fitted by applying the Boltzmann formula [35].

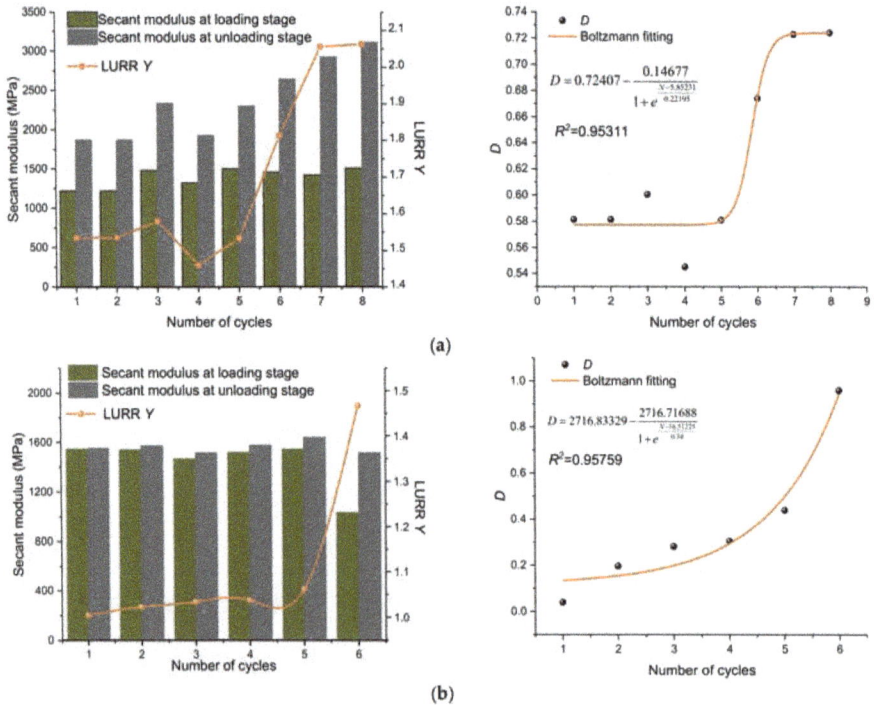

(a)

(b)

Figure 17. *Cont.*

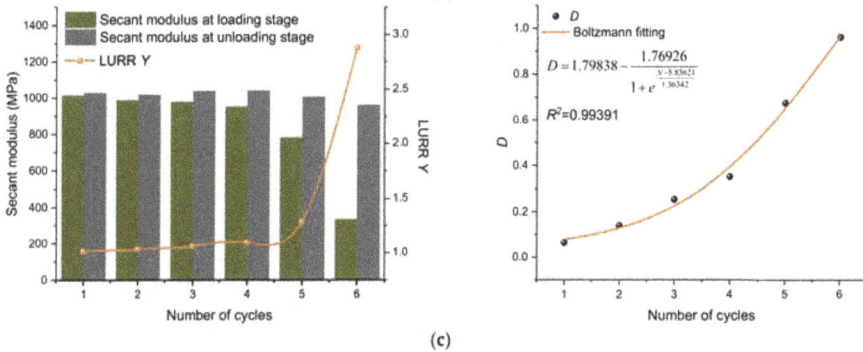

(c)

Figure 17. The changes in LURR and D of raw coal with stress under different cyclic loading–unloading paths. (**a**) Stress path 1. (**b**) Stress path 2. (**c**) Stress path 3.

6. Conclusions

The mechanical properties and seepage characteristics of gas-bearing coal evolve with changes in the loading pattern, which could reveal the evolution of permeability in a protected coal seam and allow gas extraction engineering works to be designed using the effect of mining multiple protective seams. Based on the results of the similarity experiments on the stress state of protected seam during mining of a coal seam group, the seepage and damage evolution characteristics of gas-containing coal under cyclic loading–unloading effects were explored by conducting fluid–solid–coupling tests on the coal under three simplified stress paths. On this basis, the influence of mining multiple protective seams for pressure relief on the seepage characteristics of gas-containing coal in a protected seam was investigated.

Unloading and loading stress exerted different influences on the permeability and a dual influence of stress and damage accumulation was seen on coal permeability. The pores in coal samples, before entering their yield stage, exhibited a plastic structure due to the effects of changing stress and the loading–unloading effect can reduce the permeability. Cyclic loading and unloading resulted in damage to coal samples while having a negative influence on the permeability. The permeability of coal samples cannot increase before the stress exceeds the yield stress of the coal samples. The stress paths also affect the mining-enhanced permeability of coal. The mining-enhanced permeability of coal samples in the loading stage showed a three-phase increasing trend with growing stress and number of cycles. The mining-enhanced permeability caused by the change per unit volume increased, the mining-enhanced permeability in the unloading stage showed an N-shaped growth trend during stepped-cyclic loading while growing linearly under the other two paths, and the permeability increasing effect under the cyclic loading became better with an increasing number of cycles.

Furthermore, during the stepped-cyclic loading and unloading, when the stress was less than 10 MPa, the permeability of these coal samples was sensitive to the stress loading while it was insensitive to stress unloading, which indicates that plastic deformation occurred in the pores in the coal mass. This could not be recovered after stress unloading. When the stress exceeded 10 MPa, the pores in these coal samples were compressed under load while they recovered to some degree upon unloading. This behavior indicated that the sensitivity to loading stress was reduced while indicating that the sensitivity to the unloading stress increased. With increasing peak stress and damage accumulation in the coal samples, the sensitivity of the permeability to stress gradually decreased. As loads were applied to the coal samples, the influence of stress unloading on their permeability was first strengthened and then reduced. By calculating the LURRs at various stages of cyclic loading, the evolution of D of the coal samples was obtained. Through regression analysis, it can be seen that the relationship between D and the number of cycles can be characterized by applying the Boltzmann function. The function can be used as an empirical formula for damage prediction.

These conclusions validate the observed evolution of the permeability of gas-containing coal in protected seams at different stages of the mining of multiple protective seams during the mining of a coal seam group. On this basis, the results can be used to guide the design of engineering operations such as gas extraction in protected seams (e.g., setting a borehole, extracting gas from stress unloading zones, and avoiding stress concentration zones) and the prevention and control of gas disasters.

Author Contributions: Q.L. and Y.L. conceived and designed the experiments. Q.L. and Q.Z. performed the experiments. Q.L. analyzed the data. Q.L. and Q.Z. wrote the paper.

Funding: This work is financially supported by the State Key Research Development Program of China (2017YFC0804206 and 2016YFC0801404), the National Natural Science Foundation of China (51674050 and 51704046), the National Science and Technology Major Project of China (2016ZX05043005), the Fundamental Research Funds for the Central Universities (2018CDQYZH0001 and 106112017CDJXY240001), and the Open Fund Research Project of State Key Laboratory Breeding Base for Mining Disaster Prevention and Control (MDPC201710).

Acknowledgments: The authors thank the editor and anonymous reviewers for their valuable advice.

Conflicts of Interest: The authors declare no conflict of interest.

References

1. Dou, L.; He, X.; Ren, T.; He, J.; Wang, Z. Mechanism of coal-gas dynamic disasters caused by thesuperposition of static and dynamic loads and its control technology. *J. China Univ. Min. Tech.* **2018**, *47*, 48–59. [CrossRef]
2. Lin, B.; Liu, T.; Yang, W. Solid-gas coupling model for coalseams based on dynamic diffusion and its application. *J. China Univ. Min. Tech.* **2018**, *47*, 32–39. [CrossRef]
3. Mark, C. Coal bursts that occur during development: A rock mechanics enigma. *Int. J. Min. Sci. Tech.* **2018**, *28*, 35–42. [CrossRef]
4. Zhou, A.; Wang, K.; Fan, L.; Kiryaeva, T.A. Gas-solid coupling laws for deep high-gas coal seams. *Int. J. Min. Sci. Tech.* **2017**, *27*, 675–679. [CrossRef]
5. Zhou, P.; Zhang, Y.; Huang, Z.A.; Gao, Y.; Wang, H.; Luo, Q. Coal and gas outburst prevention using new high water content cement slurry for injection into the coal seam. *Int. J. Min. Sci. Tech.* **2017**, *27*, 669–673. [CrossRef]
6. Zou, Q.; Lin, B. Fluid–solid coupling characteristics of gas-bearing coal subjected to hydraulic slotting: An experimental investigation. *Energ. Fuels* **2018**, *32*, 1047–1060. [CrossRef]
7. Ni, G.; Li, Z.; Xie, H. The mechanism and relief method of the coal seam water blocking effect (WBE) based on the surfactants. *Powder Technol.* **2018**, *323*, 60–68. [CrossRef]
8. Xu, J.; Zhai, C.; Liu, S.; Qin, L.; Dong, R. Investigation of temperature effects from LCO 2 with different cycle parameters on the coal pore variation based on infrared thermal imagery and low-field nuclear magnetic resonance. *Fuel* **2018**, *215*, 528–540. [CrossRef]
9. Ye, Q.; Jia, Z.; Zheng, C. Study on hydraulic-controlled blasting technology for pressure relief and permeability improvement in a deep hole. *J. Petrol. Sci. Eng.* **2017**, *159*, 433–442. [CrossRef]
10. Zhang, L.; Zhang, H.; Guo, H. A case study of gas drainage to low permeability coal seam. *Int. J. Min. Sci. Tech.* **2017**, *27*, 687–692. [CrossRef]
11. Zou, Q.; Lin, B.; Zheng, C.; Hao, Z.; Zhai, C.; Liu, T.; Liang, J.; Yan, F.; Yang, W.; Zhu, C. Novel integrated techniques of drilling–slotting–separation-sealing for enhanced coal bed methane recovery in underground coal mines. *J. Nat. Gas Sci. Eng.* **2015**, *26*, 960–973. [CrossRef]
12. Liu, T.; Lin, B.Q.; Yang, W. Impact of matrix-fracture interactions on coal permeability: Model development and analysis. *Fuel* **2017**, *207*, 522–532. [CrossRef]
13. Wu, Y.; Pan, Z.; Zhang, D.; Lu, Z.; Connell, L.D. Evaluation of gas production from multiple coal seams: A simulation study and economics. *Int. J. Min. Sci. Tech.* **2018**, *28*, 359–371. [CrossRef]
14. Chang, X.; Tian, H. Technical scheme and application of pressure-relief gas extraction in multi-coal seam mining region. *Int. J. Min. Sci. Tech.* **2018**, *28*, 483–489. [CrossRef]
15. Ye, Q.; Wang, G.; Jia, Z.; Zheng, C.; Wang, W. Similarity simulation of mining-crack-evolution characteristics of overburden strata in deep coal mining with large dip. *J. Petrol. Sci. Eng.* **2018**, *165*, 477–487. [CrossRef]

16. Liu, T.; Lin, B.Q.; Yang, W.; Zou, Q.L.; Kong, J.; Yan, F.Z. Cracking Process and Stress Field Evolution in Specimen Containing Combined Flaw Under Uniaxial Compression. *Rock Mech. Rock Eng.* **2016**, *49*, 3095–3113. [CrossRef]

17. Du, W.; Zhang, Y.; Meng, X.; Zhang, X.; Li, W. Deformation and seepage characteristics of gas-containing coal under true triaxial stress. *Arab. J. Geosci.* **2018**, *11*, 1–13. [CrossRef]

18. Danesh, N.; Chen, Z.; Connell, L.D.; Kizil, M.S.; Pan, Z.; Aminossadati, S.M. Characterisation of creep in coal and its impact on permeability: An experimental study. *Int. J. Coal Geol.* **2017**, *173*, 200–211. [CrossRef]

19. Peng, S.J.; Xu, J.; Yang, H.W.; Liu, D. Experimental study on the influence mechanism of gas seepage on coal and gas outburst disaster. *Saf. Sci.* **2012**, *50*, 816–821. [CrossRef]

20. Tang, Y.; Okubo, S.; Xu, J.; Peng, S. Study on the Progressive Failure Characteristics of Coal in Uniaxial and Triaxial Compression Conditions Using 3D-Digital Image Correlation. *Energies* **2018**, *11*, 1215. [CrossRef]

21. Yin, G.; Jiang, C.; Wang, J.G.; Xu, J. Geomechanical and flow properties of coal from loading axial stress and unloading confining pressure tests. *Int. J. Rock Mech. Min. Sci.* **2015**, *76*, 155–161. [CrossRef]

22. Zhang, D.; Yang, Y.; Chu, Y.; Zhang, X.; Xue, Y. Influence of loading and unloading velocity of confining pressure on strength and permeability characteristics of crystalline sandstone. *Results Phys.* **2018**, *9*, 1363–1370. [CrossRef]

23. Wang, Y.; Guo, P.; Dai, F.; Li, X.; Zhao, Y.; Liu, Y. Behavior and Modeling of Fiber-Reinforced Clay under Triaxial Compression by Combining the Superposition Method with the Energy-Based Homogenization Technique. *Int. J. Geomech.* **2018**, *18*, 04018172. [CrossRef]

24. Wu, F.; Chen, J.; Zou, Q. A nonlinear creep damage model for salt rock. *Int. J. Damage Mech.* **2018**. [CrossRef]

25. Yang, D.S.; Qi, X.Y.; Chen, W.Z.; Wang, S.G.; Yang, J.P. Anisotropic Permeability of Coal Subjected to Cyclic Loading and Unloading. *Int. J. Geomech.* **2018**, *18*, 04018093. [CrossRef]

26. Jiang, C.; Duan, M.; Yin, G.; Wang, J.G.; Lu, T.; Xu, J.; Zhang, D.; Huang, G. Experimental study on seepage properties, AE characteristics and energy dissipation of coal under tiered cyclic loading. *Eng. Geol.* **2017**, *221*, 114–123. [CrossRef]

27. Zhang, Z.; Wang, E.; Li, N.; Li, X.; Wang, X.; Li, Z. Damage evolution analysis of coal samples under cyclic loading based on single-link cluster method. *J. Appl. Geophys.* **2018**, *152*, 56–64. [CrossRef]

28. Yang, Y.J.; Duan, H.Q.; Xing, L.Y.; Deng, L. Fatigue Characteristics of Coal Specimens under Cyclic Uniaxial Loading. *Geotech. Test. J.* **2019**, *42*, 20170263. [CrossRef]

29. Song, D.; Wang, E.; Liu, J. Relationship between EMR and dissipated energy of coal rock mass during cyclic loading process. *Saf. Sci.* **2012**, *50*, 751–760. [CrossRef]

30. Ye, Z.; Zhang, L.; Hao, D.; Zhang, C.; Wang, C. Experimental study on the response characteristics of coal permeability to pore pressure under loading and unloading conditions. *J. Geophys. Eng.* **2017**, *14*, 1020–1031. [CrossRef]

31. Xie, H.; Gao, F.; Zhou, H.; Cheng, H.; Zhou, F. On theoretical and modeling approach to mining-enhanced permeability for simultaneous exploitation of coal and gas. *J. China Coal Soc.* **2013**, *38*, 1101–1108. [CrossRef]

32. Meng, Z.; Li, G. Experimental research on the permeability of high-rank coal under a varying stress and its influencing factors. *Eng. Geol.* **2013**, *162*, 108–117. [CrossRef]

33. Yin, X.; Yin, C. Precursor and Earthquake Prediction of Nonlinear System Instability—Response Ratio Theory and Its Application. *Sci. China Ser. B* **1991**, *5*, 55–60.

34. Zhang, L.; Yin, X.; Liang, N. Study on relation between load/unload response ratio and damage variable. *Chin. J. Rock Mech. Eng.* **2008**, *27*, 1874–1881.

35. Wu, L.; White, C.; Scanlon, T.J.; Reese, J.M.; Zhang, Y. Deterministic numerical solutions of the Boltzmann equation using the fast spectral method. *J. Comput. Phys.* **2013**, *250*, 27–52. [CrossRef]

![processes logo] *processes*

MDPI

Article

Effects of Pulse Interval and Dosing Flux on Cells Varying the Relative Velocity of Micro Droplets and Culture Solution

Zhanwei Wang [1] [ID], Kun Liu [1,*] [ID], Jiuxin Ning [1], Shulei Chen [1], Ming Hao [1], Dongyang Wang [1] [ID], Qi Mei [2], Yaoshuai Ba [1] and Dechun Ba [1,*]

1   School of Mechanical Engineering and Automation, Northeastern University, Shenyang 110819, China; 1700459@stu.neu.edu.cn (Z.W.); jxning@stumail.neu.edu.cn (J.N.); 1710120@stu.neu.edu.cn (S.C.); 1500445@stu.neu.edu.cn (M.H.); wdysend@gmail.com (D.W.); yshba@mail.neu.edu.cn (Y.B.)
2   Department of Oncology, Tongji Hospital, Tongji Medical College, Huazhong University of Science and Technology, Wuhan 430030, China; meiqi_tj@hust.edu.cn
*   Correspondence: kliu@mail.neu.edu.cn (K.L.); dchba@mail.neu.edu.cn (D.B.);
    Tel.: +86-24-8367-6945 (K.L. & D.B.)

Received: 28 June 2018; Accepted: 3 August 2018; Published: 7 August 2018

Abstract: Microdroplet dosing to cell on a chip could meet the demand of narrow diffusion distance, controllable pulse dosing and less impact to cells. In this work, we studied the diffusion process of microdroplet cell pulse dosing in the three-layer sandwich structure of PDMS (polydimethylsiloxane)/PCTE (polycarbonate) microporous membrane/PDMS chip. The mathematical model is established to solve the diffusion process and the process of rhodamine transfer to micro-traps is simulated. The rhodamine mass fraction distribution, pressure field and velocity field around the microdroplet and cell surfaces are analyzed for further study of interdiffusion and convective diffusion effect. The cell pulse dosing time and drug delivery efficiency could be controlled by adjusting microdroplet and culture solution velocity without impairing cells at micro-traps. Furthermore, the accuracy and controllability of the cell dosing pulse time and maximum drug mass fraction on cell surfaces are achieved and the drug effect on cells could be analyzed more precisely especially for neuron cell dosing.

Keywords: cell dosing; microdroplet; convective diffusion; interdiffusion; numerical simulation

1. Introduction

Microfluidic chip can achieve the basic function of macroscopic laboratory by manipulating the flow at the micro scale. The use of microfluidic chip for chemical analysis, drug screening and transportation, cell culture and other functions by means of microfluidic control is regarded as the miniaturization and integration of macroscopic laboratories [1–4], which has advantages as follows: (1) controllable liquid flow; (2) rare consumption of samples and reagents; (3) analysis process automation; (4) significant increased analysis efficiency [5]. In the aspect of cell culture, the microfluidic chip cell laboratory has micron-level and relatively closed three-dimensional cell culture, sorting, cracking and other operating units, which plays a unique role in cell research [6,7]. For cell dosing, the latest fourth-generation targeted dosing system can concentrate the drug to the target cell or the target tissue, which accumulates the drug in a small range to realize the improvement of therapeutic effect yet reducing the drug side-effect [8,9]. Therefore, many researchers have been focused on the application of microfluidic chip in cell dosing system. The first invented microchip based on electrochemical dissolution microcapsule experimentally demonstrated the release of drug [10]. A multilayer polymer drug handling facility was designed for oral dosing of treatment and provided a variety of unidirectional release treatments, the concentration of drug into the intestinal epithelial cells was more than 10 times that of the conventional method [11]. Three-dimensional tissue platform based on a microfluidic chip can be used to diagnose and quantify cellular heterogeneity [12]. Integrated microfluidic chip platform can simulate the microenvironment of the blood-brain barrier and glioma in vivo and conduct drug penetration experiments [13].

The conventional cell dosing methods on a chip often inject drug solution into microfluidic chip directly or use two-phase fluid in the microchannel through the semi-permeable membrane to realize drug diffusion [14–18]. The drug was wasted a lot and the precision was low for the former and it was difficult to control drug distribution for the latter. So, the jet microfluidic was used to dose cell and the control precision was improved, both the consumption of drug and the dosing time were reduced [19–23]. But the dosing pressure towards to cell was quite tough, which may impair cell severely and the dosing time was difficult to control. For overcoming these dosing problems, the microporous membrane is used to support drug microdroplets flowing and buffer the diffusion, the microdroplets velocity is controlled to adjust dosing time. Considering that polydimethylsiloxane (PDMS) has good biocompatibility, high transparence and is permeable to water and CO_2 and polycarbonate (PCTE) etching membrane also has good biocompatibility, high transparence and impact resistance, customizable micropore diameter and porosity, a novel three-layer sandwich structure of PDMS/PCTE microporous membrane/PDMS chip was designed, which supports microdroplets flow in the microchannel upon membrane, the drug carried in microdroplets could diffuse through microporous membrane into cells trapped at micro-traps. Precise pulse dosing to cells could be achieved by controlling the flow of microdroplets and culture solution. Researching the drug transfer process in this novel device is necessary for supporting the control method for cell pulse dosing. It is known that the mass transfer between deionized water droplets and continuous phases octanol can be observed that there is a vortex on the surface of the microdroplets [24] and the solute transfer at the instantaneous two-phase interface is proportional to the square of the time [25]. The drug diffusion to cells through microporous membrane from microdroplet surround by oil is influenced by these factors, especially the mechanism of mass transfer through membrane in fluid. The hydrophilic nanoporous membrane does not exhibit any pore wetting because of the dramatically increased shear viscous force of water with respect to the hydrophilic pore wall [26]. The surface tension of water decreases with the decreasing of film thickness on the surface of aqueous nano-films [27]. More investigations about drug diffusion through membrane could be expanded and more deeply discussed based on these achievements.

In current study, we studied the convective diffusion process of microdroplet cell dosing in the three-layer sandwich structure of PDMS/PCTE microporous membrane/PDMS chip. The flow velocities of microdroplets and culture medium were investigated for the convenience of studying and controlling the cells dosing pulse time ignoring other factors influence. Although the microdroplets size or surface chemical complexity and so forth also influence the diffusive dosing process. But it is not convenient to control the pulse time in drug experiments by change the microdroplets size or surface chemical character. The mathematical model was established to solve the convective diffusion process and the process of drug transfer to micro-traps was simulated. The drug concentration, convective diffusion time, pressure field and velocity field around micro-traps were analyzed. The applicability of microdroplet micropore diffusion for cell dosing was further promoted, which contributes more controllable instant pulse dosing time and flux to cells while less impact to cells.

2. Computational Model

2.1. Mathematical Model

The micropore is considered as cuboid and simplified as a one-dimensional model ignoring the influence of cuboid length as Figure 1a, which means drug convection diffusion from source to the other side directly. A rectangular coordinate system is set on the section of the model as Figure 1b. The fluid is considered incompressible and isotropic ignoring temperature changes.

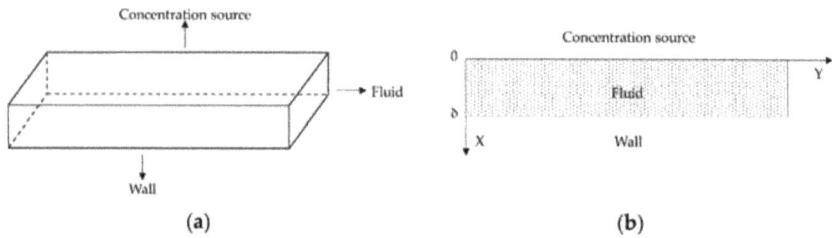

Figure 1. Convective diffusion model of microfluidic chip for drug delivery: (**a**) The simplified convective diffusion model; (**b**) The rectangular coordinate system of the section of convective diffusion model.

The concentration diffusion is distributed by Fick's second law during unsteady transmission when ignoring convection, which means that the change rate of mole concentration C in the cross-section x is equal to the change rate of diffusion flux by x.

$$\frac{\partial C}{\partial t} = \frac{\partial}{\partial x}\left(D\frac{\partial C}{\partial x}\right) \tag{1}$$

Due to the velocity along the Y direction in micropore which means convection term, the equation is altered:

$$\frac{\partial C(x,t)}{\partial t} + v\frac{\partial C(x,t)}{\partial x} = D\frac{\partial^2 C(x,t)}{\partial x^2}) \tag{2}$$

where C is mole concentration of diffusate, V is convection coefficient, D is diffusion coefficient, t is diffusion time.

The generalized equation of convective diffusion is modified for calculation:

$$\frac{\partial C(x,t)}{\partial t} = -a\frac{\partial C(x,t)}{\partial x} + b^2\frac{\partial^2 C(x,t)}{\partial x^2} \tag{3}$$

The boundary between diffusion source and microchannel is free diffusion boundary, the wall below microchannel is the boundary without mass and energy exchange. The upper and lower boundaries satisfy the first boundary condition and the second boundary condition and the math equation is shown:

$$\begin{cases} C(0,t) = C_0, & t \geq 0 \\ C(\delta,t) = 0, & t \geq 0 \end{cases} \tag{4}$$

$$\frac{\partial C(0,t)}{\partial x} = 0, \ t \geq 0 \tag{5}$$

$$\frac{\partial C(\delta,t)}{\partial x} = 0, \ t \geq 0 \tag{6}$$

The initial concentration in microchannel is constant:

$$C(x,0) = C_0, \ 0 < x \leq \delta \tag{7}$$

The fundamental equation is equivalently transformed for the convenience to solve the problem and the new function $L(x, t)$ is introduced:

$$C(x,t) = L(x,t)e^{\frac{a}{2b^2}x - \frac{a^2}{4b^2}t} \tag{8}$$

Separating variables is the fundamental method to solve the mathematical physical equation, which transforms the multi-variable partial differential equation to ordinary differential equation with several univariates:

$$L(x,t) = H(x) \cdot K(t) \tag{9}$$

The fundamental equation is shown:

$$
\begin{aligned}
b^2 K(t) \frac{\partial^2 H(x)}{\partial x} &= H(x) \frac{\partial K(t)}{\partial t} \\
\frac{1}{H(x)} \frac{\partial^2 H(x)}{\partial x} &= \frac{1}{b^2 K(t)} \frac{\partial K(t)}{\partial t} \\
\frac{H''(x)}{H(x)} &= \frac{K'(t)}{b^2 K(t)}
\end{aligned}
\tag{10}
$$

The t and x are separately independent variables, so the equation is established when they are constant. The equation is shown assuming the separation constant as $-\lambda^2$:

$$\frac{H''(x)}{H(x)} = \frac{K'(t)}{b^2 K(t)} = -\lambda^2 \tag{11}$$

The equation is obtained by substituting boundary conditions:

$$
\begin{cases}
H(0) \cdot K(t) = L_0 \cdot e^{\frac{a^2}{4b^2}} \\
H'(0) \cdot K(t) = -\frac{a^2}{2b^2} L_0 e^{\frac{a^2}{4b^2}}
\end{cases}
\tag{12}
$$

The eigenvalue constant λ^2 must be obtained by the boundary conditions, otherwise the fundamental equation would have no non-zero solution. The eigenvalue is discussed as below.

Firstly, assuming that:

$$H(x) = e^{\gamma x} \tag{13}$$

If $\lambda^2 = 0$, the general solution is shown:

$$H(x) = (c_1 + c_2 x) \cdot e^{0 \cdot x} \tag{14}$$

Substituting boundary conditions, the eigenvalue is obtained:

$$\lambda^2 = \left(\frac{\frac{\pi}{2} + n\pi}{\delta} \right)^2, \quad n = 0,1,2,3 \cdots \tag{15}$$

The general solution of $K(t)$ is obtained by solving the total differential equation:

$$K_n(x) = c_1 e^{-b^2 \left[\frac{(n+\frac{1}{2})\pi}{\delta} \right]^2 t}, \quad n = 0,1,2,3 \cdots \tag{16}$$

Assuming that:

$$\varphi(x,t) = \sin \left[\frac{\left(n+\frac{1}{2}\right)\pi}{\delta} x \right] e^{-b^2 \left[\frac{(n+\frac{1}{2})\pi}{\delta} \right]^2 t} \tag{17}$$

The solution of $C(x,t)$ is shown:

$$C(x,t) = 2L_0 \sum_{0}^{\infty} \frac{1}{\frac{a^2\delta}{4b^4} + \frac{(n+1)^2\pi^2}{\delta}} \left[1 - \frac{\delta a}{(2n+1)\pi b^2} e^{-\frac{a\delta}{2b^2}} \sin\left(n+\frac{1}{2}\right)\pi \right] \cdot \varphi(x,t) e^{\frac{a}{2b^2}x - \frac{a^2}{4b^2}t} \tag{18}$$

2.2. Numerical Simulation

The microdroplet dosing chip model was established and numbered as Figure 2. The simulation zone is a part of the chip with the overall size of 350 μm in length, 150 μm in width, 90 μm in height, which consists of microdroplet channel, microporous membrane and micro-traps array considering the repeatability of the whole structure. A microdroplet channel is 350 μm long, 150 μm wide and 50 μm high. A microporous membrane is 350 μm long, 200 μm wide and 20 μm thick. A double-slit micro-trap array zone is 350 μm long, 200 μm wide and 20 μm high, which contains 7 micro-traps and every double-slit micro-trap is 50 μm long, 50 μm wide and 20 μm high. The material of microdroplet channel and micro-traps is PDMS and the material of microporous membrane is PCTE.

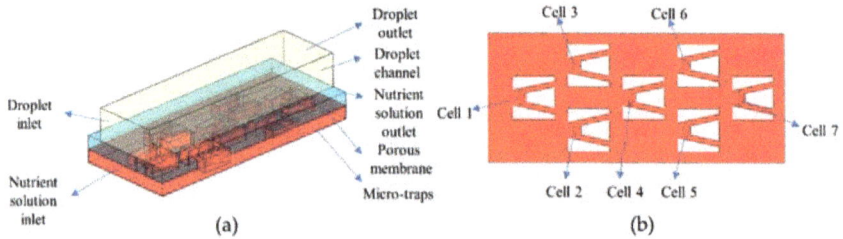

Figure 2. Three-layer sandwich structure of PDMS/PCTE microporous membrane/PDMS microfluidic chip: (**a**) computational zone of cell dosing microfluidic chip; (**b**) the number of cells trapped in micro-traps.

The diffusion process of rhodamine solution microdroplet from droplet channel to cells through microporous membrane is shown as Figure 3. In this study, we focused on controlling the drug delivery process by changing fluids velocity, so the rhodamine concentration effects were neglected for controlling variate. It is applicable in multi drug solution microdroplets dosing experiments and the evaluation accuracy under specific conditions could be further improved by adding relevant initial drug concentration. The rhodamine solution microdroplet is regarded as a component, considering the self-diffusion, inter-diffusion and convection diffusion in microdroplet and culture medium. This microfluidic system promotes the microdroplet diffusion in culture medium mainly by natural convection and especially the effect of forced convection. The rhodamine spread phenomenon in culture medium is regarded as the characterization of rhodamine molecular diffusion effect. The microdroplet is surrounded by oil and flows from inlet to outlet with oil in droplet channel upon microporous membrane. In the meanwhile, the culture medium is flowing from inlet to outlet in micro-traps channel. The rhodamine solution diffuses from microdroplet into culture medium through microporous membrane. The cell surface maximum rhodamine mass fraction and pulse time are controlled by changing the culture medium velocity and microdroplet relative velocity.

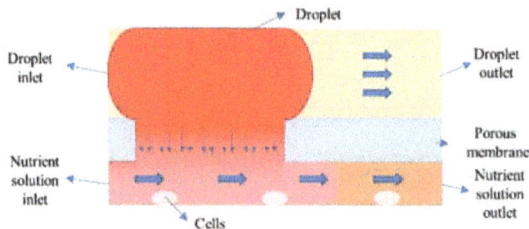

Figure 3. Diffusion process of rhodamine solution from microdroplet to cells.

In the microfluidic system, fluid is laminar and is mainly influenced by capillary force. Capillary force is represented by dimensionless number Ca. Ca = $\mu U/\gamma$, where γ is the surface tension coefficient between the two phases, μ and U is the fluid viscosity and velocity of the continuous phase respectively.

In this paper, the volume of fluid method and the transport of species transport without chemical reaction method was used to simulate the dynamics of microdroplet dosing to cells trapped at micro-traps in 3-D microfluidic chip. Water and oil were chosen as dispersed phase and continuous phase respectively. The fluids were modeled as incompressible Newtonian flow. The continuity equation and momentum equation were used in the application mode.

$$\nabla \cdot u = 0 \tag{19}$$

$$\rho \, du/dt = F - \nabla p + \nabla^2 u \tag{20}$$

where P, u, ρ and μ is pressure, velocity field, the volume-fraction-weighted density and viscosity respectively. F is the surface tension force which needs special treatment to be computed.

The Young–Laplace equation was used to obtain the contact angle of silicon oil-water on PDMS. The surface tension of silicon oil-water system was taken as 42.6 mN/m [28], the air-silicon oil system as 20.8 mN/m (provided by the silicon oil manufacturer Dow Corning), the air-water system as 72.1 mN/m and the contact angle for air-silicon oil system was taken as 15° [29], the air-water system as 98° [29]. The contact angle of oil-water system in these simulations was obtained as 135° [29]. The contact angle of oil-water on PCTE is not necessary to be calculated, because the capillary force between microdroplet and culture solution would impact the shape of microdroplet apparently.

Numerical simulations were performed in the boundary condition with $Q_{water} \leq Q_{oil} \leq 60$ μL/min or $U_{water} \leq U_{oil} \leq 0.1$ m/s according to our former experiments [29]. The interface tension between the two phases was taken as 0.06 N/m, because the microdroplet fully fills the microchannel so as to cling to microporous membrane to prompt convective diffusion. The culture medium was blended by RPMI 1640, Fatal bovine serum, Penicillin-Streptomycin Solution, dextran and so forth were all purchased from Sangon Biotech Co., Ltd. (Shanghai, China), and the dynamic viscosity was measured by PND401a intelligent Kinematic viscosity testers, Puruite Instrument, Jilin, China. The velocity field around cells, static pressure and mass fraction of rhodamine B on cell surface were monitored every 0.5 ms. The simulation parameters are set as Table 1.

Table 1. Numerical simulation parameters.

Parameters	Value
Culture medium density ρ_1	1038 kg/m^3
Culture medium dynamic viscosity μ_1	3.07×10^{-3} Pa·s
Rhodamine B solution density ρ_2	1.053 kg/m^3
Rhodamine B solution dynamic viscosity μ_2	1×10^{-3} Pa·s
Oil density ρ_2	960 kg/m^3
Oil density dynamic viscosity μ_3	0.012–0.096 Pa·s
Inlet velocity v_0	≤ 0.1 m/s
Porosity ρ	0.1

3. Results and Discussion

3.1. Area-Weighted Rhodamine Mass Fraction on Cell Surfaces

The advantages of microdroplet dosing to cell on a chip is tiny waste of drug, controllable pulse dosing and less impact to cell. So, the area-weighted mass fraction on cell surfaces that describes the rhodamine molecules mass reaching cell surfaces, the method for controlling pulse time and the impact on cells especially about cell surfaces pressure are significant parameters in this microfluidic system. More rhodamine mass fraction on cell surfaces, more mild pressure impact on cell surfaces and less pulse dosing time would contribute better cell pulse dosing effect.

The rhodamine mass fraction profiles at Cell 1–7 cell surfaces except at Cell 3 and Cell 6 cells are shown as Figure 4, because Cell 3 and Cell 6 cells are symmetric along the plane at middle of micro-traps perpendicular to Y axis with Cell 4 and Cell 5 cells. The diffusion effect is uniform and the maximum rhodamine mass fraction is invariable when fluid velocities in droplet channel and culture solution channel are same. The maximum rhodamine mass fraction on Cell 1–7 cell surfaces is 0.38 and all cells have attained it. It is obvious that the maximum rhodamine mass frication on cell surfaces is independent with velocity change when microdroplet velocity is equal to culture solution velocity.

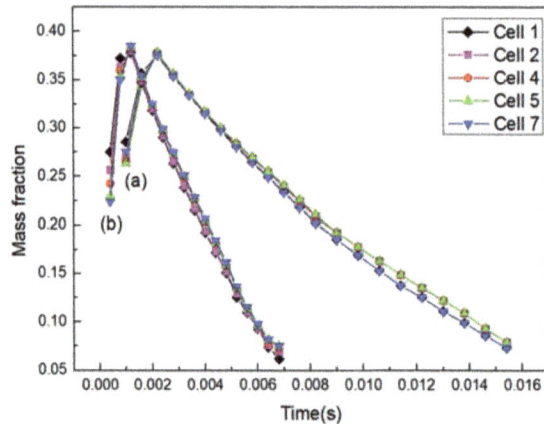

Figure 4. Change of cell surfaces mass fraction with time at Cell 1–7 micro-trap: (a) the microdroplet and culture solution flow velocities are 0.05 m/s; (b) the microdroplet and culture solution flow velocities are 0.1 m/s.

As the rhodamine molecules last Brownian motion, they would collide with other molecules. The rhodamine molecules would keep a certain speed and would move a distance after the collision at certain moments causing self-diffusion [30]. The multi-component inter-diffusion happens when concentration gradient exists in mixture, rhodamine molecules diffuse to culture solution while culture solution diffuses to droplet. But in this dosing system, there are convective diffusion and capillary force particularly influencing the rhodamine transport process besides self-diffusion and inter-diffusion. These factors play different important roles as the microdroplet and culture solution inlet velocity change.

More drug delivery efficiency could be realized by controlling the relative velocity of microdroplet to culture solution. The maximum mass fraction at cell surfaces when the relative velocity of microdroplet to culture solution changes from 0–0.09 m/s is shown as Figure 5. As culture solution velocity is 0.01 m/s, the maximum mass fraction at cell surfaces improved with the increase of the relative velocity of microdroplet and it almost all exceeds 0.5 when relative velocity is 0.03 m/s but the maximum mass fraction decreases gradually when relative velocity exceeds 0.03 m/s.

The trend of convective diffusion of rhodamine from microdroplet to culture solution was aroused by the flow of high velocity fluid to low velocity fluid zone and was enhanced gradually when the relative velocity of microdroplet increased. However, large difference between fluid velocities also promotes impacts of oil on membrane and microdroplet, a little two phases mixing caused by a part of oil break the surface tension between water phase and oil phase. The vortex between microdroplet and culture solution caused by the relative velocity was aggravated, so the mass diffusion was hindered [31].

Thus, the rhodamine convective diffusion is weakened when the vortex and mix effect are stronger as microdroplets flow velocity increases.

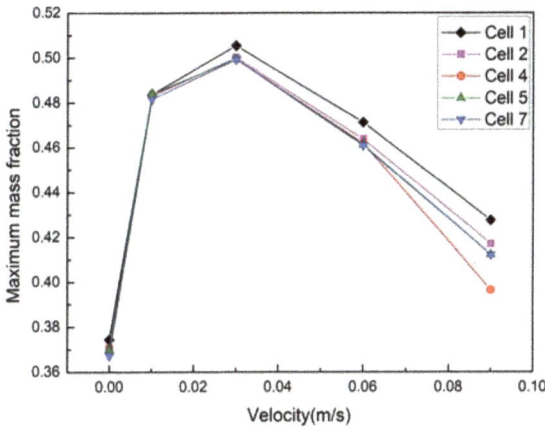

Figure 5. Change of maximum rhodamine mass fraction on cell surfaces with the relative velocity of microdroplet to culture solution when culture solution velocity is 0.01 m/s.

3.2. Pulse Time

Controllable and instant pulse drug delivery is the most advantage of this microfluidic chip. The time is assumed as effective dosing time when the rhodamine solution mass fraction at cell surfaces is larger than 0.01. The drug delivery pulse time changing with microdroplet and culture solution inlet velocity is shown as Figure 6. The pulse time decreases rapidly as all fluids inlet velocity increases to 0.03 m/s but then decreases slowly when culture solution inlet velocity is over 0.03 m/s. The pulse time decreases rapidly when the relative velocity of microdroplet to culture solution varies from 0.01 m/s to 0.06 m/s but decreases slightly under other relative velocities.

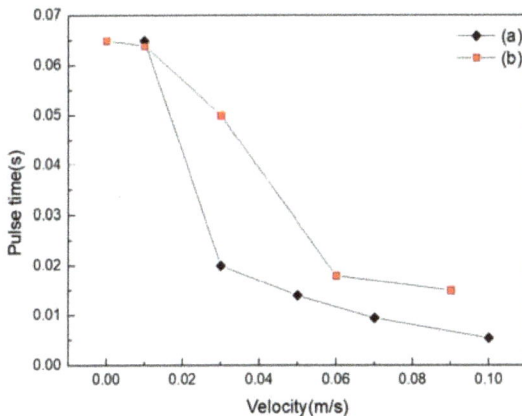

Figure 6. Change of the cell surfaces dosing pulse time with fluids flow velocity: (a) the relative velocity between microdroplet and culture solution is 0 m/s and microdroplet and culture solution velocity change from 0.01 m/s to 0.1 m/s; (b) change of the relative velocity of microdroplet to culture solution from 0 to 0.09 m/s.

The inter-diffusion effect is predominant when the concentration gradient is pretty large and culture solution inlet velocity is relatively slow, so culture solution takes more time to dilute and carry rhodamine to outlet direction. The rhodamine mass fraction distribution in micro-traps at 2 ms is shown as Figure 7. The rhodamine mass fraction diffuses from microdroplet to culture solution through microporous membrane once the microdroplet attains membrane zone and it dramatically decreased from inlet to outlet at 4 ms when both microdroplet and culture solution velocity are 0.03 m/s. However, the rhodamine mass fraction is quite uniform in 7 micro-traps just at 2 ms when the relative velocity of microdroplet to culture solution is 0.06 m/s.

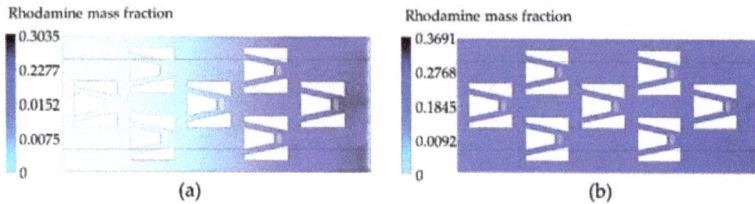

Figure 7. Distribution of rhodamine mass fraction in micro-traps: (**a**) microdroplet and culture solution velocity are 0.03 m/s; (**b**) relative velocity of the microdroplet to culture solution is 0.06 m/s.

Inter-diffusion velocity along X direction is nearly equal to culture solution velocity when little rhodamine diffuses to inlet direction and it is around 0.03 m/s. Thus, the time of rhodamine staying at cell surfaces almost only depends on culture solution inlet velocity, because inter-diffusion velocity is far less than culture solution velocity, the inter-diffusion direction was changed mainly towards outlet. The rhodamine was carried away by culture solution rapidly once they diffused to cell surfaces. Moreover, the microdroplet is extruded to micro-traps zone through porous zone when relative velocity is larger than 0.01 m/s. As a result, the mass fraction of rhodamine on micro-traps grows more rapidly than the situation when the microdroplet velocity is the same as culture solution velocity and the convective diffusion effect is much enhanced when the relative velocity increases. But the time of microdroplet extruded into culture solution decreases slightly when the relative velocity of microdroplet to culture solution is larger than 0.06 m/s, the fluid velocity located under microdroplet zone could not be improved significantly.

3.3. Pressure Field

The fluid pressure on cell surfaces is directly related with cell culture and dosing analysis, high pressure on cell surfaces should be avoided in this system. Other than that, the fluid pressure around microdroplet also influences the rhodamine diffusion efficiency and the impact to cells. The microdroplet was affected by capillary force and especially fluid pressure, its shape is changed to trapezoid from slug as shown in Figure 8. The contact angle and area between microdroplet and culture solution are expanded when microdroplet velocity and culture solution velocity increase and the microdroplet is extruded to culture solution more rapidly when relative velocity enlarges.

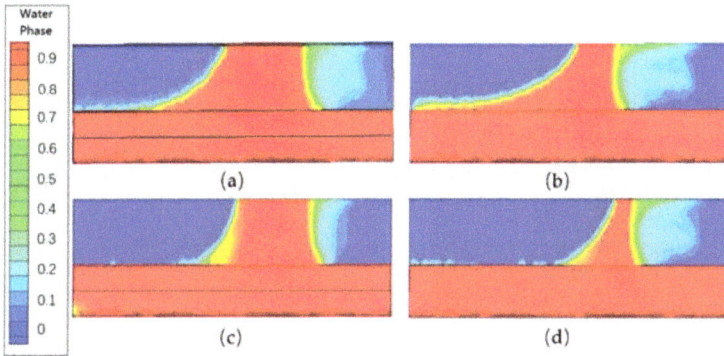

Figure 8. Distribution of water phase: (**a**) both microdroplet and culture solution flow velocities are 0.05 m/s; (**b**) microdroplet and culture solution flow velocities are 0.1 m/s; (**c**) the relative velocity of microdroplet to culture solution is 0.03 m/s when culture solution velocity is 0.01 m/s; (**d**) the relative velocity of microdroplet to culture solution is 0.09 m/s when culture solution velocity is 0.01 m/s.

From the pressure change around microdroplet with microdroplet and culture solution velocities shown as Figure 9. The pressure increase on the inlet side is 1200 Pa when the microdroplet and culture solution velocities increase from 0.05 m/s to 0.1 m/s; meanwhile, the pressure on the outlet side only increases by 200 Pa. From Figures 8 and 9 it can be concluded that the pressure around the microdroplet is mainly responsible for the much larger contact angle and area on the outlet side than those on inlet side. Because the pressure difference between the two ends of the droplet will increase when the relative velocity of the droplet increases, regardless of whether the medium velocity increases or not. The interface between microdroplet and oil close to inlet suffers from much severer fluid pressure, which pushes microdroplet to flow and maintain original shape. However, the interface between microdroplet and oil towards outlet side not only lacks sufficient pressure support but also is dragged by faster culture solution below the membrane. The pressure difference between the two ends of the droplet is mainly affected by the oil phase velocity. The contact angle of between microdroplets and microporous membrane that close the outlet increases with the increase of the flow velocity of the culture medium, so that the contact area between the droplet and the medium also increases.

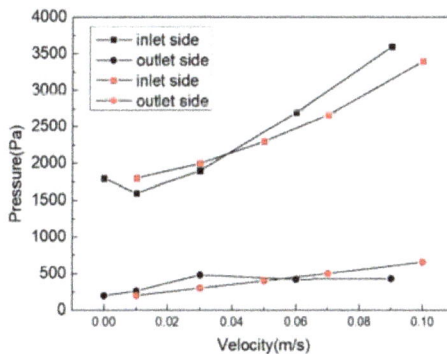

Figure 9. Pressure around microdroplet influenced by velocity: black symbol means the pressure change with the relative velocity of microdroplet; red symbol means the pressure change with microdroplet velocity when there is no relative velocity between microdroplet and culture solution.

The capillary force at microporous membrane zone plays an important role in rhodamine diffusion when microdroplet flow velocity is equal to culture solution flow velocity. The capillary force drags the microdroplet, arousing a tendency to flow into culture solution. What's more, the microporous membrane surface is quite rough in microscale, which causing droplet spreading and absorption phenomenon [32]. As a result, the contact area microdroplet between culture solution is relatively extended, so rhodamine inter-diffusion effect is improved and the number of cells dosing at a moment is increased too.

In addition, microdroplet would be extruded to culture solution channel through porous membrane gradually because of relative velocity and it is extruded more rapidly as the relative velocity becomes larger. The cell surfaces pressure caused by fluids flow field change may influence cell culture environment, culture solution velocity and the relative velocity of microdroplet and should be focused. The cell surfaces pressure influenced by culture velocity when the relative velocity of microdroplet is 0 and as shown in Figure 10a, the cell surfaces pressure influenced by the relative velocity of microdroplet when culture velocity is 0.01 m/s as shown in Figure 10b. All the cell surfaces pressure increases especially for Cell 1–4 when the flow velocity of microdroplet and culture solution increase at equal rates. All the cell surfaces pressure increased generally when microdroplet and culture velocity or the relative velocity of microdroplet enlarged but it is only Cell 4 cell surface pressure always keeps growing tendency.

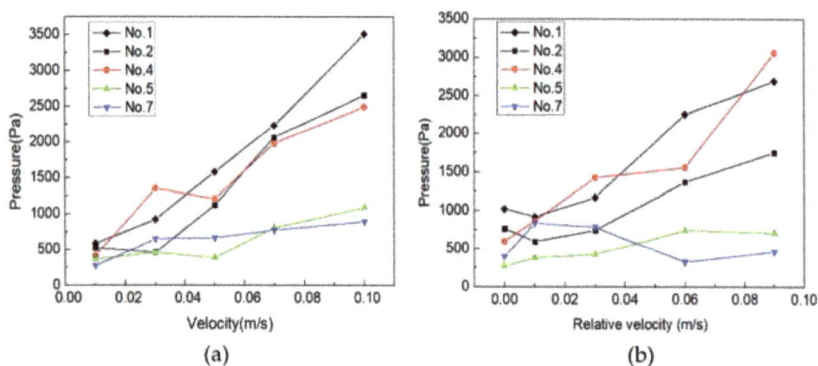

Figure 10. Simulation results of cell surfaces suffer pressure: (**a**) relation of cell surfaces pressure with culture solution inlet velocity when microdroplet relative velocity is 0; (**b**) relation of cell surfaces pressure with the relative velocity of microdroplet when culture solution inlet velocity is 0.01 m/s.

Cells located at micro-traps close to the culture solution inlet suffer more dramatic fluid impact force when microdroplet velocity increases, while the cells far from the inlet suffer only slight impact. The microdroplet velocity does not influence cells because the laminar situation, the convective diffusion between microdroplet and culture solution is tiny. Because the microdroplet diffusion process always covers Cell 4 cell, it causes more pressure as it diffuses through microporous membrane faster. The cells behind Cell 4 would not suffer too much from microdroplet diffusion pressure because of most microdroplets diffused to culture solution and buffered located around Cell 1–4 cells zone. In general, this microfluidic system could further reduce the drug consumption in cells dosing experiments and promote experiments efficiency and observability. The impairment of cells caused by fluid pressure would be reduced a lot compared to jet dosing and the dosing efficiency would be improved, as well as an instant controllable dosing time is realized [19–23].

3.4. Combined Influence on Cell Dosing

The cell pulse dosing should be carried out in condition of low pressure, as high as possible drug mass fraction on cell surfaces and controllable pulse time. From above results and discussion, the change of rhodamine mass fraction with relative velocity and pressure is shown in Figure 11. The diffusion process of rhodamine to cells attains best effect when the relative velocity of microdroplet to culture solution is 0.03 m/s and cells suffer more pressure when relative velocity increases.

Convective velocity of Rhodamine molecules to micro-traps increases when the relative velocity of microdroplet enhanced. The pressure around the microdroplets would not only extrude microdroplets into culture medium but also increase the friction between the microdroplets and the wall surface or the surface of the microporous membrane. And the vortex between the microdroplets and the microporous membrane would be enhanced because of high fluid pressure, which may hinder the diffusion process. In the meanwhile, cells would be impaired in high fluid pressure environment should be considered. As a result, high efficiency cell pulse dosing could be achieved at the relative velocity of microdroplet is 0.03 m/s when culture solution velocity is 0.01 m/s and shorter pulse time could be achieved by increasing microdroplet and culture solution velocity at the same time or increasing the relative velocity of microdroplet to culture solution.

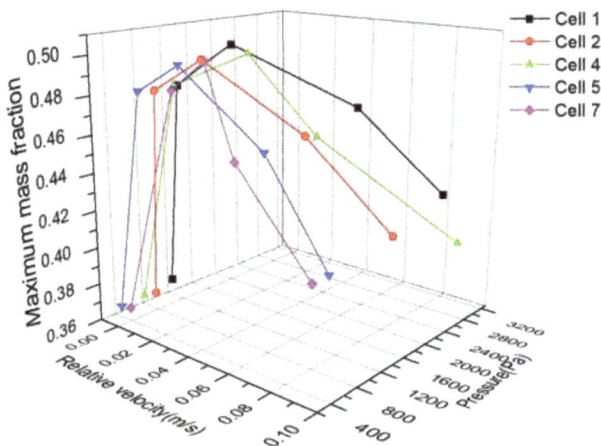

Figure 11. Simulation results of cell surfaces maximum rhodamine mass fraction influenced by the relative velocity of microdroplet and cell surfaces pressure.

4. Conclusions

In this paper, we analyzed the convective diffusion process of microdroplet cell dosing in the three-layer sandwich structure of PDMS/PCTE microporous membrane/PDMS chip, which mainly focused on the drug delivery mechanism, pulse time, mass fraction on cell surfaces, the effect of the change on microdroplet velocity and culture solution velocity. We found that cells close to culture solution suffer more fluid pressure impact when culture solution inlet velocity or the relative velocity of microdroplet to culture solution increased, otherwise cells are affected slightly. The mechanism of the changes of cell dosing pulse time with microdroplet and culture solution velocity is obtained. Inter-diffusion and convective diffusion play different degrees of importance during drug delivery through microporous membrane to cells. In condition of the relative velocity of microdroplet to culture solution is 0, the cell dosing pulse time decreases notably until flow velocity increases to 0.03 m/s, while it decreased slowly when flow velocity is faster than 0.03 m/s. In condition of constant culture solution velocity, the cell dosing pulse time decreases notably until the relative velocity of microdroplet increases to 0.06 m/s, while it decreases slowly when flow velocity is faster than 0.06 m/s. The pulse time could be controlled by changing microdroplet and culture solution velocities. For better cell dosing efficiency, the rhodamine mass fraction on cell surfaces attains the maximum value when the relative velocity of microdroplet to culture solution is 0.03 m/s. And the cells dosing pulse time or drug concentration could be controlled by changing the flow velocity of microdroplets and culture medium, which is convenient for studying the pulse time influence. Compared to traditional dosing in living body or jet dosing in microfluidic chip, this microdroplet cell dosing chip could achieve more controllable pulse dosing and observe the instant drug effects on cells. The knowledge obtained in this work may provide useful information for microfluidic chip applied to cell culture and dosing.

Author Contributions: K.L. and D.B. proposed the idea of single-cell trapping model; J.N. and M.H. designed the structure of the chip and performed the draft simulation; Z.W. and S.C. proposed the numerical model and performed the simulation; Z.W. and D.W. analyzed the data; Z.W. wrote the paper; Y.B. joined in the discussion and inspected the accuracy; Q.M. contributed reagents and materials and joined in the discussion of experimental data; K.L. and D.B. organized the entire teamwork and supported this work with their funding.

Funding: This research was jointly funded by the Fundamental Research Funds for the Central Universities of China (N160302001), National Natural Science Foundation of China (51376039), the CAST-BISEE Fund Project (CAST-BISEE2017-011), the 13th Five-Year National Science and Technology Major Project (2017ZX02201005-001).

Acknowledgments: This research has been supported by Key Laboratory of Vibration and Control of Aeronautical Equipment, Ministry of Education, Northeastern University (VCAME201604).

Conflicts of Interest: The authors declare no conflict of interest.

References

1. Millet, L.J.; Gillette, M.U. New perspectives on neuronal development via microfluidic environments. *Trends Neurosci.* **2012**, *35*, 752–761. [CrossRef] [PubMed]
2. Lion, N.; Rohner, T.C.; Dayon, L.; Arnaud, I.L.; Damoc, E.; Youhnovski, N.; Wu, Z.Y.; Roussel, C.; Josserand, J.; Jensen, H.; et al. Microfluidic systems in proteomics. *Electrophoresis* **2003**, *24*, 3533–3562. [CrossRef] [PubMed]
3. Wang, P.; DeVoe, D.L.; Lee, C.S. Integration of polymeric membranes with microfluidic networks for bioanalytical applications. *Electrophoresis* **2001**, *22*, 3857–3867. [CrossRef]
4. Benjamin, R.K.; Hochberg, F.H.; Fox, E.; Bungay, P.M.; Elmquist, W.F.; Stewart, C.F.; Gallo, J.M.; Collins, J.M.; Pelletier, R.P.; de Groot, J.F.; et al. Review of microdialysis in brain tumors, from concept to application: First annual Carolyn Frye-Halloran symposium. *Neuro Oncol.* **2004**, *6*, 65–74. [CrossRef] [PubMed]
5. Fair, R.B. Digital microfluidics: Is a true lab-on-a-chip possible? *Microfluid. Nanofluid.* **2007**, *3*, 245–281. [CrossRef]
6. Chengfu, C.; Drew, K.L. Droplet-based microdialysis—Concept, theory, and design considerations. *J. Chromatogr. A* **2008**, *1209*, 29–36. [CrossRef]
7. Lee, T.Y.; Hyun, K.A.; Kim, S.I.; Jung, H.I. An integrated microfluidic chip for one-step isolation of circulating tumor cells. *Sens. Actuators B Chem.* **2016**, *238*, 1144–1150. [CrossRef]
8. Mao, H.; Cremer, P.S.; Manson, M.D. A sensitive, versatile microfluidic assay for bacterial chemotaxis. *Proc. Natl. Acad. Sci. USA* **2003**, *100*, 5449–5454. [CrossRef] [PubMed]
9. Dertinger, S.K.; Jiang, X.; Li, Z.; Murthy, V.N.; Whitesides, G.M. Gradients of substrate-bound laminin orient axonal specification of neurons. *Proc. Natl. Acad. Sci. USA* **2002**, *99*, 12542–12547. [CrossRef] [PubMed]
10. Schiff, N.D.; Giacino, J.T.; Kalmar, K.; Victor, J.D.; Baker, K.; Gerber, M.; Fritz, B.; Eisenberg, B.; O'connor, J.; Kobylarz, E.J.; et al. Behavioural improvements with thalamic stimulation after severe traumatic brain injury. *Nature* **2007**, *448*, 600–603. [CrossRef] [PubMed]
11. Asaad, W.; Eskandar, E. The movers and shakers of deep brain stimulation. *Nat. Med.* **2008**, *14*, 17–18. [CrossRef] [PubMed]
12. Fytagoridis, A.; Åström, M.; Wårdell, K.; Blomstedt, P. Stimulation-induced side effects in the posterior subthalamic area: Distribution, characteristics and visualization. *Clin. Neurol. Neurosurg.* **2013**, *115*, 65–71. [CrossRef] [PubMed]
13. Lopes, C.D.; Gomes, C.P.; Neto, E.; Sampaio, P.; Aguiar, P.; Pêgo, A.P. Microfluidic-based platform to mimic the in vivo peripheral administration of neurotropic nanoparticles. *Nanomedicine* **2016**, *11*, 3205–3221. [CrossRef] [PubMed]

14. Wu, L.; Tsutahara, M.; Kim, L.; Ha, M. Numerical simulations of droplet formation in a cross-junction microchannel by the lattice Boltzmann method. *Int. J. Numer. Methods Fluids* **2007**, *57*, 793–810. [CrossRef]
15. Wu, L.; Tsutahara, M.; Kim, L.S.; Ha, M. Hree-dimensional lattice Boltzmann simulations of droplet formation in a cross-junction microchannel. *Int. J. Multiph. Flow* **2008**, *34*, 852–864. [CrossRef]
16. Raj, R.; Mathur, N.; Buwa, V.V. Numerical simulations of liquid–liquid flows in microchannels. *Ind. Eng. Chem. Res.* **2010**, *49*, 10606–10614. [CrossRef]
17. Raimondi, N.D.M.; Prat, L. Numerical study of the coupling between reaction and mass transfer for liquid–liquid slug flow in square microchannels. *AICHE J.* **2011**, *57*, 1719–1732. [CrossRef]
18. Raimondi, N.M.; Prat, L.; Gourdon, C.; Cognet, P. Direct numerical simulations of mass transfer in square microchannels for liquid–liquid slug flow. *Chem. Eng. Sci.* **2008**, *63*, 5522–5530. [CrossRef]
19. Pfaff, D.; Waters, E.; Khan, Q.; Zhang, X.; Numan, M. Minireview: Estrogen Receptor-Initiated Mechanisms Causal to Mammalian Reproductive Behaviors. *Endocrinology* **2012**, *152*, 1209–1217. [CrossRef] [PubMed]
20. Serwer, L.P.; James, C.D. Challenges in drug delivery to tumors of the central nervous system: An overview of pharmacological and surgical considerations. *Adv. Drug Deliv. Rev.* **2012**, *64*, 590–597. [CrossRef] [PubMed]
21. Chen, Y.; Liu, L.H. Modern methods for delivery of drugs across the blood-brain barrier. *Adv. Drug Deliv. Rev.* **2012**, *64*, 640–665. [CrossRef] [PubMed]
22. Monopoli, M.P.; Aberg, C.; Salvati, A.; Dawson, K.A. Biomolecular coronas provide the biological identity of nanosized materials. *Nat. Nanotechnol.* **2012**, *7*, 779–786. [CrossRef] [PubMed]
23. Sugiura, S.; Hattori, K.; Kanamori, T. Microfluidic serial dilution cell-based assay for analyzing drug dose response over a wide concentration range. *Anal. Chem.* **2010**, *82*, 8278–8282. [CrossRef] [PubMed]
24. Mary, P.; Studer, V.; Tabeling, P. Microfluidic droplet-based liquid-liquid extraction. *Anal. Chem.* **2008**, *80*, 2680–2687. [CrossRef] [PubMed]
25. Nerín, C.; Salafranca, J.; Aznar, M.; Batlle, R. Critical review on recent developments in solventless techniques for extraction of analytes. *Anal. Bioanal. Chem.* **2009**, *393*, 809–833. [CrossRef] [PubMed]
26. Chua, Y.T.; Ji, G.; Birkett, G.; Lin, C.X.C.; Kleitz, F.; Smart, S. Nanoporous organosilica membrane for water desalination: Theoretical study on the water transport. *J. Membr. Sci.* **2015**, *482*, 56–66. [CrossRef]
27. Peng, T.; Li, Q.; Chen, J.; Gao, X. Quantitative analysis of surface tension of liquid nano-film with thickness: Two stage stability mechanism, molecular dynamics and thermodynamics approach. *Phys. A Stat. Mech. Appl.* **2016**, *462*, 1018–1028. [CrossRef]
28. Qu, J.; Wang, Q.; He, Z.X.; Jian, Q.U.; Qian, W.; Zhi-Xia, H.E.; Han, X.Y.; Zi-Cheng, H.U.; Tao, L. An experimental study of formation and flow charactristics of droplets in a rectangular microchannel. *J. Shanghai Jiaotong Univ.* **2015**, *49*, 86–90. [CrossRef]
29. Chen, S.; Liu, K.; Liu, C.; Wang, D.; Ba, D.; Xie, Y.; Du, G.; Ba, Y.; Lin, Q. Effects of surface tension and viscosity on the forming and transferring process of microscale droplets. *Appl. Surf. Sci.* **2016**, *388*, 196–202. [CrossRef]
30. Fujita, H.; Einaga, Y. Self-Diffusion and Viscoelasticity in Entangled Systems I. Self-Diffusion Coefficients. *Polym. J.* **1985**, *17*, 1131–1139. [CrossRef]
31. Dunca, A.A.; Neda, M.; Rebholz, L.G. A mathematical and numerical study of a filtering-based multiscale fluid model with nonlinear eddy viscosity. *Comput. Math. Appl.* **2013**, *66*, 917–933. [CrossRef]
32. Espin, L.; Kumar, S. Droplet spreading and absorption on rough, permeable substrates. *J. Fluid Mech.* **2015**, *784*, 465–486. [CrossRef]

processes

MDPI

Article

Numerical Simulation of a New Porous Medium Burner with Two Sections and Double Decks

Zhenzhen Jia, Qing Ye *, Haizhen Wang, He Li and Shiliang Shi

School of Resource, Environment and Safety Engineering, Hunan University of Science and Technology, Xiangtan 411201, China; jiazhenzhen1982@126.com (Z.J.); yelongtian@hotmail.com (H.W.); lihecumt@126.com (H.L.); sslhnust@yeah.net (S.S.)
* Correspondence: 1010068@hnust.edu.cn; Tel.: +86-0731-5829-0040

Received: 11 July 2018; Accepted: 26 September 2018; Published: 6 October 2018

Abstract: Porous medium burners are characterized by high efficiency and good stability. In this study, a new burner was proposed based on the combustion mechanism of the methane-air mixture in the porous medium and the preheating effect. The new burner is a two-section and double-deck porous medium with gas inlets at both ends. A mathematical model for the gas mixture combustion in the porous medium was established. The combustion performance of the burner was simulated under different equivalence ratios and inlet velocities of premixed gas. The methane combustion degree, as well as the temperature and pressure distribution, was estimated. In addition, the concentrations of emissions of NO_x for different equivalence ratios were investigated. The results show that the new burner can not only realize sufficient combustion but also save energy. Furthermore, the emission concentration of NO_x is very low. This study provides new insights into the industrial development and application of porous medium combustion devices.

Keywords: two sections and double decks; porous medium; combustion; equivalence ratio; gas velocity in inlet; NO_x

1. Introduction

Nowadays, a lot of low-grade combustible gases are not effectively used. Direct emission of these gases will result in severe energy dissipation and environment pollution. Porous media have many advantages such as good storage, conduction and radiation of heat, and explosion suppression, etc. These features help the porous medium to achieve super-adiabatic combustion [1,2]. Porous medium combustion technology presents a new type of combustion. Compared with traditional combustion technology, it has a larger combustion limit, higher combustion efficiency, and lower pollutant emission [3–6]. Therefore, porous medium combustion technology is suitable for the utilization of low-calorific-value gases such as low-concentration gas, blast furnace gas, biomass gas, and landfill gas. Based on the chemical kinetic model of the Gibbs minimum free energy theory, Slimane et al. established a mathematical model to simulate the combustion of hydrogen sulphide in porous medium. The composition of combustion products and combustion limit were numerically determined [7]. The combustion of the propane-air mixture was comprehensively investigated by using the ceramic material of honeycomb foam as the porous medium [8]. The temperature of the gas was much higher than that of the adiabatic flame, which was attributed to the internal heat recirculation. Hayashi simulated gas mixture combustion in a porous medium using a three-dimensional mathematical model [9]. The results showed that the flame tended to spread to the combustion zone with lowest air proportion. Hayashi suggested that the pore size of the porous medium should be properly reduced to avoid tempering. Kotani et al. proposed a two-layer porous medium burner [10]. The results showed that the space structure mutation of the junction plane of two different porous media can stabilize the main body of the flame in the flame support layer. Pereira et al. studied the energy focus-ability of a gas mixture in a burner with two-layer porous media [11]. The results showed that the excess temperature is a function of the modified Lewis number, the porosity of the porous medium, and the gas thermal conductivity ratio of solid to gas. Mital et al. concluded that the radiant heat efficiency in the outlet of the burner can reach 30% [12]. According to the one-step chemical reaction of the methane-air mixture, the one-dimensional concentration and flame velocity field were calculated, and the temperature distribution of the flame under different optical thicknesses was obtained [13]. Based on the multistep reaction model, the gas-solid energy equation was established, and the combustion reaction was simulated [14]. Also, the thermal conductivity, volume heat transfer coefficient, and outlet boundary condition of the porous medium were studied.

The influences of the combustion equivalence ratio of the gas mixture and steady airflow velocity on the combustion efficiency were experimentally studied by Wang Enyu [15]. The results showed that the combustion and flow of gas can be improved by using a gradient structure. The characteristics of combustion, pollutant emission, heat diffusion, and resistance in gradually varied porous media (GVPM) were studied, and a GVPM gas-fired boiler was designed [16]. The results revealed that the GVPM burner has the merits of lower NO_x and CO emission, higher heat diffusion, and lower resistance. A "volume mean transport equation" of premixed combustible gas flow in isotropic porous medium was proposed by Du [17,18]. Using this equation, he simulated the super-adiabatic combustion of the gas flow in a porous medium burner and found that the reciprocating flow of premixed combustible gas in porous medium has greater advantages than the single-direction flow. The effect of the characteristic parameters of the porous medium on the combustion temperature and pressure can be found in the literature [19–21]. The industrial applications and numerical simulation of premixed gas combustion in a porous medium were studied in the literature [22]. The results showed that a square burner can achieve a stable combustion of premixed gases, but it is more demanding in terms of working conditions, and there is a problem of uneven combustion. Circular burners are recommended for industrial applications. The excess enthalpy flames stabilized in a radial multichannel as a model of a cylindrical radial-flow porous medium burner were experimentally and theoretically studied [23]. The result showed that the multichannel flames had excess combustion velocities due to heat recirculation via the channel walls. The stabilization diagram of premixed methane-air flames in finite porous medium under a uniform ambient temperature was numerically studied [24]. The results showed that both stable and unstable solutions, for upper and lower flames, either exist on the surface or submerged in the porous matrix. The conversion of nitric oxide inside a porous medium was investigated in the literature [25]. The effects of equivalence ratio and flow velocity on the flame stabilization were investigated. At the same time, the NO_x and TFN (total fixed nitrogen) conversion ratios and temperature profiles along the burner were obtained. The flame propagation and stationary mechanism in single/double-layer porous medium burners were studied by Zhao [26]. The results showed that the double-layer porous medium burner can realize stable standing in the vicinity of the material interface, which can effectively prevent tempering and stripping. It was also found that the burner with multiple sections or double layers would more easily achieve the above objectives than the burner with a single direction or single layer. The combustion characteristics of gaseous fuels with low calorific value in a two-layer porous burner were numerically simulated [27]. The results showed that OH and O radicals are most actively involved in combustion. OH-involved elementary reactions are significantly faster than O-involved elementary reactions. NO is the only NO_x species which is significantly presented in the exhaust gases. The concentrations of NO_2 and N_2O never exceeded 1 mg/m^3. The performance characteristics of premixed inert porous burners were numerically simulated [28]. The results indicated that the flow, temperature, and concentration fields of species appear to be approximately one-dimensional. Compared with the traditional burner, the chemical reaction zone is extended in the porous medium burner, while the gas temperature gradient near the flame zone is reduced. Premixed natural gas combustion in porous inert media was numerically studied [29]. The results showed that the radiation of the solid and the heat exchange between gas and solid are the two main factors impacting the combustion performance of porous inert media burners.

Based on the wide application of porous medium burners with a two-section structure ("i.e., ordinary burner") at present and the advantages of the porous medium burner, the combination of porous media is optimized in this paper. Finally, a new two-section and double-deck porous medium burner with gas inlets at both ends ("i.e., new burner") is proposed and designed, and the combustion effects of the two kinds of burners are numerically simulated. It is expected that it is simpler and more effective to realize regenerative heat from the combustion zone to the preheating zone, which can effectively solve the problem of low temperature in the vicinity of the burner inlet.

2. Mechanism of Gas Combustion in a Porous Medium

The physical-chemical reaction process of gas combustion is very complicated. Free radicals, ions, electrons molecules, and other transient/intermediate products will be produced during combustion. The flame features of the combustion reaction of CH_4-air can be reflected by study of the oxidation reaction mechanism of CH_4 [30]. Experimental study on the combustion reaction of N_2 and CH_4 showed that the oxidation process of CH_4 includes a series of free radicals involved in the reaction. According to the element composition of the gas mixture, ions and molecules containing C, H, O, and N are the vast majority of the intermediate products and transient products. These products, especially free radicals, become the activation center in the reaction process, which greatly promotes the continuation, acceleration, and termination of the reaction. The conversion of methane to hydrogen in a porous medium reactor was investigated using fuel with equivalence ratios ranging from 1.5 to 5. The wave velocity, peak combustion temperature, flame structure, volumetric heat release, wave thickness, and hydrogen yield were obtained. The parameters affecting these characteristics included inlet velocity, equivalence ratio, and the thermal conductivity, and the specific heat of the porous medium [31–35].

Generally, when the temperature is high, the oxidation of CH_4 can be simplified to $CH_4 \rightarrow CH_3 \rightarrow CH_2O \rightarrow CO \rightarrow CO_2$. A continuous combustion reaction increases the temperature, which in turn accelerates the combustion; as a result, the temperature rapidly rises. In this process, hot molecules, ions, and electrons are sharply generated. They carry a huge impulse, momentum, and kinetic energy, and collide with each other. So, the N_2 molecules are bound to be impacted by these active particles, and then be decomposed and involved in the oxidation of CH_4, which also explains why the pollutants produced by gas combustion in porous media are NO_x mostly, in addition to CO_2. The design of the new burner fully considers the material structure and the outstanding regenerative properties which can promote chain break and reduce the activation energy in gas combustion reactions. The combustion efficiency is increased, and the combustion of lean gas is promoted.

3. Methodology

3.1. Numerical Model

At present, burners with two-section structures have been widely used, as shown in Figure 1. This burner is composed of two porous media with different porosities. The smaller-pored medium is regarded as the preheating zone, and the other is the combustion zone. As illustrated in Figure 2, the new burner is divided into two sections: the preheating zone and the combustion zone. Our previous research suggested an optimal ratio of 1.2:1 for the burner. The exterior of the new burner is a cylinder with a 0.025 m radius and 0.22 m length. The lengths of the porous media of the preheating and combustion zones are 0.12 m and 0.1 m, respectively. The new burner consists of double decks, both of which are composed of two porous media with different porosities. The material is silicon carbide. The media with different pores of the inner and outer deck are staggered, and the small-pore medium is used as the preheating zone of the combustion; the other is the main combustion zone. The cross-sectional radius of the inner deck is 0.0125 m. The gases simultaneously enter into the preheating zone of the small-pore medium from two ends of the new burner. In addition to convection heating, the premixed gas can be heated by the solid heat conduction and heat radiation from the surrounding combustion zone with the big-pore medium. The heat reflux will further enhance the heat storage effect, and the combustion stability is thus improved. The contact area between the small-pore preheating zone and the big-pore combustion zone of the new burner is larger. Therefore, the fully premixed gas will be burned in a flow process, which completes the chemical reaction and releases energy. The equivalence ratio of premixed gas in the burner inlet was set to 0.65, the corresponding volume ratio of methane/air was 6%, the inlet flow velocity was 0.85 m/s, and the temperature of premixed gas was 300 K.

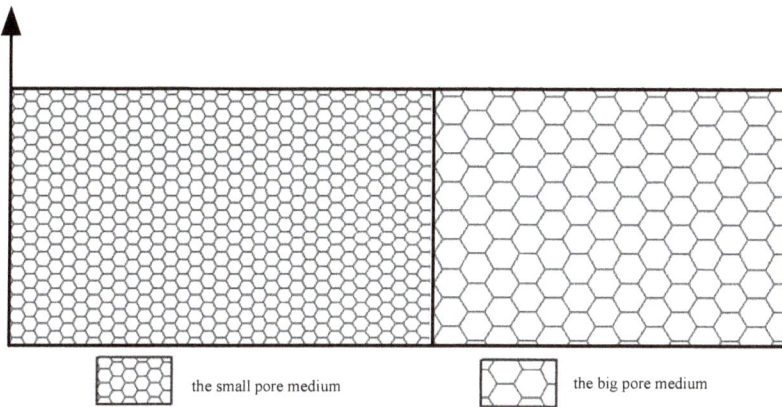

Figure 1. The ordinary burner.

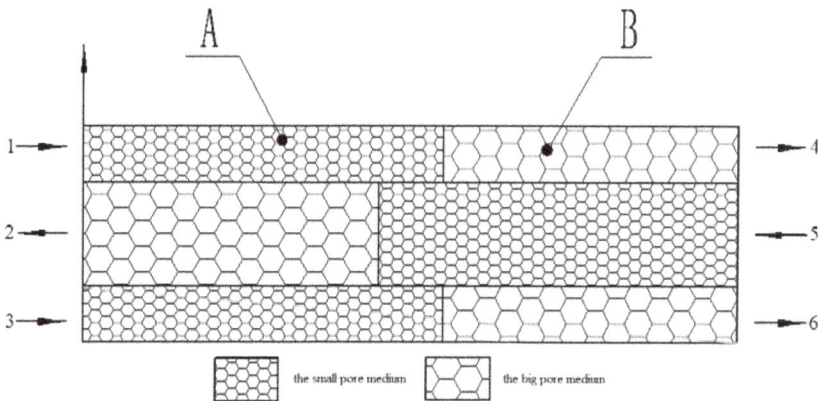

Figure 2. Sectional view of the physical model of the new burner. 1, premixed gas inlet; 2, exhaust gas outlet; 3, premixed gas inlet; 4, exhaust gas outlet; 5, premixed gas inlet; 6, exhaust gas outlet.

Meshing enables the discretization of the geometry into small units of simple shapes, referred to as mesh elements. Because the new burner has double decks of a nested type, the unstructured mesh can adapt to the structural requirements. The partition method was the T mesh and the element form was the regular tetrahedral element. Due to the obvious symmetry of the model, the axisymmetric model was adopted in the mesh. To ensure the accuracy of the calculation, the mesh spacing was 0.1 mm; finally, the mesh with 16,028 nodes was selected. According to the research theory [36], the reaction mechanism has little effect on the temperature distribution of porous medium burners. Therefore, a simple one-step reaction mechanism was adopted in the numerical model, namely,

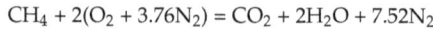

$$CH_4 + 2(O_2 + 3.76N_2) = CO_2 + 2H_2O + 7.52N_2$$

3.2. Fundamentals

Because the structure of the porous medium burner is simple, the pores of the porous material are small, and the vortex is also small, it is assumed that the turbulence effect increases the flame front thickness and heat transfer and that the flame propagation is one-dimensional. The chemical source can be calculated by the Arrhenius formula. To simplify the calculation, the following assumptions were made:

①The gravity effect is ignored, and the gradient of each physical quantity is zero in the cross section of the circular cylinder.

②The porous medium is considered a homogeneous optical material and dispersion structure.

③The outer wall of the burner is adiabatic and nonslip, whereas the inner wall is set to have grey body radiation. The heat process between different parts is an unsteady-state process.

④Gas phase radiation is neglected.

⑤The potential high-temperature solid catalysis effect is ignored.

⑥The flame surface is at the coordinate origin of the X axis during steady combustion; the flame front is located near the interface of the two-deck porous medium.

⑦The premixed methane-air gas is regarded as an ideal gas.

Based on the above assumptions, the balance equations are as follows:

(1) Continuity equation:

$$\frac{\partial(\rho\varepsilon)}{\partial t} + \frac{\partial(\varepsilon\rho u_i)}{\partial x_i} = 0 \tag{1}$$

where ρ is the density of the premixed gas, ε is the porosity of the porous medium, and u_i is the velocity vector.

(2) Momentum conservation equation:

$$\frac{\partial(\varepsilon\rho u_j)}{\partial t} + \frac{\partial}{\partial x_i}(\varepsilon\rho u_j u_i) - \frac{\partial}{\partial x_i}\left(\rho\varepsilon v \frac{\partial u_j}{\partial x_i}\right) =$$
$$-\frac{\partial(\varepsilon P)}{\partial x_j} + \frac{\partial}{\partial x_i}\left[\rho\varepsilon v\left(\frac{\partial u_i}{\partial x_j} - \frac{2}{3}\frac{\partial u_k}{\partial x_k}\delta_{ij}\right)\right] - \frac{\mu\varepsilon^2}{K_P}u_j \tag{2}$$

where v is the viscous resistance coefficient, μ is the inertia resistance coefficient, and K_p is the permeability of the porous medium.

(3) Component conservation equation:

$$\frac{\partial(\rho_g Y_n)}{\partial t} + \nabla \cdot (\rho_g u_i Y_n) = -\nabla \cdot (\rho_g Y_n V_i) + \omega_n W_n \tag{3}$$

where Y_n is the mass fraction of the nth component, V_i is the diffusion velocity, u_i is the velocity with respect to a stationary coordinate system, W_n is the molecular weight of the nth component, and ω_n is the molar formation rate of the nth component.

(4) Gas energy equation:

$$\frac{\partial(\varepsilon\rho_g C_g T_g)}{\partial t} + \nabla \cdot [\varepsilon u(\rho_g C_g T_g + \rho)] =$$
$$h_v(T_s - T_g) + \nabla \cdot \left[\varepsilon\lambda_g \Delta T_g - \left(\varepsilon\sum_n h_n\rho Y_n(u_i - u)\right) + \varepsilon(\tau u)\right] + \varepsilon Q \tag{4}$$

where Q is the heat efficiency of the chemical reaction, λ_g is the thermal conductivity of the premixed gas, h_v is the volumetric convective heat transfer coefficient, h_n is the molar enthalpy of formation of the nth component, C_g is the specific heat capacity of the mixed gas at constant pressure, and T_g is the gas phase temperature.

Because the gas content is low, the one-step chemical reaction mechanism of methane/air is used. The gas viscosity is significantly related to the temperature but is almost independent of the pressure. When the temperature T_g is less than 2000 K, the gas viscosity can be calculated by the Sutherland formula.

(5) Solid state energy equation:

$$\frac{\partial[(1-\varepsilon)\rho_s C_s T_s]}{\partial t} = \nabla \cdot (\lambda_e^s \Delta T_s) + hv(T_g - T_s) \tag{5}$$

where $\lambda_e^s = \lambda_e^c + \lambda_e^r$ is the effective heat transfer coefficient of the solid porous medium, λ_e^c is the effective thermal conductivity of the porous medium, and λ_e^r is the heat transfer coefficient; β_s is the radiation attenuation coefficient of the porous medium, $\beta_s = m(1-\varepsilon)/d_p$, and the coefficient m reduces with the increase in the average pore size of the porous medium, but increases with the increase in porosity. According to the porosity of the porous medium and the results of the experiment, m is 4, ρ_s is the density of the porous medium, T_s is the porous medium temperature, and C_s is the heat capacity at constant pressure.

(6) State equation of ideal gas:

$$\rho_g = \frac{WP}{RT_g} \tag{6}$$

where W is the average molecular weight of the mixed gas, and R is the gas constant. Without consideration of the rich combustion condition, the single-step chemical reaction mechanism of methane/air is used.

(7) Determination of solid radiation parameters:

The radiation absorption coefficient of the foam ceramic material is 0.4 in the literature [34]. Based on the literature data, the radiation scattering rate of foam ceramics is defined as a function of porosity:

$$\omega = 0.251 + 0.635\varphi \tag{7}$$

The radiation scattering rate can be calculated by the following formula [35]:

$$\omega = \frac{\sigma_s}{\beta} = \frac{\sigma_s}{\alpha + \sigma_s} \tag{8}$$

After conversion, it can be concluded that the calculation formulae of the scattering coefficient and absorption coefficient of the foam ceramic are as follows:

$$\sigma_s = \omega m(1-\varepsilon)/d_p \tag{9}$$

$$\alpha = \omega m(1-\varepsilon)(1-\omega)/d_p \tag{10}$$

The mechanism of methane oxidation (GRI1.2) is used in the chemical reaction, which consists of 32 components and 177 basic elements; the mechanism has been verified by many scholars. The thermal physical properties of the porous media are assumed to be constant [37].

3.3. Boundary Condition

The energy and component conservation equations are boundary value problems, and the relevant boundary conditions are required for the solution. The inlet of the gas mixture was set as the velocity inlet, and the inlet temperature and ambient temperature of the premixed mixture were

300 K. The external wall heat loss includes two parts: heat transfer to the environment and thermal radiation. The specific settings are as follows:

$$\text{Inlet}: T_g = T_0 = 300\,K,\ Y_{CH_4} = Y_{CH_4,in},\ Y_{O_2} = Y_{O_2,in},\ U_g = U_{g,in}$$
$$\text{Outlet}: \frac{\partial Y_n}{\partial x} = \frac{\partial T_g}{\partial x} = 0,\ P_{out} = 0.1\ \text{MPa}$$

(11)

At the burner outlet, there are large radiation heat losses; the outlet boundary condition for a given solid temperature is as follows:

$$\lambda_s \frac{\partial T_s}{\partial x} = -\varepsilon_T \sigma (T_s - T_0)$$

(12)

where ε_T is the surface radiation coefficient of the porous medium.

3.4. Numerical Scheme

The equations were discretized and solved using Fluent software. When the energy equation of the porous medium is solved, only the single-temperature model can be used in the software. However, the high-temperature heat in the flame zone of a porous medium burner will be transferred to a solid medium. At the same time, the solid medium can accept the energy of convection, heat conduction, and heat radiation. Therefore, two energy equations need to be used. To ensure good convergence of the calculation results, sub-relaxation iteration was adopted. The residual error of the energy equation was set to 1.0×10^{-6}, and the residual error of all other variables was set to 1.0×10^{-5}. When the initialization was finished, the initial velocity of the whole flow field was set to the flow velocity of the premixed gas in the inlet, and steady-state iteration was carried out. When chemical reactions were calculated, to ensure the stability of the calculation results and accelerate the convergence, a high-temperature heat source for the ignition needed to be set up, so the temperature in the calculation area was initially set to 1600 K. The same setting also was found in the literature [28]. Eight turbulence models were set up in Fluent software according to the different situations of turbulent fluid flow. The standard k–ε model was used to simulate the combustion in this study. To obtain the characteristics of premixed gas combustion in a porous medium with different equivalence ratios, the equivalence ratios simulated were 0.85, 0.75, 0.45, and 0.35, with inlet flow velocities of 1.15 m/s, 1 m/s, 0.7 m/s and 0.55 m/s, respectively. Parameters of the porous medium burner are listed in Table 1.

Table 1. Physical parameters of the porous media.

	Medium with Small Pores (Upstream)	Medium with Big Pores (Downstream)
Length (m)	0.12	0.1
Pore diameter (PPcm)	25.6	3.9
Porosity	0.835	0.87
Thermal conductivity (W/m·k)	0.2	0.1
Radiation attenuation coefficient (m^{-1})	1707	257
Scattering albedo	0.8	0.8

4. Results and Discussion

4.1. Analysis of the Effect of the Premixed Gas Combustion

Figure 3 shows the stable temperature of the premixed gas combustion in the ordinary burner and the new burner when the equivalence ratio is 0.65 and inlet flow velocity is 0.85 m/s. In Figure 3a, the temperature change is not obvious in the small-pore-medium zone, indicating that the combustion has not been fully developed. A lot of the premixed gas is burned at the interface of the porous medium, and the stationary flame front is formed in the big-pore-medium zone. When the equivalence ratio is

0.65 and the flow velocity in the inlet is 0.85 m/s, the temperature peak value in the ordinary burner reaches about 1900 K, and the temperature in the outlet is about 1850 K; this result is consistent with previous work in the literature [19–21]. There is a difference between the results in this simulation and the literature; the reason for this difference is that a three-dimensional reaction model is used in this study, which is different from the two-dimensional model [21]. In addition, the simulation results are sensitive to the model.

Under the same conditions, the temperature peak of the new burner is up to approximately 2600 K, and the temperature in the outlet zone is higher than 2400 K. The temperature peak is higher than the corresponding maximum temperature of the ordinary burner, which indicates that premixed gas combustion in the new burner is carried out more thoroughly, and the heat storage and heat transfer capacity of the porous medium in the new burner can be better realized. In the ordinary burner, the temperature in the outlet zone is slightly lower than that in the inner zone. The porous medium with high temperature in the outlet transfers a part of heat upstream; therefore, the temperature in the outlet will decrease. Among the released heat from premixed gas combustion in the porous medium, a considerable part is from high-temperature radiation of the porous medium; the radiation heat can directly heat materials, and has high heat transfer efficiency.

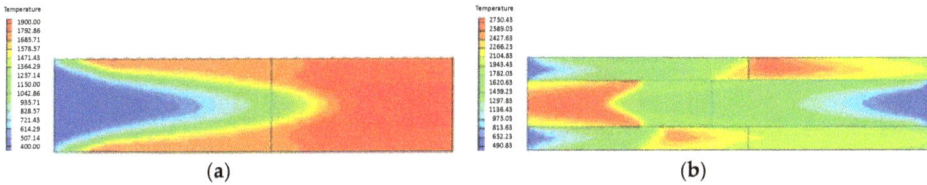
(a) (b)

Figure 3. Temperature contours in the ordinary burner and the new burner when the equivalence ratio is 0.65 and the inlet flow velocity is 0.85 m/s. (**a**) Temperature contours in the ordinary burner; (**b**) Temperature contours in the new burner.

Figure 4 illustrates the gas concentration contours in the two kinds of burners after steady combustion. The gas mass fractions of the axis outlets in the two kinds of burners stabilize at 0.002. The advantages of premixed combustion in a porous medium are fully reflected; in consequence, the efficient use of gaseous fuels can be achieved using the new burner.

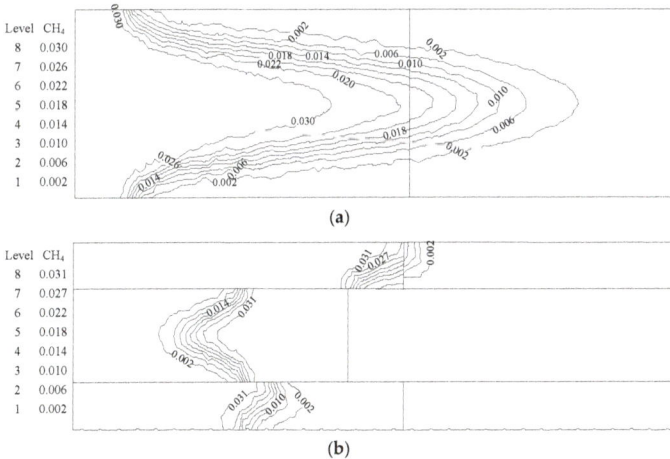
(a)
(b)

Figure 4. CH_4 concentration contours in the ordinary burner and the new burner when the equivalence ratio is 0.65 and the inlet flow velocity is 0.85m/s. (**a**) CH_4 concentration contours in the ordinary burner; (**b**) CH_4 concentration contours in the new burner.

At a certain flow velocity, the new burner can obtain a higher combustion temperature and higher combustion efficiency, because the contact area between the preheating zone and the combustion zone is larger in the new burner. Thus, the heat transfer effect of the reverse heat reflux, heat conduction, and radiation heat in the combustion zone is more significant, and the reaction proceeds more fully.

4.1.1. Influence of the Equivalence Ratio of Premixed Gas on Combustion Degree

Gas concentration is often described using the concentration equivalence ratio. The greater the ratio, the higher the proportion of methane in the mixture. Figures 5 and 6 show the temperature contours of gas mixtures with equivalence ratios of 0.75 and 0.85, respectively, during combustion in the ordinary and new burners. Figures 7 and 8 illustrate the gas concentration contours correspondingly.

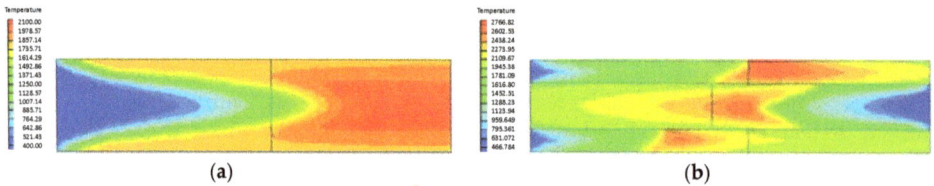

Figure 5. Temperature contours in the ordinary burner and the new burner when the equivalence ratio is 0.75 and the inlet flow velocity is 0.85 m/s. (**a**) Temperature contours in the ordinary burner; (**b**) Temperature contours in the new burner.

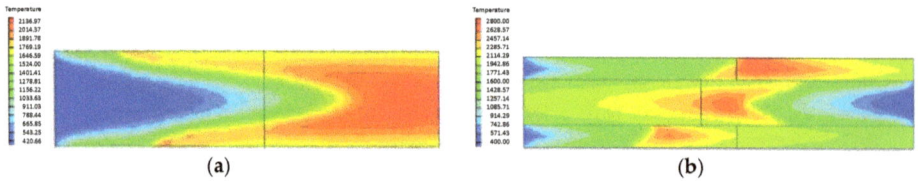

Figure 6. Temperature contours in the ordinary burner and the new burner when the equivalence ratio is 0.85 and the inlet flow velocity is 0.85 m/s. (**a**) Temperature contours in the ordinary burner; (**b**) Temperature contours in the new burner.

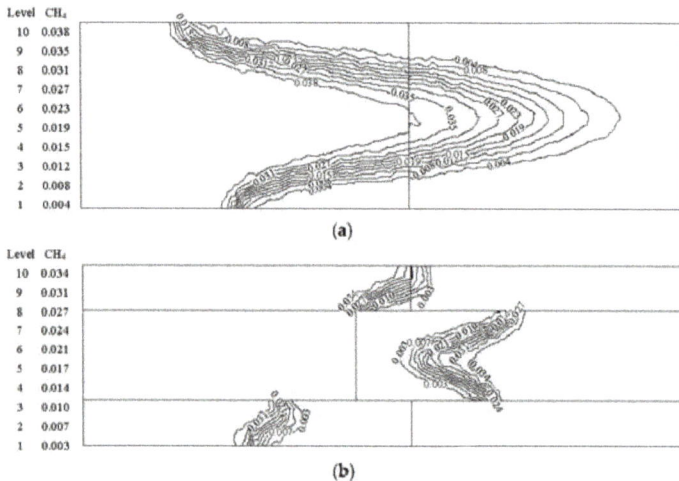

Figure 7. CH_4 concentration contours in the ordinary burner and the new burner when the equivalence ratio is 0.75 and the inlet flow velocity is 0.85 m/s. (**a**) CH_4 concentration contours in the ordinary burner; (**b**) CH_4 concentration contours in the new burner.

(a)

(b)

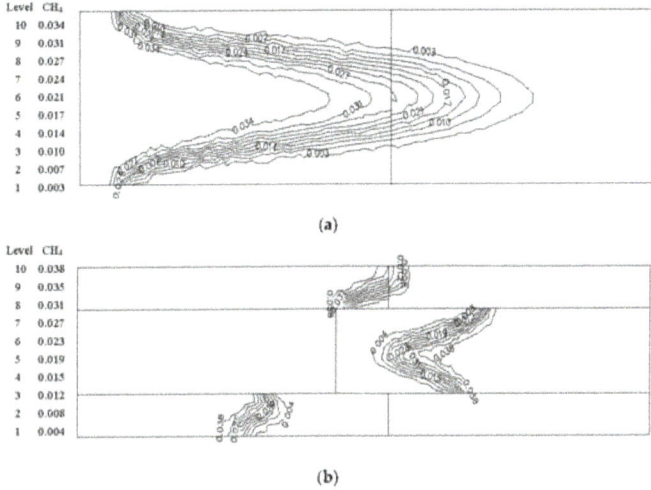

Figure 8. CH₄ concentration contours in the ordinary burner and the new burner when the equivalence ratio is 0.85 and the inlet flow velocity is 0.85 m/s. (**a**) CH₄ concentration contours in the ordinary burner; (**b**) CH₄ concentration contours in the new burner.

When the equivalence ratio is high and the excess air coefficient is small, it is hard to complete the combustion. Under these two equivalence ratios, the location of the flame surface and the gas mixture combustion consumption are similar. When the equivalence ratio is 0.85, the inner temperature of the new burner can reach 2700 K. However, the outlet temperature is not less than 2000 K, which is higher than the temperature peak in the ordinary burner under the same conditions. This shows that the combustion degree and the temperature in the new burner are higher, and the new burner's advantage of saving energy is very obvious.

Figures 9 and 10 show the temperature contours of gas mixture combustion in the two kinds of burners with equivalence ratios of 0.45 and 0.35. Figures 11 and 12 are the gas concentration contours of gas mixture combustion in the two kinds of burners.

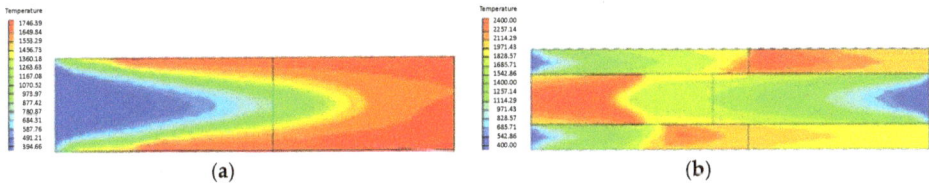

(a) (b)

Figure 9. Temperature contours in the ordinary burner and the new burner when the equivalence ratio is 0.45 and the inlet flow velocity is 0.85 m/s. (**a**) Temperature contours in the ordinary burner; (**b**) Temperature contours in the new burner.

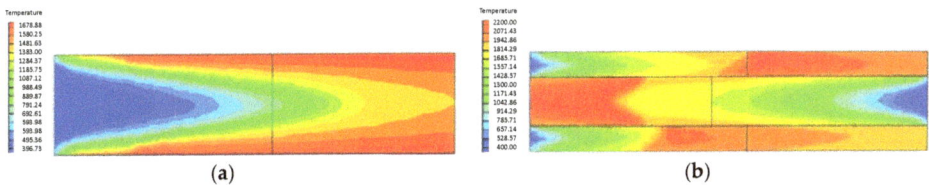

(a) (b)

Figure 10. Temperature contours in the ordinary burner and the new burner when the equivalence ratio is 0.35 and the inlet flow velocity is 0.85 m/s. (**a**) Temperature contours in the ordinary burner; (**b**) Temperature contours in the new burner.

(a)

(b)

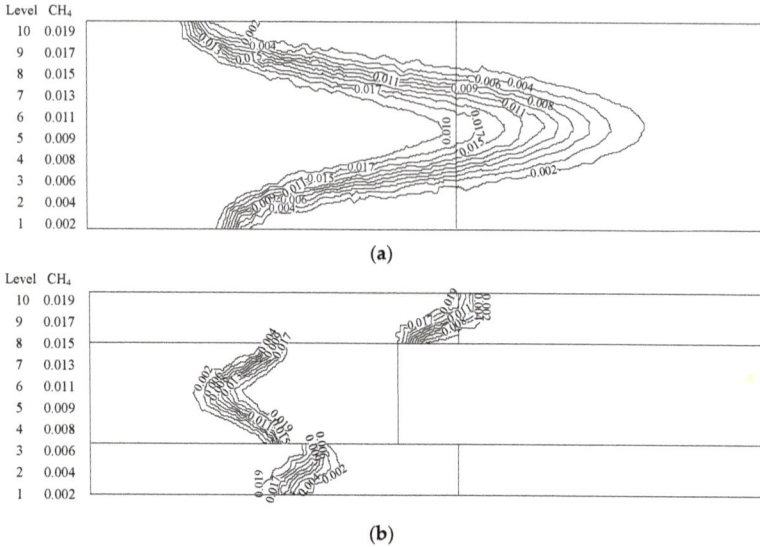

Figure 11. CH$_4$ concentration contours in the ordinary burner and the new burner when the equivalence ratio is 0.45 and the inlet flow velocity is 0.85 m/s. (**a**) CH$_4$ concentration contours in the ordinary burner; (**b**) CH$_4$ concentration contours in the new burner.

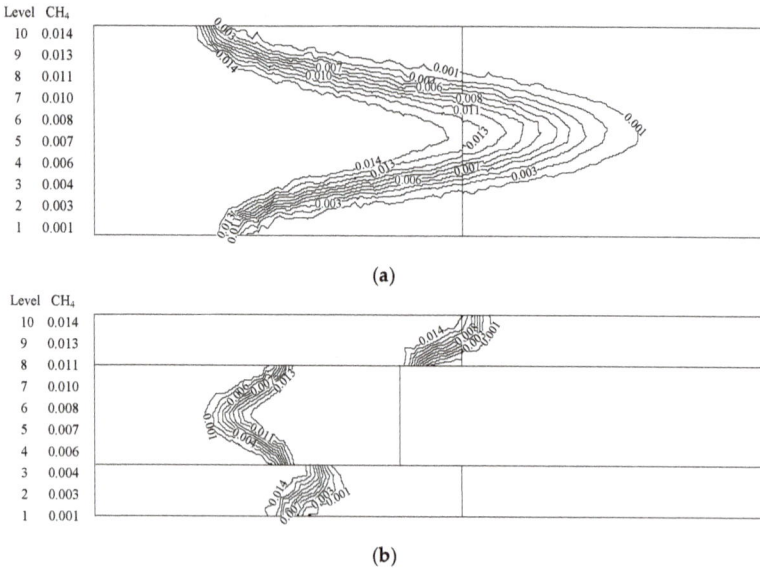

(a)

(b)

Figure 12. CH$_4$ concentration contours in the ordinary burner and the new burner when the equivalence ratio is 0.35 and the inlet flow velocity is 0.85 m/s. (**a**) CH$_4$ concentration contours in the ordinary burner; (**b**) CH$_4$ concentration contours in the new burner.

With the reduction of the equivalence ratio, the combustion flame is oriented towards the rear and becomes closer to the outlet position; this is reflected in both kinds of burners and is consistent with the laws from the literature [19–21,30]. The reliability of the simulation results is also proved. By making this comparison, it can be concluded that when low-concentration gas combustion is carried out in the ordinary burner, the combustion flame is oriented towards the rear and dispersed.

At a low equivalence ratio, the outlet temperature of the new burner is not lower than 2000 K, and the temperature peak of the new burner is up to 2400 K (equivalence ratio 0.45) or 2200 K (equivalence ratio 0.35); the new burner not only has a higher temperature, but also has better stability of the flame surface.

4.1.2. Influence of the Inlet Flow Velocity of Premixed Gas on Combustion Intensity

To analyze the inner temperature distribution of the burner and the gas temperature of the outlet at different mass flow velocities, the same equivalence ratio was set up to investigate the influence of the inlet flow velocity of premixed gas on the combustion state. Although the porous medium has the characteristics of high temperature resistance and corrosion resistance, the excessive temperature of the material may cause damage to the bearing capacity, affecting the working ability of the burner and threatening production. On the contrary, if the premixed gas temperature is too low, it is difficult to achieve effective heating for the material. Numerical simulation can predict the adjustment range of the inlet flow velocity of the new burner. According to the related literature [22], the combustion intensity should not be higher than 1500 kw/m^2 (namely, when the equivalence ratio is 0.65, the corresponding inlet flow velocity is about 0.85 m/s), so the four flow velocities near to the flow velocity in the literature were calculated and compared. Figures 13–16 are the inner temperature distributions of the two kinds of burners when the equivalence ratio is 0.65 and the inlet flow velocity is 0.65 m/s, 0.7 m/s, 0.95 m/s, or 1.15 m/s, respectively. Figures 17–20 show the gas concentration contours of the two kinds of burners regarding these four flow velocities, respectively.

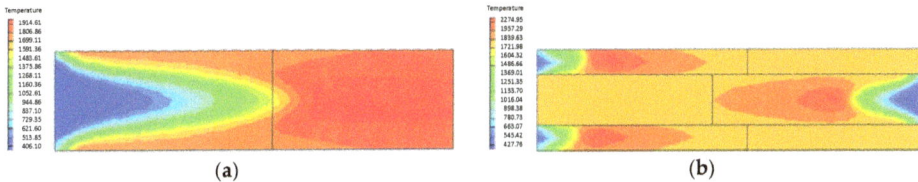

Figure 13. Temperature contours in the ordinary burner and the new burner when the equivalence ratio is 0.65 and the inlet flow velocity is 0.65 m/s. (**a**) Temperature contours in the ordinary burner; (**b**) Temperature contours within the new burner.

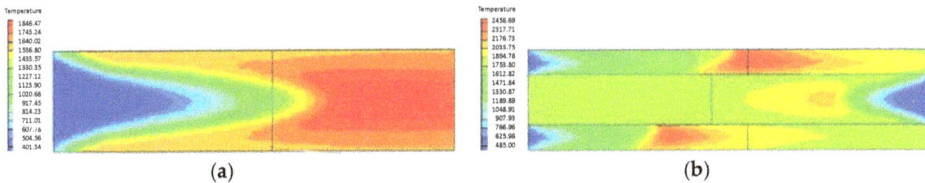

Figure 14. Temperature contours in the ordinary burner and the new burner when the equivalence ratio is 0.65 and the inlet flow velocity is 0.7 m/s. (**a**) Temperature contours in the ordinary burner; (**b**) Temperature contours in the new burner.

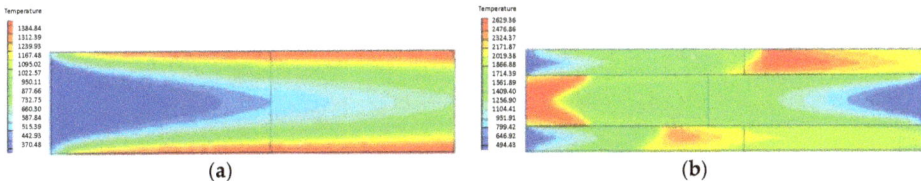

Figure 15. Temperature contours in the ordinary burner and the new burner when the equivalence ratio is 0.65 and the inlet flow velocity is 0.95 m/s. (**a**) Temperature contours in the ordinary burner; (**b**) Temperature contours in the new burner.

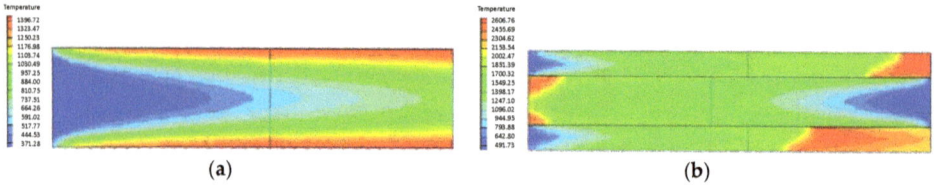

Figure 16. Temperature contours in the ordinary burner and the new burner when the equivalence ratio is 0.65 and the inlet flow velocity is 1.15 m/s. (**a**) Temperature contours in the ordinary burner; (**b**) Temperature contours in the new burner.

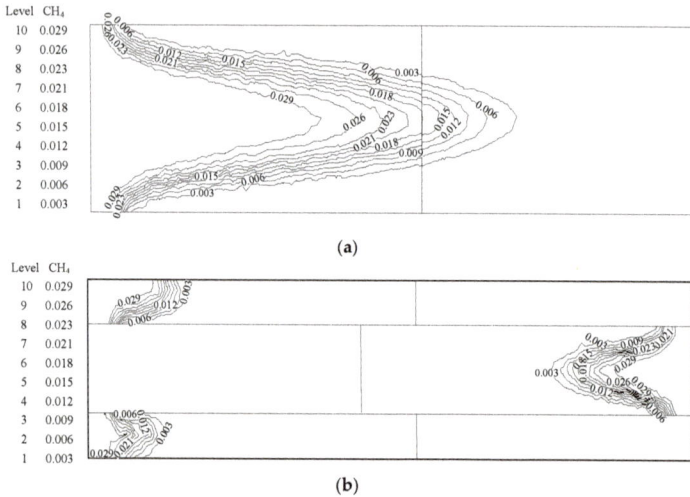

Figure 17. CH$_4$ concentration contours in the ordinary burner and the new burner when the equivalence ratio is 0.65 and the inlet flow velocity is 0.65 m/s. (**a**) CH$_4$ concentration contours in the ordinary burner; (**b**) CH$_4$ concentration contours in the new burner.

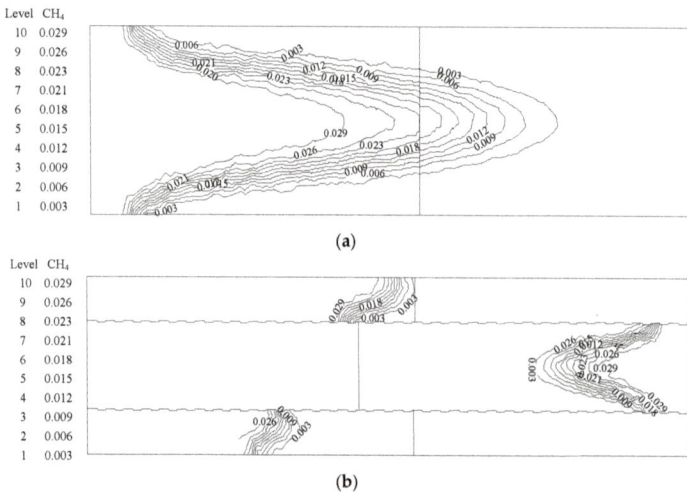

Figure 18. CH$_4$ concentration contours in the ordinary burner and the new burner when the equivalence ratio is 0.65 and the inlet flow velocity is 0.75 m/s. (**a**) CH$_4$ concentration contours in the ordinary burner; (**b**) CH$_4$ concentration contours in the new burner.

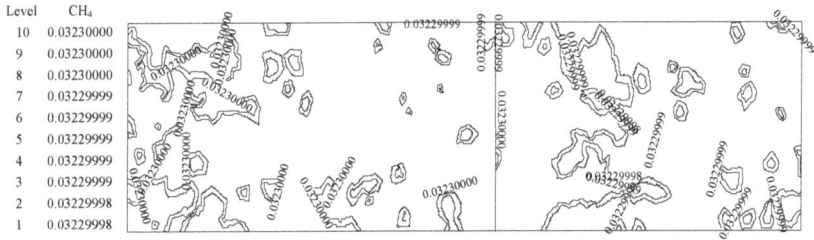

(a)

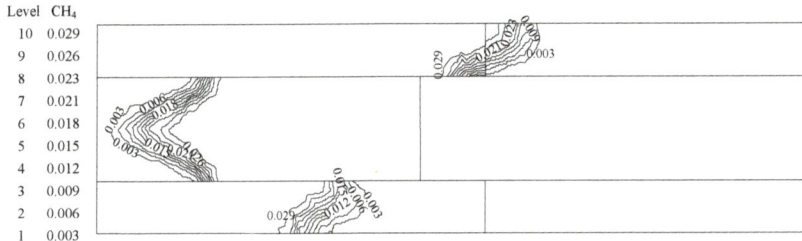

(b)

Figure 19. CH$_4$ concentration contours in the ordinary burner and the new burner when the equivalence ratio is 0.65 and the inlet flow velocity is 0.95 m/s. (**a**) CH$_4$ concentration contours in the ordinary burner; (**b**) CH$_4$ concentration contours in the new burner.

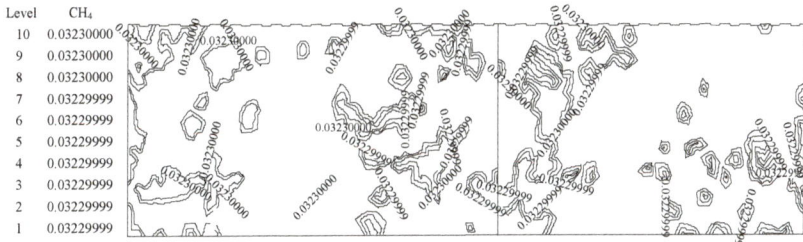

(a)

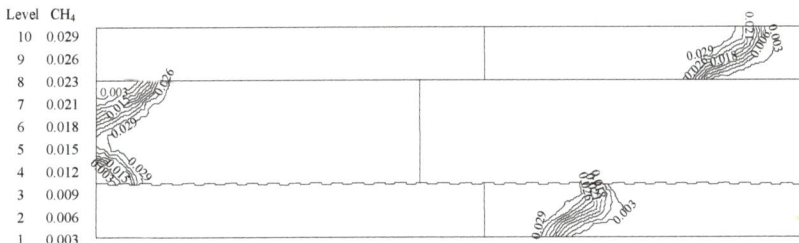

(b)

Figure 20. CH$_4$ concentration contours in the ordinary burner and the new burner when the equivalence ratio is 0.65 and the inlet flow velocity is 1.15 m/s. (**a**) CH$_4$ concentration contours in the ordinary burner; (**b**) CH$_4$ concentration contours in the new burner.

As can be seen from Figures 13–16, the temperature in the vicinity of the burner inlet at low flow velocity is higher than that at high flow velocity. The combustion reaction begins earlier when the flow velocity is low. The main reason for this is that due to the low flow velocity, the contact time between the gas and the porous medium is prolonged, and the preheating effect is more durable. The temperature near the outlet of the burner is higher. The reason for this is that due to the complete reaction in the porous medium burner, the mass flow velocity of the premixed gas increases with the increase of the inlet flow velocity, i.e., the reaction amount of gas in unit time increases, which produces more heat. Figures 13–16 show that the outlet temperature is 1800–2000 K, according to the high temperature resistance of the porous medium. This temperature range is acceptable for practical industrial applications. At the same time, at the velocity of 0.65~1.15 m/s, the temperature peak of the new burner is always higher than that of the ordinary burner.

When the inlet flow velocity of the premixed gas is less than the minimum flame propagation velocity, the flame will move toward the inlet and produce tempering, but premixed gas can be burned in the porous medium. If the inlet flow velocity of the premixed gas is more than the maximum flame propagation velocity, the flame will move toward the outlet and produce combustion extinguishment. In the flame stabilization zone, each equivalence ratio corresponds to a velocity range that can stabilize flame in the support layer. With the increase of the equivalence ratio, the flame stability zone increases. As the premixed gas has been effectively preheated, the flame propagation velocity is significantly improved. With the increase of the equivalence ratio, the flame stability zone becomes wider.

As can be seen from the Figures 17–20, at the lower inlet flow velocity of the premixed gas, the premixed gas with the equivalence ratio of 0.65 can realize effective combustion in the two kinds of burners, and the combustion consumption degree is similar. When the inlet flow velocity reaches 0.95 m/s and 1.15 m/s, the temperature cloud figure and gas concentration contour show that gas combustion is not effective in the ordinary burner (the concentration experiences almost no change); however, in the new burner, the high-temperature flame moves toward the rear with the increase in premixed gas mass flow, the temperature peak is maintained over 2500 K, the attenuation of the gas concentration gradient is normal, and the combustion state is good.

4.2. NO_X Emission Characteristics of Different Equivalence Ratios in the New Burner

When the combustion reaction occurs, the nitrogen in the air/fuel is heated and decomposed under the oxidizing conditions of combustion, and toxic nitrogen oxides are then produced. Its compositions are mainly NO and NO_2, which are collectively called NO_x. The harm from emission of NO_x cannot be ignored, and many countries have strict regulations concerning their emissions [38]. For this portion of the study, the flow velocity of the premixed gas in the new burner was fixed at 0.65m/s. Figure 21 is the schematic diagram of NO_x emission, which depicts the NO_x emission states under equivalence ratios of 0.35, 0.45, 0.65, 0.75, and 0.85, respectively.

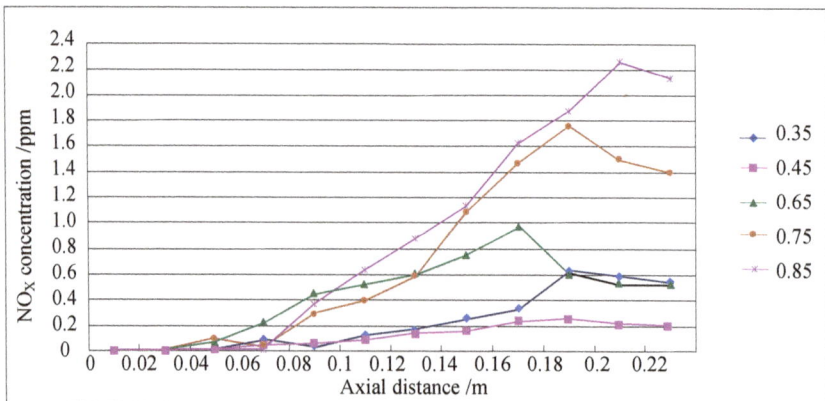

Figure 21. NO_x emissions schematic diagram for the new burner at different equivalence ratios.

As can be seen from Figure 21, in the porous medium combustion, the emission of pollutant NO_x is very low. When the equivalence ratio of premixed gas is 0.65 and inlet flow velocity is 0.85 m/s, the NO_x concentration in the new burner outlet is only 1 ppm approximately. When the equivalence ratio is 0.85, the highest concentration of NO_x in the ordinary burner is about 2.3 ppm. Because NO_x has a strong dependence on temperature, with the increase of mass flow velocity of the premixed gas, the combustion temperature becomes higher and higher, and the NO_x emission consequently shows a trend of increase.

The NO_x emission for an equivalence ratio of 0.45 is lower than that for an equivalence ratio of 0.35, which shows that NO_x emission will decrease with the increase in the complete reaction degree(the emission concentration is lower than the minimum standard (100 ppm) in Beijing, China). However, the total emission amount of NO_x will show an upward trend with the increase in the equivalence ratio. The reason for this is that with the increase in the equivalence ratio, the mass flow velocity of the combustible gas increases, the incomplete reaction degree increases, the combustion temperature is higher, and the NO_x relatively increases. However, the emission is still lower than that from free combustion.

5. Conclusions

Based on the combustion mechanism of methane-air premixed gas in a porous medium, a new burner was proposed. A mathematical model for the premixed gas combustion in porous medium was established. The combustion performance for different equivalence ratios and premixed gas velocities in inlets was simulated. The methane combustion degree and distribution of temperature and pressure in the porous medium were analyzed. In addition, the concentrations of NO_x emission for different equivalence ratios were studied. The following results were obtained:

(1) Compared with the ordinary porous medium burner, the new burner can gather heat more effectively, thereby promoting gas combustion. As a result, the gas temperature in the new burner is much higher than that in the ordinary one.

(2) At higher flow velocity, the short preheating time results in poor reaction of the premixed gas in the porous medium of the ordinary burner. On the contrary, under the condition of high flow velocity, the gas concentration contour gradually fades along the gas flow direction in the new burner. The combustion state is good, and the combustion heat is large.

(3) The change of flow velocity of the premixed gas in the inlet exerts an important influence on the combustion in the porous medium. In the case of low flow velocity, the temperature in the inlet increases significantly. In contrast, the temperature in the outlet increases dramatically in the case of high flow velocity. The inlet flow velocity should be set within a reasonable range to ensure the safety and normal operation of the burner.

(4) The emission concentration of major pollutants (NO_x) from the new burner is much lower than the minimum standard ((100 ppm) in Beijing, China).

(5) The radiation attenuation coefficient of the small-pore medium in the new burner increases, but the radiation attenuation coefficient of the big-pore medium in the new burner remains unchanged, which increases the stable operating range of the burner.

(6) The thermal conductivity coefficient of the small-pore medium in the new burner increases, but the thermal conductivity coefficient of the big-pore medium in the new burner remains unchanged or reduces slightly, which increases the temperature peak and the stable operating range. The aperture of the porous medium should not be too large. A reasonable range of aperture size can maintain the stability of flame front and obtain the ideal temperature.

(7) Compared with the traditional burner, the new porous medium burner has advantages in improving combustion efficiency, expanding flammability limits, and improving environmental friendliness. Therefore, the application prospects of the new porous medium burner are very broad. The influences of various factors on the combustion process of the new porous medium

burner were simulated by CFD (computational fluid dynamics) and the basic characteristics were mastered. The numerical results can provide a theoretical basis and data support for the design of the new porous medium burner. If the design and manufacturing process of the new porous medium burner can be further strengthened, better application value will surely be attained.

Author Contributions: All the authors designed research, performed research, analyzed data, wrote and revised the paper, all authors contributed equally.

Funding: This research was funded by the National Natural Science Foundation Project of China, grant number (51704111, 51374003, 51774135 and 51604110).

Conflicts of Interest: The authors declare no conflicts of interest.

References

1. Zhang, Y.; Lu, C.; Li, T. A Practical Countercurrent Fluid Catalytic Cracking Regenerator Model for in Situ Operation Optimization. *AICHE J.* **2012**, *58*, 2810–2819. [CrossRef]
2. Jia, Z.; Feng, T. Gas Explosion Characteristics and Its Control Measures in Closed Fire Zone. *Comput. Modell. New Technol.* **2014**, *18*, 21–25.
3. Kamal, M.M.; Mohamad, A.A. Combustion in Porous Media. *Proc. Inst. Mech. Eng. Part A J. Power Energy* **2006**, *220*, 487–508. [CrossRef]
4. Valerio, G.; Rajnish, N.S.; Robert, R.R. Premixed Combustion of Methane-Air Mixture Stabilized over Porous Medium: A 2-D Numerical Study. *Chem. Eng. Sci* **2016**, *152*, 591–605.
5. Mujeebu, M.A.; Abdullah, M.Z.; Bakar, M.A.; Mohamad, A.A.; Abdullah, M.K. Applications of Porous Media Combustion Technology Review. *Appl. Energy* **2009**, *86*, 1365–1375. [CrossRef]
6. Wood, S.; Harris, A.T. Porous Burners for Lean-Burn Applications. *Prog. Energy Combust. Sci.* **2008**, *34*, 667–684. [CrossRef]
7. Slimane, R.B.; Lau, F.S.; Khinkis, M.; Bingue, J.P.; Saveliev, A.V.; Kennedy, L.A. Conversion of Hydrogen Sulphide to Hydrogen by Super Adiabatic Partial Oxidation: Thermodynamic Consideration. *Int. J. Hydrogen Energy* **2004**, *29*, 1471–1477. [CrossRef]
8. Min, D.K.; Shin, E.D. Laminar Premixed Flame Stabilized inside a Honeycomb Ceramic. *Int. J. Heat Mass Transf.* **1991**, *34*, 341–356.
9. Hayashi, T.C.; Malico, I.; Pereira, J.C.F. Three-dimensional Modeling of Two-Layer Porous Burner for Household Application. *Comput. Struct.* **2004**, *82*, 1543–1550. [CrossRef]
10. Kotani, Y.; Takeno, T. An Experimental Study of Stability and Combustion Characteristics of an Excess Enthalpy Flame. *Symp. (Int.) Combust.* **1982**, *19*, 1503–1509. [CrossRef]
11. Pereira, F.M.; Oliveira, A.A.M. Analysis of the Combustion with Excess Enthalpy in Porous Media. In Proceeding of the European Combustion Meeting, Orléans, France, 2003; pp. 388–392.
12. Mital, R.; Gore, J.P.; Viskanta, R. A Study of the Structure of Submerged Reaction Zones in Porous Ceramics Radiant Burners. *Combust. Flame* **1997**, *111*, 175–184. [CrossRef]
13. Sathe, S.B.; Kulkarni, M.R.; Peck, R.E.; Tong, T.W. An Experimental and Theoretical Study of Porous Radiant Burner Performance. *Symp. (Int.) Combust.* **1990**, *23*, 1011–1018. [CrossRef]
14. Chen, Y.K.; Matthews, R.D.; Howell, J.R. *The Effect of Radiation on the Structure of Premixed Flame within a Highly Porous Inert Medium*; American Society of Mechanical Engineers, Heat Transfer Division, (Publication) HTD: Boston, MA, USA, 1987; pp. 35–41.
15. Wang, E.; Chen, L.; Wu, J. Application and Existing Problems of Porous Ceramics in the Combustion Field. *Foshan Ceram.* **2005**, *100*, 35–39.
16. Chu, J. Study and Development of a Gradually-varied Porous Media Burner. Master's Thesis, Zhejiang University, Hangzhou, China, 2005.
17. Du, L.; Xie, M. Numerical Investigation on Effects of Porous Media in Premixed Combustion. *J. Dalian Univ. Technol.* **2004**, *1*, 70–75.
18. Du, L. Investigation on Super-adiabatic Combustion of Lean Premixed Gases in Porous Media. Ph.D. Thesis, Dalian University of Technology, Dalian, China, 2003.
19. Liu, H. Experimental Study and Numerical Simulation of Premixed Gas Combustion in Porous Media. Ph.D. Thesis, Northeastern University, Shengyang, China, 2010.

20. Liu, H.; Dong, S.; Xing, N.; Li, B. Numerical Simulation of the Influence of Porous Medium Characteristics on Combustion Temperature and Pressure. *Appl. Energy Technol.* **2009**, *10*, 31–37.
21. Barra, A.J.; Diepvens, G.; Ellzey, J.L.; Henneke, M.R. Numerical Study of the Effects of Material Properties on Flame Stabilization in a Porous Burner. *Combust. Flame* **2003**, *134*, 369–379. [CrossRef]
22. Wang, H. Industrial Application and Numerical Simulation of Premixed Gas Combustion in Porous Media. Master's Thesis, Northeastern University, Shengyang, China, 2009.
23. Lee, D.K.; Noh, D.S. Experimental and Theoretical Study of Excess Enthalpy Flames Stabilized in a Radial Multi-Channel as a Model Cylindrical Porous Medium Burner. *Combust. Flame* **2016**, *170*, 79–90. [CrossRef]
24. Mendes, M.A.A.; Pereira, J.M.C.; Pereira, J.C.F. A Numerical Study of the Stability of One-Dimensional Laminar Premixed Flame in Inert Porous Media. *Combust. Flame* **2008**, *153*, 525–539. [CrossRef]
25. Afsharvahid, S.; Ashman, P.J.; Dally, B.B. Investigation of NO_x Conversion Characteristics in a Porous Medium. *Combust. Flame* **2008**, *152*, 604–615. [CrossRef]
26. Zhao, P. Study on Premixed Combustion in Inert Porous Media. Ph.D. Thesis, University of Science and Technology of China, Hefei, China, 2007.
27. Huang, R. Combustion Characteristics of Multi-Species Low Calorific Gaseous Fuels in a Two-Layer Porous Burner. Master's Thesis, Zhejiang University, Hangzhou, China, 2016.
28. Dong, S. Preliminary Experimental and Numerical Study of Premixed Combustion in Porous Inert Media. Master's Thesis, Northeastern University, Shengyang, China, 2008.
29. Zheng, Z. Numerical Study of Natural Gas Premixed Combustion in Porous Inert Media. Master's Thesis, Shanghai Jiaotong University, Shanghai, China, 2007.
30. Jia, Z.; Lin, B.; Ye, Q. Analysis of Influencing Factors and Flame Acceleration Mechanism of Gas Explosion Propagation in Tube. *Min. Eng. Res.* **2009**, *24*, 57–62.
31. Gao, Z.; Cui, G.; Zuo, D. Discussion on the Reaction Mechanism of Carbon Combustion. *J. Shandong Polytech. Univ.* **2002**, *16*, 39–42.
32. Yang, S.; Tao, W. *Heat Transfer*, 3rd ed.; Higher Education Press: Beijing, China, 1998.
33. Han, Z. *Fuel and Combustion*; Metallurgical Industry Press: Beijing, China, 2007.
34. Wang, P. Theoretical and Experimental Study on Gas Thermal Counter Current Oxidation of Low Concentration Methane in Coal Mine. Ph.D. Thesis, Central South University, Changsha, China, 2012.
35. Dhamrat, R.S.; Ellzey, J.L. Numerical and Experimental Study of the Conversion of Methane to Hydrogen in a Porous Media Reactor. *Combust. Flame* **2006**, *144*, 698–709. [CrossRef]
36. Mohammadi, I.; Hossainpour, S. The Effects of Chemical Kinetics and Wall Temperature Performance of Porous Media Burners. *Heat Mass Transf.* **2013**, *49*, 869–877. [CrossRef]
37. Younis, L.B.; Viskanta, R. Experimental Determination of the Volumetric Heat Transfer Coefficient between Stream of Air and Ceramic Foam. *Int. J. Heat Mass Transf.* **1993**, *36*, 1425–1434. [CrossRef]
38. Li, Z.; Cheng, Y. *Fire Engineering*; China University of Mining and Technology Press: Xuzhou, China, 2002.

processes

MDPI

Article

Dispersion Performance of Carbon Nanotubes on Ultra-Light Foamed Concrete

Jing Zhang [1,2] and Xiangdong Liu [1,*]

1 School of Materials Science and Engineering, Inner Mongolia University of Technology, Hohhot 010051, China; Jingzhang602@126.com
2 Department of Mechanical and Electrical Engineering, Inner Mongolia Technical College of Construction, Hohhot 010070, China
* Correspondence: liuxd@imut.edu.cn

Received: 29 September 2018; Accepted: 12 October 2018; Published: 17 October 2018

Abstract: This study investigates the effect of carbon nanotube (CNT) dispersion on the mechanical properties and microstructures of ultra-light foamed concrete. A type of uniform and stable CNT dispersion solution is obtained by adding nano-$Ce(SO_4)_2$. Results show that CNT dispersion increases the compressive and breaking strengths of foamed concrete. CNTs play a nuclear role in the crystallization of C–S–H, and CNT dispersion effectively promotes the grain growth of C–S–H. The effect of CNT dispersion on the compressive and breaking strengths of foamed concrete is predicted through simulation.

Keywords: ultra-light foamed concrete; carbon nanotubes (CNTs); nano-$Ce(SO_4)_2$; CNT dispersion; simulation; compressive strength; breaking strength

1. Introduction

Foamed concrete is typically composed of cement, fly ash, foam, and water. The foam in the concrete generates a large number of air pores with diameters ranging from 1 mm to 3 mm. Consequently, the density of foamed concrete generally ranges from 200 kg/m^3 to 2000 kg/m^3 [1]. In particular, low-density foamed concrete ($\leq$300 kg/m^3) is regarded as ultra-light foamed concrete. Substantial theoretical research and engineering experience suggest that ultra-light foamed concrete is unstable [2]. The inevitable instability of ultra-light foamed concrete usually results in poor mechanical properties, such as compressive and breaking strengths, which limit the application of this material.

Several experimental investigations evaluated the compressive and breaking strengths of foamed concrete. Furthermore, an increasing number of studies attempted to enhance the compressive and breaking strengths of ultra-light foamed concrete by introducing various fibers [3–6]. In recent years, researchers have proven that the introduction of carbon nanotubes (CNTs) effectively increases the compressive and breaking strengths of concrete/foamed concrete, and this effect is influenced by CNT dispersion [7–10]. Achieving a uniform distribution of CNTs in cement paste, which is key to enhancing the compressive and breaking strengths of foamed concrete, is complicated by their tendency to agglomerate [7].

Previous studies have attempted to increase CNT dispersion by using various methods, including mechanical processing and surface modification. An increasing number of researchers have employed surface modification methods [11–14], such as HCl/H_2SO_4/HNO treatment and thermal chemical vapor deposition. However, the strong acids used in these CNT dispersion methods can damage concrete/foamed concrete. Conversely, reactions between the organic matter in CNT dispersion and the chemical substances in foamed concrete hinder the growth of the foamed concrete particles. Consequently, the applications of these traditional methods of CNT dispersion in concrete/foamed concrete are limited. This limitation explains the lack of investigations on concrete and foamed concrete materials.

In the present study, we introduced inorganic matter nano-Ce(SO$_4$)$_2$ to realize high-performance CNT dispersion. The effect of nano-Ce(SO$_4$)$_2$ on the dispersion, structure, and surface of the prepared CNTs was determined. This study aimed to develop a new type of CNT dispersion method for enhancing the mechanical properties and microstructure of ultra-light foamed concrete. Moreover, modeling and simulation issues should be considered in order to develop a thorough approach, and thus make well-informed decisions [15,16]. We also simulated the effect of CNT dispersion on the compressive and breaking strengths of foamed concrete.

2. Materials and Test Methods

2.1. Materials

The following materials were used in the experiments: cement with a compressive strength of 42.5 MPa after 28 days (Hohhot, China), fly ash (L2 dry ash), naphthalene water reducer (industrial grade, Shanghai, China), FeCl$_3$ (industrial grade, Hangzhou, China), H$_2$O$_2$ (27.5%, Beijing, China), calcium stearate (analytical reagent, Beijing, China), multi-walled CNTs (diameter of 10–30 nm and length of 5–15 μm, Beijing, China), nano-Ce(SO$_4$)$_2$ (analytical reagent, Shanghai, China), and water (tap water, Hohhot, China).

Transmission electron microscopy (TEM, Hillsboro, OR, USA) images of the CNTs are shown in Figure 1. Figure 1a shows that the CNTs present serious agglomerations, and Figure 1b illustrates that the CNTs are indeed multi-walled. The density of foamed concrete and the content (0.20–0.24 g) of CNTs added were varied. The physical and mechanical properties of the multi-walled CNTs are listed in Table 1.

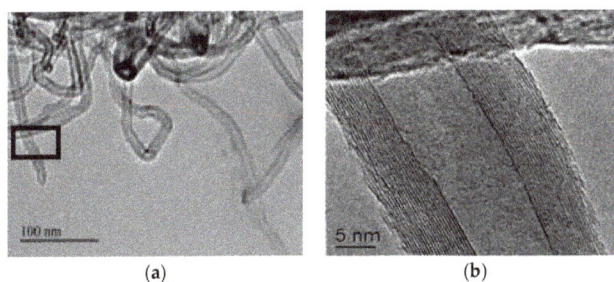

(a)   (b)

Figure 1. TEM of CNTs. (a) aggregate of CNTs (b) multi-walled CNT.

Table 1. Physical and mechanical properties of multi-walled carbon nanotubes.

Type	Inner Diameter (nm)	External Diameter (nm)	Length (μm)	Tensile Strength (GPa)	Modulus of Elasticity (TPa)
Multi-walled carbon nanotubes	8–12	25–30	5–15	10–60	1

2.2. Preparation of CNT Dispersion

In a beaker, a certain quantity of CNTs and tap water were mixed and stirred with a glass rod for 5 min until complete dissolution. The solution was ultrasonically cleaned, and the temperature was controlled at 40 °C, considering that solution temperature affects CNT dispersion. Specifically, the structure of CNTs is damaged by high temperatures [17–20]. After 20 min of ultrasound, different contents of nano-Ce(SO$_4$)$_2$ were added, and ultrasonic was performed at 99 W for 1 h to prepare a homogeneous CNT dispersion solution [21]. The different contents of CNT dispersion solution and the corresponding foamed concretes are shown in Table 2.

Table 2. CNT dispersion and the corresponding foamed concretes.

CNT (mass%)	Nano-Ce(SO$_4$)$_2$ (mass%)	CNT Dispersion	Foamed Concrete
0.0%	0.0%	S0-0	FC0-0
0.1%	0.3%	S1-1	FC1-1
0.1%	0.6%	S1-2	FC1-2
0.2%	0.3%	S2-1	FC2-1
0.2%	0.6%	S2-2	FC2-2
0.3%	0.3%	S3-1	FC3-1
0.3%	0.6%	S3-2	FC3-2
0.4%	0.3%	S4-1	FC4-1
0.4%	0.6%	S4-2	FC4-2
0.5%	0.3%	S5-1	FC5-1
0.5%	0.6%	S5-2	FC5-2

A type of CNT dispersion solution is prepared in Table 3. Three beakers (A, B, and C), each containing 50 mL of water, were prepared. Then, 0.3 wt% CNTs were added into each of the three beakers [22]. Finally, 0.3 wt% nano-Ce(SO4)$_2$ was added to beaker B, and 0.6 wt% nano-Ce(SO$_4$)$_2$ was added to beaker C.

Table 3. CNT dispersion solution.

Solution	A	B	C
Water	50 mL	50 mL	50 mL
Nano-Ce(SO$_4$)$_2$	0%	0.3 wt%	0.6 wt%

2.3. Characteristics of CNT Dispersion

CNT dispersion solution was prepared by dissolving CNTs (0.1, 0.2, 0.3, 0.4, and 0.5 wt%) and nano-Ce(SO$_4$)$_2$ (0.3 and 0.6 wt%) in water at room temperature. After approximately 6–7 h, CNT particles settled in the bottom of the beaker completely, based on visual inspection. The dispersion property of the CNTs in the solution was observed under a FEI Talos200X transmission electron microscope (Hillsboro, OR, USA). In addition, CNT dispersion was analyzed by obtaining the test absorption spectra using a UV-1700 spectrophotometer (Tokyo, Japan). Absorbance spectra refer to the distribution of CNTs or to the settlement of CNT particles in the solution.

2.4. Preparation of Foamed Concrete and CNT/Foamed Concrete

Ordinary Portland cement, fly ash, and calcium stearate were mixed with self-made foam concrete mixing equipment for 2–3 min. The water/CNT dispersion solution was added and then stirred for 2–3 min to form foamed concrete as a thick liquid, and the water temperature was controlled at approximately 35 °C. Finally, FeCl$_3$ and H$_2$O$_2$ were added successively and then stirred at a high speed for 30 s. After stirring, the liquid was poured in a mold and then demolded after 24 h. The sample was prepared into two standard test pieces with dimensions of 100 mm × 100 mm × 100 mm and 400 mm × 100 mm × 100 mm, which were then stored in the standard curing room for 7 and 28 days, respectively.

2.5. Characteristic of CNT Dispersion in Foamed Concrete

A CNT foamed concrete block was prepared by adding the CNT dispersion solution into foamed concrete paste (Table 2). The microstructure of the CNT foamed concrete was observed under a QUANTA FEG 650 electron scanning microscope (SEM) to evaluate CNT dispersion.

2.6. Measurement Methods

(1) Dry density test was performed in accordance with the provisions in foamed concrete of JG/T266-2011 (the construction industry standard of the People's Republic of China), which is promulgated by the ministry of housing and urban-rural development of the People's Republic of China.

(2) Compressive strength test was performed in accordance with the provisions in the foamed concrete of JG/T266-2011 (the construction industry standard of the People's Republic of China), which is promulgated by the ministry of housing urban-rural development of the People's Republic of China.

(3) Breaking strength test was performed in accordance with the provisions in the foamed concrete of JG/T266-2011 (the construction industry standard of the People's Republic of China), which is promulgated by the ministry of housing and urban-rural development of the People's Republic of China.

(4) QUANTA FEG 650 electron scanning microscope was used for morphological observation (Hillsboro, OR, USA).

(5) The solution was sonicated using a SK250LH ultrasonic cleaner (Shanghai, China).

(6) CNT dispersion was observed using a FEI Talos200X transmission electron microscope (Hillsboro, OR, USA).

(7) Absorbance spectra were tested using a UV-1700 spectrophotometer (Tokyo, Japan).

(8) Pro/Engineer (Pro-E) software (Parametric Technology Corporation, Boston, MA, USA) is an entity modeling system that uses parametric design based on physical characteristics. It was used for the modeling and simulation of foamed concrete.

3. Results and Discussion

3.1. Uniformity of CNT Dispersion

The effects of different contents of nano-$Ce(SO_4)_2$ on CNT dispersion are reported in this section. Figures 2–4 display the effects of the nano-$Ce(SO_4)_2$ values on CNT dispersion and surface.

An important parameter influencing CNT dispersion is the degree of uniformity of particle distribution in solution. The degree of CNT dispersion properties, including particle size, degree of uniformity of particle distribution in solution, and dispersion stability, can be determined based on the absorbance [23,24]. CNT dispersion is analyzed by detecting absorbance. Ten types of CNT dispersion solutions (S1-1, S1-2, S2-1, S2-2, S3-1, S3-2, S4-1, S4-2, S5-1, and S5-2) are prepared (Table 2). These CNT dispersion solutions are subjected to ultrasound for 0.5 h and then allowed to stand for 2 h. The absorbance the solutions is then obtained with a UV-1700 ultraviolet spectrophotometer (Figure 2). As reported in previous studies [23,24], higher absorbance values indicate better dispersion. Figure 2 shows that the increase in absorbance spectra with increasing CNT content from 0.1 wt% to 0.5 wt% is due to the dispersion of CNT particles by 0.3% nano-$Ce(SO_4)_2$ in the solution, approximately 0.31–0.46%. The histograms of 0.3% nano-$Ce(SO_4)_2$ and 0.6% nano-$Ce(SO_4)_2$ show the same patterns, and the increasing range of absorbance spectra with 0.6% nano-$Ce(SO_4)_2$ is 0.72–0.96%. This result can be explained by the presence of large CNT aggregates in the solution when 0.3% nano-$Ce(SO_4)_2$ is added. When nano-$Ce(SO_4)_2$ is continuously added into the solution, the newly increased 0.3% nano-$Ce(SO_4)_2$ has a greater tendency to enhance the dispersion of CNT particles in the solution. The absorbance spectra of CNT dispersion increase by 132.26% (0.1% CNTs), 97.30% (0.2% CNTs), 88.10% (0.3% CNTs), 86.67% (0.4% CNTs), and 108.7% (0.5% CNTs) after adding 0.6% nano-$Ce(SO_4)_2$ relative to 0.3% nano-$Ce(SO_4)_2$. This behavior indicates that the remaining aggregate CNTs are dispersed effectively by the newly increased 0.3% nano-$Ce(SO_4)_2$.

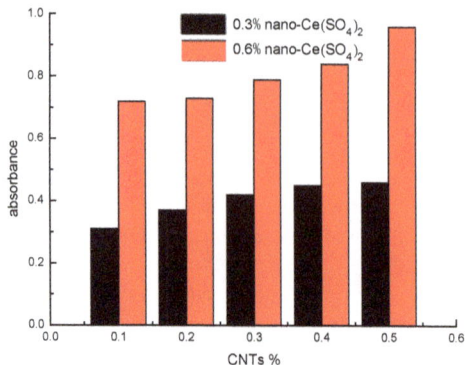

Figure 2. Absorbance of CNT dispersion.

Another important parameter of the CNT dispersion is the dispersion stability. The absorbance spectra of the CNT dispersions with 0.3% nano-Ce(SO$_4$)$_2$ and 0.6% nano-Ce(SO$_4$)$_2$ are shown in Figure 3a,b. Figure 3a shows that the absorbance spectra decrease slowly from 1 h to 6 h, indicating that the CNT particles are setting slowly. Then, the spectra almost become constant from 7 h to 12 h. In particular, the absorbance spectra of all the specimens are almost overlapping from 7 h to 12 h, indicating that the CNT particles are completely set. This result is consistent with the visual inspection result that the CNT particles are completely set at about 6 h later. The absorbance spectra of the CNT dispersion solutions with 0.3% nano-Ce(SO$_4$)$_2$ and 0.6% nano-Ce(SO$_4$)$_2$ almost coincide, except for two results described below. First, the CNT particles with 0.6% nano-Ce(SO$_4$)$_2$ set completely about 1 h later than those with 0.3% nano-Ce(SO$_4$)$_2$. Second, all of the absorbance spectra of 0.6% nano-Ce(SO$_4$)$_2$ are higher than those of 0.3% nano-Ce(SO$_4$)$_2$ at the same time. Consequently, the setting behavior and dispersion stability of the CNT particles are prevented by the addition of 0.3% nano-Ce(SO$_4$)$_2$ into the CNT dispersion solution.

Figure 3a,b show that the absorbance spectra are almost overall declined (from S1-1, S2-1 to S5-1, from S1-2, S2-2, to S5-2). Nevertheless, the absorbance spectra of S1-1 basically overlap with those of S2-1, and the absorbance spectra of S1-1 at 3, 5, and 9 h are below those of S2-1. This result indicates that the absorbance spectra (0.3% nano-Ce(SO$_4$)$_2$) are unstable when 0.1% and 0.2% CNTs are added.

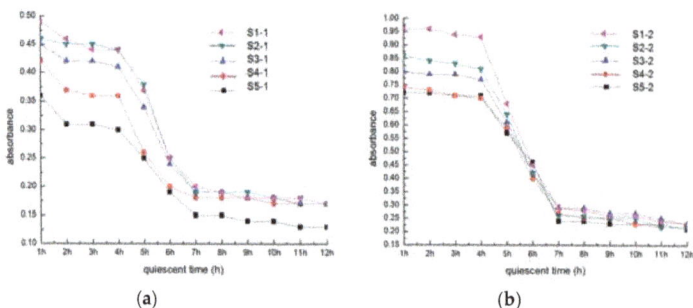

(a)

(b)

Figure 3. Impact of time on the absorbance of CNT dispersion. (**a**) 0.3% nano-Ce(SO$_4$)2, (**b**) 0.6% nano-Ce(SO$_4$)$_2$.

A type of CNT dispersion solution is prepared as described in Table 3 to investigate the impact of different contents of nano-Ce(SO$_4$)$_2$ on CNT dispersion. Figure 4 shows the TEM images of the CNTs obtained with FEI Talos200X (Hillsboro, OR, USA). Figure 4a,b show that the increased 0.3% nano-Ce(SO$_4$)$_2$ can disperse the aggregates of CNTs. Figure 4b,c displays that the newly increased 0.3% nano-Ce(SO$_4$)$_2$ can disperse the aggregates of CNTs in beaker B, which correspond to the results in

Figures 2 and 3. Figure 4d shows that single CNTs can be obtained from aggregates of CNTs when the content of nano-Ce(SO$_4$)$_2$ is increased to 0.6%. These results indicate that nano-Ce(SO$_4$)$_2$ is a significant factor affecting the CNT dispersion of foamed concrete.

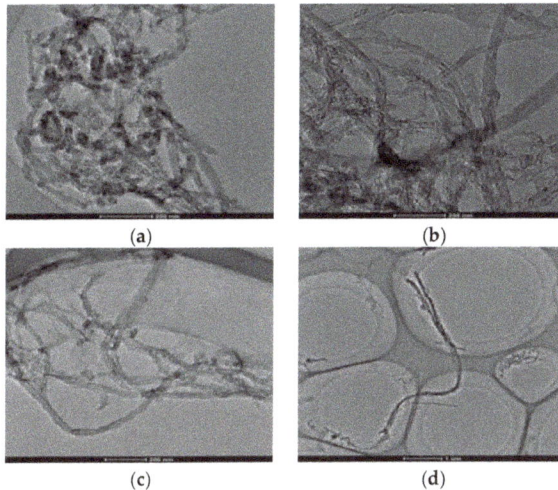

Figure 4. CNT dispersion. (a) without Ce(SO4)$_2$, (b) 0.3 wt.% Ce(SO4)$_2$, (c) 0.6 wt.% Ce(SO$_4$)$_2$, (d) 0.6 wt.% Ce(SO$_4$)$_2$ (single).

CNT dispersion is determined by the species, diameter, and morphology of CNTs, and by the dispersing agent [25]. The results above show that nano-Ce(SO$_4$)$_2$, which can be adsorbed onto the surface of CNTs by ultrasound, plays a major role in CNT dispersion. Thus, nano-Ce(SO$_4$)$_2$ shows "superactivity" with the action of ultrasonic wave and prevents the aggregation of CNTs. Once nano-Ce(SO$_4$)$_2$ is adsorbed onto the surface of CNTs, it can also adhere stably onto the surface of CNTs (Figure 5). A detailed description of nano-Ce(SO$_4$)$_2$ distribution on the surface of CNTs is shown in the two TEM images in Figure 5. As illustrated in Figure 5a,b, the surface of CNTs forms a saw-tooth shape with non-uniform thickness because nano-Ce(SO$_4$)$_2$ is adsorbed irregularly onto the surface of CNTs. In addition, the excellent binding force between CNTs and foamed concrete is obtained due to the saw-tooth shape of the surface of CNTs, and it is the most important factor determining the improvement of the mechanical performance of foamed concrete.

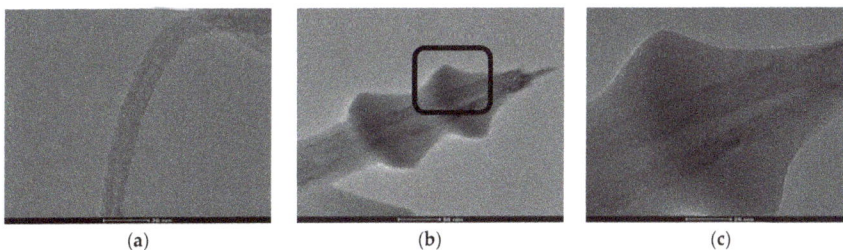

Figure 5. Surface of CNTs. (a) without nano-Ce(SO$_4$)$_2$, (b) with nano-Ce(SO$_4$)$_2$, (c) with nano-Ce(SO$_4$)$_2$.

Ce^{2+} is easily adsorbed onto the surface of materials [26,27]. Microbubbles in the solutions are exploded instantly under the effect of ultrasound, and then a large amount of energy is produced in the following. The large amount of energy can impact the CNT aggregates seriously and damage the Van der Waals force among them. Meanwhile, the temperature of some solutions is increased rapidly due to the large amount of thermal energy released. On the basis of the two reasons discussed above, the connection among the aggregates of CNTs is interrupted. Nano-Ce(SO$_4$)$_2$ is adsorbed onto the surface of CNTs, and then an electrostatic layer is developed on the surface of CNTs, leading to the destruction of the Van der Waals force among the CNTs. Steric hindrance of the reactions is increased by the adsorption of nano-Ce(SO$_4$)$_2$ onto the surface of CNTs, and then the access between CNTs is disabled. Thus, the stability of CNT dispersion is achieved upon the complete dispersion of the CNTs (Figure 6). Figure 6 describes the process of adsorption from nano-Ce(SO$_4$)$_2$ to CNTs through Pro/Engineer (Pro-E) software. Figure 6a presents the large amount of aggregative CNT. Figure 6b shows that nano-Ce(SO$_4$)$_2$ is added into CNT dispersion. Figure 6c indicates that nano-Ce(SO$_4$)$_2$ is adsorbed onto the surface of CNTs under the effect of ultrasound.

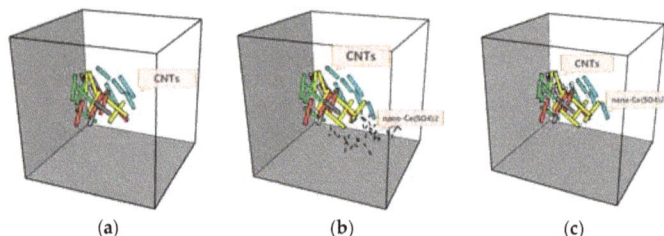

(a)   (b)   (c)

Figure 6. Adsorption of nano-Ce(SO$_4$)$_2$ on the surface of CNTs. (**a**) without nano-Ce(SO$_4$)$_2$, (**b**) with nano-Ce(SO$_4$)$_2$, (**c**) nano-Ce(SO$_4$)$_2$ is adsorbed.

3.2. Compressive and Breaking Strength

The compressive and breaking strengths of foamed concrete described in Table 1 are presented in Figure 7. Figure 7a shows the following results:

- The compressive strengths of FC1-2 and FC1-1 are 17.5% and 5% greater than that of FC0-0, respectively; those of FC2-2 and FC2-1 are 35% and 15% greater than that of FC0-0, respectively; those of FC3-2 and FC3-1 are 40% and 22.5% greater than that of FC0-0, respectively; those of FC4-2 and FC4-1 are 55% and 35% greater than that of FC0-0, respectively; and those of FC5-2 and FC5-1 are 65% and 47.5% greater than that of FC0-0, respectively. Better CNT dispersion corresponds to higher compressive strength of foamed concrete.
- The compressive strength of foamed concrete is increased by 11.9% from FC1-2 to FC1-1, by 17.3% from FC2-2 to FC2-1, by 14.2% from FC3-2 to FC3-1, by 14.8% from FC4-2 to FC4-1, and by 11.9% from FC5-2 to FC5-1. The results indicate that the increases in compressive strength of foamed concrete are almost the same due to the same contents of 0.3% nano-Ce(SO$_4$)$_2$ added in the CNT dispersion solution although the contents of CNTs are different. The result is the same as described in Figure 2, above. Thus, CNT dispersion is an important factor that determines the compressive strength of foamed concrete.
- Cementing material C–S–H affects the compressive strength of foamed concrete [28,29]. Homogeneous CNT dispersion can promote the formation of the cementing material C–S–H, similar to the results of Figure 8a–c described below.

Figure 7b shows the following results:

- The breaking strengths of FC1-2 and FC1-1 are 63.6% and 36.4% greater than that of FC0-0, respectively; those of FC2-2 and FC2-1 are 90.9% and 54.5% greater than that of FC0-0, respectively; those of FC3-2 and FC3-1 are 118.2% and 81.8% greater than that of FC0-0, respectively; those of FC4-2 and FC4-1 are 181.8% and 136.4% greater than that of FC0-0, respectively and those of FC5-2 and FC5-1 are 218.2% and 163.6% greater than that of FC0-0, respectively. Better CNT dispersion corresponds to higher breaking strength of foamed concrete.

- The breaking strength of foamed concrete is increased by 20% from FC1-2 to FC1-1, by 23.5% from FC2-2 to FC2-1, by 20% FC3-2 to FC3-1, by 19.2% from FC4-2 to FC4-1, and by 20.7% from FC5-2 to FC5-1. The results indicate that the increases in breaking strength of foamed concrete are almost the same due to the same contents of 0.3% nano-Ce(SO$_4$)$_2$ added in the CNT dispersion solution although the contents of CNTs are different. The result is the same as described in Figure 2, above. Thus, CNT dispersion is an important factor that determines the breaking strength of foamed concrete.
- Interface bonding strength between the particles of foamed concrete affects the breaking strength of foamed concrete [30,31]. Homogeneous CNT dispersion can reinforce the interface bonding strength, similar to the results shown in Figure 8d,e below.
- Meanwhile, the crack propagation of foamed concrete is inhibited by CNTs. A large area of the crack propagation of foamed concrete can be inhibited due to the homogeneous CNT dispersion, and the breaking strength of foamed concrete is enhanced.

Figure 7a,b show that (i) CNT dispersion has a greater contribution to increasing the compressive strength than the breaking strength and (ii) the effect of the same content of weight of 0.3% nano-Ce(SO$_4$)$_2$ on the compressive and breaking strengths of foamed concrete is almost the same, i.e., 11.9–17.3% increase rate of compressive strength (5.4% maximum difference) and 19.2–23.5% increase rate of breaking strength (4.3% maximum difference).

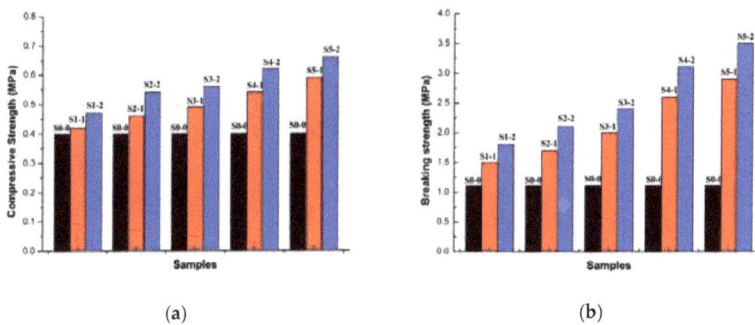

(a) (b)

Figure 7. Mechanical property (28 days). (**a**) Compressive strength, (**b**) breaking strength.

3.3. Microstructure

The microstructure of the CNT/foamed concretes can be observed in Figure 8. The effect of CNTs on the C–S–H growth and interface bonding force is discussed. CNT fibers are distributed uniformly between the foamed concrete and combined in the foamed concrete particles closely, as illustrated in Figure 8a. Figure 8b,c shows the relationship between CNTs and C–S–H. CNTs and C–S–H are interlaced with each other, forming a compact grid. On the basis of nucleation mechanism, CNTs play a nuclear role in the crystallization of C–S–H, signifying that the effective dispersion of CNTs promotes the grain growth of C–S–H. Thus, CNT dispersion is the most important factor influencing the increase in compressive strength.

In addition to the mean CNTs values, the other values regarding the interface bonding force between foamed concrete particles and C-S-H are reported. As shown in Figure 8d,e, CNTs are distributed not only between foamed concrete particles and particles but also between foamed concrete particles and C–S–H. Figure 7b illustrates that foamed concrete with well-dispersed CNTs exhibits an improved behavior, suggesting that CNTs can improve the interface bonding force.

Figure 8. Microstructure of CNTs/foamed concretes with different magnification times. (**a**) 10 μm, (**b**) 5 μm, (**c**) 2 μm, (**d**) 10 μm, (**e**) 5 μm.

3.4. CNT Dispersion Model and Mechanical Properties of the Simulation

Despite the importance of CNT dispersion in the performance of foamed concrete, CNT dispersion models are established scarcely possibly because the mass of pores in the foamed concrete is difficult to model. Foamed concrete is a typical porous material with characteristics similar to those of a pore structure unit and with a certain distribution of pores at a macroscopic state. This study focuses on the influence of aggregated and dispersed CNTs on the mechanical properties of foamed concrete. A type of isotropic foamed concrete is established, and the pores in the foamed concrete are not considered when modeling. In the process modeling, CNTs form the micro unit, whereas foamed concrete is the macro unit.

Pro/Engineer (Pro-E) software is an entity modeling system that uses parametric design based on physical characteristics. The CNT dispersion of the foamed concrete model is established, and the mechanical performance is simulated, using the Pro-E software. The effects of the CNT dispersion of foamed concrete on the compressive and breaking strengths are studied.

3.4.1. Modeling

The boundary representation model, the decomposition model, and the construct solid geometry model are widely used at present. Precision is difficult to achieve with the boundary representation model due to the large number of curves and surfaces in the CNT/foamed concrete material [31]. If the decomposition model is used, two scales of macro (foamed concrete) and micro(CNTs) are present in the foamed concrete [32]. In addition, not only the size and scale of the macro structure, but also the microstructure of the model, should be established, and a quantity of voxels should be used [33].

Therefore, the constructive solid geometry model is adopted in this study, which is combined through the Boolean operation method.

The model of foamed concrete features a complicated structure unit and a high degree of freedom. Thus, the calculating quantity of grid computation is large if the grid is generated directly. In particular, long computation time is expended if the finite element simulation method is used for the simulation directly, and this can lead to computer crash due to insufficient computing ability of the computer. To reduce the complexity of the grid generation and improve the grid quality, a previous study suggested simplifying the models before simulation [34]. Some researchers have introduced a method of model simplification based on rules and realized the development on the plug-in of model that is simplified based on the Pro-E software (Parametric Technology Corporation, Boston, MA, USA) [35]. Figure 8 shows that that the dispersed CNTs are distributed evenly in the foamed concrete. If the CNT unit is defined as the micro unit while the foamed concrete unit is defined as the macro unit, then the CNT unit is defined as spherical by the finite element method.

Based on the mechanical property test, two models with dimensions of 100 mm × 100 mm × 100 mm (Figure 9) and 400 mm × 100 mm × 100 mm (Figure 10) are established to simulate the compressive and breaking strength tests, respectively. Modeling parameters are shown in Table 4. A microstructure scale that describes the CNTs inside the foamed concrete and a macrostructure scale that describes the external size of foamed concrete are obtained in the two models.

Table 4. Modeling parameters.

Material	Mechanical Properties	Size	Force	Modulus of Elasticity	Poisson Ratio
Concrete	Compressive strength	100 mm × 100 mm × 100 mm	2 kN	30,000 MPa	0.2
	Breaking strength	400 mm × 100 mm × 100 mm	0.2 kN	30,000 MPa	0.2

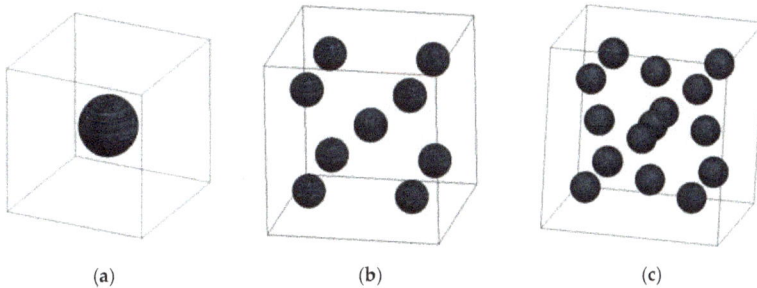

(a)   (b)   (c)

Figure 9. Model of compressive strength (Strain). (**a**) Aggregated completely, (**b**) dispersion ratio of 50%, (**c**) dispersion ratio of 100%.

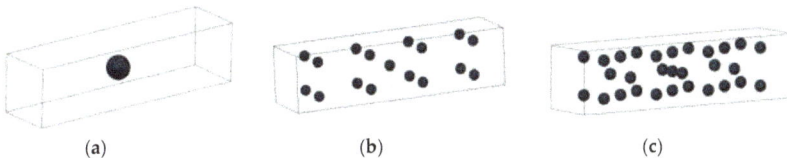

(a)   (b)   (c)

Figure 10. Model of breaking strength (Strain). (**a**) Aggregated completely, (**b**) dispersion ratio of 50%, (**c**) dispersion ratio of 100%.

3.4.2. Simulation

In this section, we report the simulation results regarding the influence of CNT dispersion on compressive and breaking strengths. To improve the accuracy of the models, we define the CNT dispersion as three conditions, namely, aggregated completely (Figures 11a and 12a), dispersion ratio of 50% (Figures 11b and 12b), and dispersion ratio of 100% (Figures 11c and 12c).

As expected, Figure 11a shows that the CNTs are aggregated completely and the compressive strength is 0.6123 MPa. Figure 11b illustrates that the dispersion ratio of CNTs is 50% and the compressive strength is 0.8765 MPa. Figure 11c displays that the dispersion ratio of CNTs is 100% and the compressive strength is 1.231 MPa. These results indicate that better CNT dispersion corresponds to increased compressive strength. This noticeable result is fully in line with the experimental findings (Figure 7a) and is useful to explain the previous quantification methods on a theoretical basis.

Figure 12 compares the same target breaking strength obtained from the three situations of CNT dispersion. Figure 12a shows that the CNTs are aggregated completely and the breaking strength is 2.124 MPa. Figure 12b shows that the dispersion ratio of CNTs is 50% and the breaking strength is 3.413 MPa. Figure 12c shows that dispersion ratio of CNTs is 100% and the breaking strength is 4.54678 MPa. Similarly, better CNT dispersion corresponds to remarkably increased breaking strength, which agrees with the experimental results (Figure 7b).

The experimental findings presented in the previous sections are analyzed to determine simulation results for prediction purposes. The simulation results are based on the idealized assumptions of foamed concrete. Therefore, the simulation results are valid only for foamed concrete specimens with similar characteristics.

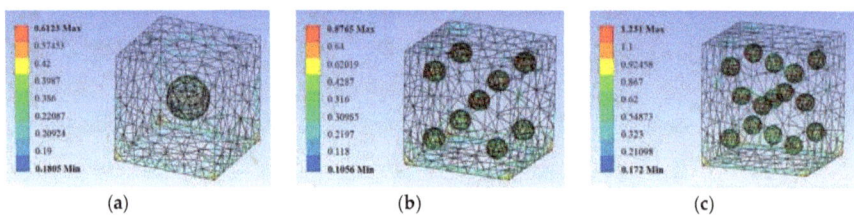

Figure 11. Simulation model of compressive strength (Strain). (**a**) Aggregated completely, (**b**) dispersion ratio of 50%, (**c**) dispersion ratio of 100%.

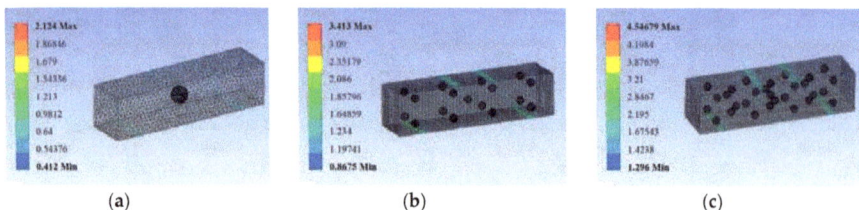

Figure 12. Simulation model of breaking strength (Strain). (**a**) Aggregated completely, (**b**) dispersion ratio of 50%, (**c**) dispersion ratio of 100%.

4. Conclusions

The effect of CNT dispersion on the microstructure and mechanical properties of foamed concrete is investigated. The effect of CNT dispersion on compressive and breaking strengths is simulated. Results show that the addition of nano-Ce(SO$_4$)$_2$ improves CNT dispersion by increasing the degree of distribution and dispersion of CNT particles in solution. In particular, the structure of CNTs is not destroyed due to the addition of nano-Ce(SO$_4$)$_2$ to the dispersion. The enormous uniformity and stability observed are ascribed to the properties, including activity, adsorption, and structure, of nano-Ce(SO$_4$)$_2$, which has modified the surface of CNTs.

CNT dispersion increases the compressive and breaking strengths of foamed concrete. CNTs play a nuclear role in the crystallization of C–S–H. The dispersion of CNTs effectively promotes the grain growth of C–S–H, forming a compact grid. Finally, the effect of CNT dispersion on the compressive and breaking strengths of foamed concrete is predicted through simulation to provide a theoretical basis for the experimental findings.

Author Contributions: J.Z. and X.L. conceived and designed the experiments; J.Z. performed the experiments; J.Z. and X.L. analyzed the data; J.Z. wrote the paper.

Funding: This research was funded by Inner Mongolia autonomous region education department grant number [NJZC13344 and NJZY17487].

Conflicts of Interest: The authors declare no conflict of interest.

References

1. Falliano, D.; de Domenico, D.; Ricciardi, G.; Gugliandolo, E. Experimental investigation on the compressive strength of foamed concrete: Effect of curing conditions, cement type, foaming agent and dry density. *Constr. Build. Mater.* **2018**, *165*, 735–749. [CrossRef]

2. She, W.; Du, Y.; Miao, C.; Liu, J.; Zhao, G.; Jiang, J.; Zhang, Y. Application of organic- and nanoparticle-modified foams in foamed concrete: Reinforcement and stabilization mechanisms. *Cem. Concr. Res.* **2018**, *106*, 12–22. [CrossRef]

3. Gunawana, P.; Setionob. Foamed lightweight concrete tech using galvalum Az 150 fiber. *Procedia Eng.* **2014**, *95*, 433–441. [CrossRef]

4. Vesova, L.M. Disperse Reinforcing Role in Producing Non-autoclaved Cellular Foam Concrete. *Procedia Eng.* **2016**, *150*, 1587–1590. [CrossRef]

5. Mahzabin, M.S.; Hock, L.J.; Hossain, M.S.; Kang, L.S. The influence of addition of treated kenaf fibre in the production and properties of fibre reinforced foamed composite. *Constr. Build. Mater.* **2018**, *178*, 518–528. [CrossRef]

6. Mastali, M.; Kinnunen, P.; Isomoisio, H.; Karhu, M.; Illikainen, M. Mechanical and acoustic properties of fiber-reinforced alkali-activated slag foam concretes containing lightweight structural aggregates. *Constr. Build. Mater.* **2018**, *187*, 371–381. [CrossRef]

7. Eftekhari, M.; Ardakani, S.H.; Mohammadi, S. An XFEM multiscale approach for fracture analysis of carbon nanotube reinforced concrete. *Theor. Appl. Fract. Mech.* **2014**, *72*, 64–75. [CrossRef]

8. Zhang, S.; Luo, J.; Li, Q.; Wei, X.; Sun, S. Physico-mechnical and damping performances of carbon nanotubes modified foamed concrete. *Concrete* **2015**, *306*, 78–81.

9. Luo, J.; Hou, D.; Li, Q.; Wu, C.; Zhang, C. Comprehensive performances of carbon nanotube reinforced foam concrete with tetraethyl orthosilicate impregnation. *Constr. Build. Mater.* **2017**, *131*, 512–516. [CrossRef]

10. Vaganov, V.; Popov, M.; Korjakins, A.; Šahmenko, G. Effect of CNT on Microstructure and Minearological Composition of Lightweight Concrete with Granulated Foam Glass. *Procedia Eng.* **2017**, *172*, 1204–1211. [CrossRef]

11. Ferreira, F.V.; Francisco, W.; Menezes, B.R.C.; Brito, F.S.; Coutinho, A.S.; Cividanes, L.S.; Coutinho, A.R.; Thim, G.P. Correlation of surface treatment, dispersion and mechanical properties of HDPE/CNT nanocomposites. *Appl. Surf. Sci.* **2016**, *389*, 921–929. [CrossRef]

12. Kim, G.M.; Yang, B.J.; Cho, K.J.; Kim, E.M.; Lee, H.K. Influences of CNT dispersion and pore characteristics on the electrical performance of cementitious composites. *Compos. Struct.* **2017**, *164*, 32–42. [CrossRef]

13. Arboleda-Clemente, L.; Ares-Pernas, A.; García, X.; Dopico, S.; Abad, M.J. Influence of polyamide ratio on the CNT dispersion in polyamide 66/6 blends by dilution of PA66 or PA6-MWCNT masterbatches. *Synth. Met.* **2016**, *221*, 134–141. [CrossRef]

14. Liu, L.; Bao, R.; Yi, J.; Li, C.; Tao, J.; Liu, Y.; Tan, S.; You, X. Well-dispersion of CNTs and enhanced mechanical properties in CNTs/Cu-Ti composites fabricated by Molecular Level Mixing. *J. Alloy. Compd.* **2017**, *726*, 81–87. [CrossRef]

15. Eftekhari, M.; Mohammadi, S.; Khanmohammadi, M. A hierarchical nano to macro multiscale analysis of monotonic behavior of concrete columns made of CNT-reinforced cement composite. *Constr. Build. Mater.* **2018**, *175*, 134–143. [CrossRef]

16. Eftekhari, M.; Karrech, A.; Elchalakani, M.; Basarir, H. Multi-scale Modeling Approach to Predict the Nonlinear Behavior of CNT reinforced Concrete Columns Subjected to Service Loading. *Structures* **2018**, *14*, 301–312. [CrossRef]

17. Zou, B.; Chen, S.J.; Korayem, A.H.; Collins, F.; Wang, C.M.; Duan, W.H. Effect of ultrasonication energy on engineering properties of carbon nanotube reinforced cement pastes. *Carbon* **2015**, *85*, 212–220. [CrossRef]
18. Mendoza, O.; Sierra, G.; Tobón, J.I. Influence of super plasticizer and Ca(OH)2 on the stability of functionalized multi-walled carbon nanotubes dispersions for cement composites applications. *Constr. Build. Mater.* **2013**, *47*, 771–778. [CrossRef]
19. Kang, L.; Du, H.L.; Zhang, H.; Ma, W.L. Systematic research on the application of steel slag resources under the background of big data. *Complexity* **2018**. [CrossRef]
20. Fu, H.; Li, Z.; Liu, Z.; Wang, Z. Research on Big Data Digging of Hot Topics about Recycled Water Use on Micro-Blog Based on Particle Swarm Optimization. *Sustainability* **2018**, *10*, 2488. [CrossRef]
21. Yang, A.M.; Yang, X.L.; Chang, J.C.; Bai, B.; Kong, F.B.; Ran, Q.B. Research on a fusion scheme of cellular network and wireless sensor networks for cyber physical social systems. *IEEE Access* **2018**, *6*, 18786–18794. [CrossRef]
22. Zhang, Q.; Jin, B.; Wang, X.; Lei, S.; Shi, Z.; Zhao, J.; Liu, Q.; Peng, R. The mono (catecholamine) derivatives as iron chelators: Synthesis, solution thermodynamic stability and antioxidant properties research. *R. Soc. Open Sci.* **2018**, *5*, 171492. [CrossRef] [PubMed]
23. Liang, X.; Li, W. Dispersion Properties of Aligned Multi-Walled Carbon Nanotubes. *J. Dispers. Sci. Technol.* **2016**, *37*, 1360–1367. [CrossRef]
24. Sindu, B.S.; Sasmal, S. Properties of carbon nanotube reinforced cement composite synthesized using different types of surfactants. *Constr. Build. Mater.* **2017**, *155*, 389–399. [CrossRef]
25. Chaichi, A.; Sadrnezhaad, S.K.; Malekjafarian, M. Synthesis and characterization of supportless Ni-Pd-CNT nanocatalyst for hydrogen production via steam reforming of methane. *Int. J. Hydrogen Energy* **2018**, *43*, 1319–1336. [CrossRef]
26. Chen, J.; Luo, W.; Guo, A. Preparation of novel carboxylate-rich palygorskite as an adsorbent for Ce^{3+} for aqueous solution. *J. Colloid Interface Sci.* **2018**, *512*, 657–664. [CrossRef] [PubMed]
27. Ogata, T.; Narita, H.; Tanaka, M. Adsorption behavior of rare earth elements on silica gelmodified with diglycol amic acid. *Hydrometallrugy* **2015**, *152*, 178–182. [CrossRef]
28. Pichler, B.; Hellmich, C.; Eberhardsteiner, J.; Wasserbauer, J.; Termkhajornkit, P.; Barbarulo, R.; Chanvillard, G. Effect of gel–space ratio and microstructure on strength of hydrating cementitious materials: An engineering micromechanics approach. *Cem. Concr. Res.* **2013**, *45*, 55–68. [CrossRef]
29. Eftekhari, M.; Mohammadi, S. Molecular dynamics simulation of the nonlinear behavior of the CNT-reinforced calcium silicate hydrate (C–S–H) composite. *Compos. Part A Appl. Sci. Manuf.* **2016**, *82*, 78–87. [CrossRef]
30. Hlobil, M.; Šmilauer, V.; Chanvillard, G. Micromechanical multiscale fracture model for compressive strength of blended cement pastes. *Cem. Concr. Res.* **2016**, *83*, 188–202. [CrossRef]
31. Liu, Z.; Cheng, K.; Li, H.; Cao, G.; Wu, D.; Shi, Y. Exploring the potential relationship between indoor air quality and the concentration of airborne culturable fungi: A combined experimental and neural network modeling study. *Environ. Sci. Pollut. Res.* **2018**, *25*, 3510–3517. [CrossRef] [PubMed]
32. Jiang, S.; Lian, M.; Lu, C.; Gu, Q.; Ruan, S.; Xie, X. Ensemble Prediction Algorithm of Anomaly Monitoring Based on Big Data Analysis Platform of Open-Pit Mine Slope. *Complexity* **2018**, *2018*, 1048756. [CrossRef]
33. Liu, T.; Liu, H.; Chen, Z.; Lesgold, A.M. Fast Blind Instrument Function Estimation Method for Industrial Infrared Spectrometers. *IEEE Trans. Ind. Inform.* **2018**. [CrossRef]
34. Li, M. Review on Engineering Analysis Reliable CAD Model Simplification. *J. Comput.-Aided Des. Comput. Graph.* **2015**, *27*, 1363–1375.
35. Papadopoulos, V.; Impraimakis, M. Multiscale modeling of carbon nanotube reinforced concrete. *Compos. Struct.* **2017**, *182*, 251–260. [CrossRef]

MDPI
St. Alban-Anlage 66
4052 Basel
Switzerland
Tel. +41 61 683 77 34
Fax +41 61 302 89 18
www.mdpi.com

Processes Editorial Office
E-mail: processes@mdpi.com
www.mdpi.com/journal/processes

www.ingramcontent.com/pod-product-compliance
Lightning Source LLC
Chambersburg PA
CBHW051726210326
41597CB00032B/5617